国家双语教学示范课程配套教材

生物技术概论

Essentials of Biotechnology

王 武 主编

科学出版社
北 京

内 容 简 介

本书为国家双语教学示范课程配套教材，全书分为基因克隆、细胞工程、酶技术、工业发酵四个部分，共21章，分别介绍了上述四大生物技术领域的基础知识、技术关键以及发展前沿与动态，并对有关技术的正负面影响提出思辨。附录部分提供了关键词汇的中英文对照；列出了本领域诺贝尔奖获得者名单及其主要贡献；提供了近十年来出版的英文原版专业书目，以利于读者自学、参考。

本书主要供相关专业本科生双语课程使用，也适合作为研究生或高职高专优秀生的选用教材。对于生物技术产业的技术人员，以及准备出国深造的青年学子而言，本书也具有一定的参考价值。

图书在版编目(CIP)数据

生物技术概论＝Essentials of Biotechnology/王武主编. —北京：科学出版社，2012

国家双语教学示范课程配套教材

ISBN 978-7-03-035239-2

Ⅰ.①生…　Ⅱ.①王…　Ⅲ.①生物工程-双语教学-高等学校-教材　Ⅳ.①Q81

中国版本图书馆CIP数据核字(2012)第177049号

责任编辑：席　慧/责任校对：刘小梅

责任印制：赵　博/封面设计：迷底书装

科学出版社 出版

北京东黄城根北街16号

邮政编码：100717

http://www.sciencep.com

保定市中画美凯印刷有限公司印刷

科学出版社发行　各地新华书店经销

*

2012年8月第　一　版　开本：787×1092　1/16

2025年1月第三次印刷　印张：22 1/4

字数：513 000

定价：89.00元

(如有印装质量问题，我社负责调换)

编委会名单

主　　编　王　武

编写人员　唐　蕾　杨海麟　辛　瑜
夏小乐　张　玲

审 稿 人　王南新　杨　昊　李珊珊　李　莉

Preface

For years, these authors have been teaching an undergraduate course named "Notes in Biotechnology" in a bilingual English-Chinese format. The National Education Authority has encouraged to build bilingual teaching systems for all undergraduates. This approach will help them improve their comprehension of scientific English, since large amounts of information in the field of biotechnology is constantly emerging and is often expressed in English. The younger Chinese generation has been taking English courses from the Primary Grade 3 through their Ph. D studies, but I have to say that many of them still have difficulty of comprehending academic articles and books with many modern scientific English terms. I'm glad to see that these authors have written this new textbook—*Essentials of Biotechnology* for upgrading their teaching.

How biotechnology should be presented in this new edition is a crucial task that the authors have to face. With the ever changing and increasingly blurred boundaries across Biology, Biochemistry, Microbiology, Physics, Chemistry, Genetics, Logics, and Information Technology, etc., biotechnology can no longer be regarded as a clearly defined subject. Since it is an enormously complicated and rapidly growing field, I understand the comprehensive approach is to collect and compile the knowledge referring to important discoveries, key theories, broadly applied techniques and classical examples of application. Four parts, including *Gene Cloning*, *Cell Engineering*, *Enzyme Technology*, and *Industrial Fermentation*, are considered as the most significant domains to be presented in this textbook, and each part is divided into several chapters which are subsequently composed of many sections. I appreciate that the authors have made every effort to limit the content to a reasonable length, in spite of the massive information that must be included in a modern, up-to-date text in biotechnology. Therefore, some of the topics are discussed briefly, yet the major significance and the main points are essentially kept. At the beginning of each Chapter, there is an abstract written in Chinese to help students catch the main idea at the first glance. Beyond the main text, an index of term translations, a list of Nobel Laureates in the related field, and the suggested further readings are provided in the Appendix section.

I believe that writing this textbook, especially in English, was no doubt a time consuming and energy demanding task. The authors must have carried a heavy burden of responsibility. Fortunately they have completed this work. I have no doubt that "Essentials of Biotechnology" will be of great benefit to biotechnology teachers and students especially those who wish to further studies abroad or pursue careers in bio-industry.

Lun Shiyi
Professor
Academician of the Chinese Academy of Engineering
June, 2012

Contents

Part I Gene Cloning

Chapter 1 Introduction to Gene Cloning

自 1970 年，科学家已实现了对生物体内的基因组进行切分，提取出含有基因的片段，将供体和载体片段连接成重组 DNA 分子，最终转化入合适的宿主细胞，表达出一种新的遗传性状。所谓的基因克隆是生物技术领域最具创新性的工作，借此，人类有可能培育出全新设计的植物和动物，并根据需要调整它们的遗传特性。这一技术也使得人类可以透彻地研究遗传物质的本质和功能，鉴别染色体中某个基因的位点。基因克隆技术为现代社会带来了各种全新的生物产品和克隆品种，但是对那些因不良因子或危险性遗传特性导入生物体而产生的负面效应，人类还必须保持高度的警惕。

1.1 Brief Overview

Gene cloning, as one of the most innovative technologies, allows scientists to discover the sophisticated essence of life and to manipulate genes in life. For a long time, **biologists** had been seeking to understand how **genetic factors** are organized and how they work in living organisms. Several generations of biologists had contributed their understanding to life issues, especially since the behavior of genes was revealed by molecular geneticists. The efforts they made led to an almost incredible technology—gene cloning, which arose in the early 1970s. **Mendel**, for example, having counted the **offspring** of pea plants, suggested that there exist some **hereditary factors** (later called **genes**) by inferring from the traits and numbers of **progeny** of the parent plants crossed. Later, people took almost a hundred years to know in detail the real genetic elements and how they are expressed. Now, based upon the powerful means of gene cloning, one could easily extract the **genetic materials**, cut the essential fragments from a genome, manipulate *in vitro* the genes, and **transform** the cloned piece into living cells to get totally novel **recombinants**.

Genes are at the very heart of life, in terms of information technology (IT), they are the master program of life. Collectively, genes constitute the **life's blueprint** and determine all the properties and capabilities of an organism. In biological terms, this master program is called the **hereditary substance**, the **chromosomes**. Constituted by chains of so called DNA molecules (in some cases RNA molecules), chromosomes carry the genes with "**code words**" or instructions of the master program.

There is an identical set of this master program in every cell of an individual species. For example a corn plant has about a billion cells, each with a set of this master program. In different sections of the plants different parts of the program are active, giving rise to differentiated tissues such as leaves, seeds and root. The living organism is like a huge computer network, much more complicated than any man-made one. Still, science has a very incomplete understanding on how this master programs is able to make bio-system cooperated in a very harmoniously and effectively coordinated way.

Although the knowledge about the master program in some **prokaryotes** are well known, but in some **eukaryotic genome** only 1～3 percent of it are carrying "code words". The existing purpose and function of the remaining 97～99 percent are very little known. But this doesn't obstruct the progress on gene cloning, since the general roles of the DNA, RNA and protein molecules are gradually revealed in detail by molecular biologists.

Just as any IT expert knows that adding just one "code Syllable" (binary code) may be disastrous to a computer program, so too it is well known in genetics that changing even a little code word in the master program could mean a big difference between being healthy and carrying deadly hereditary diseases. Insertion of genes, as done by genetic **engineers**, does not add just one syllable, but many thousands of code syllables. The results that come from the gene cloning are not always predictable, which should remind scientists of their responsibility for the whole biological society.

"**Clone**", a term derived from an ancient Greek word "Klon", means the germ line from asexual propagation, for example in the case of grape **vegetative propagation**. Later, cloning was also defined as changing an organism's DNA to make it more beneficial. In this way, the history of cloning can be traced back to very early times when people started to perform **selective breeding**. Breeders have found ways to alter or change the genes of animals and plants to their advantage for thousands of years.

But the practical technology of gene cloning came much later. Gene cloning (also called **gene manipulation** or genetic engineering) means the molecular manipulation of genetic materials *in vitro*. In many cases, genes mostly from often totally unrelated species are recombined and inserted into another genetic "master program" by certain fascinating means. The essence of gene cloning can be considered as a series of four key steps: Generating donor DNA fragments; Joining the DNA fragments to a **vector** or **carrier molecule**; Introducing the cloned molecule into a host cell; Selecting required **clones**(recombinants). The noun "clone" comes from the **colonies** of identical host cells produced during amplification of the cloned fragments. Gene cloning is sometimes referred to as "molecular cloning" to distinguish the process from the cloning of whole organisms.

Genes from species e. g. fish, corn, bacteria and viruses have been inserted into other species in genetic engineering projects and the cloned bacterial cells producing

human hormone are not surprising to us now.

1.2 Historical Events

Tab. 1.1 lists out some of the notable events which strongly propelled the progress of gene cloning. Most contributors won the **Nobel Prices** for their beautiful works done during the last century. The British microbiologist **F. Griffith** became the first person to find the genetic material extracted from the dead bacterial cells in 1928, which could be transformed into living cells and express the related **hereditary feature. G. W. Beadle & E. L. Tatum** declared the "One Gene One Enzyme" theory to illustrate the relationship between the genetic unit and enzyme protein molecule during 1940. **Chase & Hershey** worked on virus genetics in the **Lab. of Cold Spring Harbor**, in 1951, they finally proved that **phage** DNA performed the genetic role. The discovery of the structure of **Double Helix DNA** by **J. Watson** and **F. Crick** in 1953 provided the stimulus for the development of genetics at the molecular level. The following few years saw a period of intense activities and excitements as the main features of genes and their expression were determined. In 1966, a stage was set for the appearance of the new genetics, based upon the establishment of the complete set of the **Genetic Code**.

Tab. 1.1 Some of the notable events

Year	Contributors	Events
1928	F. Griffith	DNA transformation to bacteria
1940	G. W. Beetle & E. L. Tatum	One gene one enzyme
1949	E. Chargaff	Analysis of 4 nucleotide bases
1951	H. Chase	DNA is the major gene carrier
1953	J. Watson & F. Crick	DNA double helix structure
1959	A. Kornberg	DNA replication
1960	J. Monod	Lac-operon
1964	M. W. Nirenberg & P. Leder	Genetic triplet codon
1970	H. O. Smith	DNA restriction enzyme
1972	P. Berg & B. Cohen	DNA cloned into *E. coli*
1975	H. M. Temin	Revetro-transcriptase
1977	F. Sanger & W. Gilbert	DNA sequencing
1986	T. R. Cech	Self splicing of intron, indicated existence of ribozyme

In 1967, an enzyme named **DNA ligase** was isolated, which catalyzes the joining of the **complementary ends** of two DNA pieces together (some time also works for the **blunt ends** of two DNA pieces in lower efficacy), a prerequisite for the construction of recombinant molecules, and can be regarded as a sort of **molecular glue**. In 1970, H. O.

Smith declared the discovery of the first **DNA Restriction Endonuclease**, a major milestone in the development of genetic engineering. Restriction enzymes are essentially molecular scissors which cut DNA sequence at precisely defined sites. Such enzymes can be used to produce DNA fragments that are suitable for joining to other fragments with the same cutting. Thus, by 1970, the basic tools required for the construction of **recombinant DNA** were available. Right away, P. Berg & B. Cohen eagerly performed the very primitive cloning works. The first recombinant DNA molecules were generated at **Stanford University** in 1972, utilizing the cleavage properties of restriction enzymes (scissors) and the ability of DNA ligase (glue) to join DNA fragments together. When the cloned **antibiotics resistant markers** were expressed in the *E. coli* host cells, the importance of these first **tentative experiments** can hardly be overestimated. The **methodology** was extended in 1973 by joining DNA fragments to the **plasmid** which is an **extra-chromosomal genetic element** isolated from **bacteria**. These recombinant molecules behaved as either the vectors or the **replicons**; that is, they could help the cloned fragments **transform** and **replicate** in *E. coli* host cells. Thus, by creating recombinant molecules *in vitro*, and placing the recombinant molecules in a bacterial cell where it could be replicated *in vivo*, the cloned DNA fragments could be isolated from bacterial colonies that form clones (colonies derived from a single cell, in which all cells are identical) when grown on **agar plates**. This development marked the emergence of the technology that became known as gene cloning (Fig. 1.1).

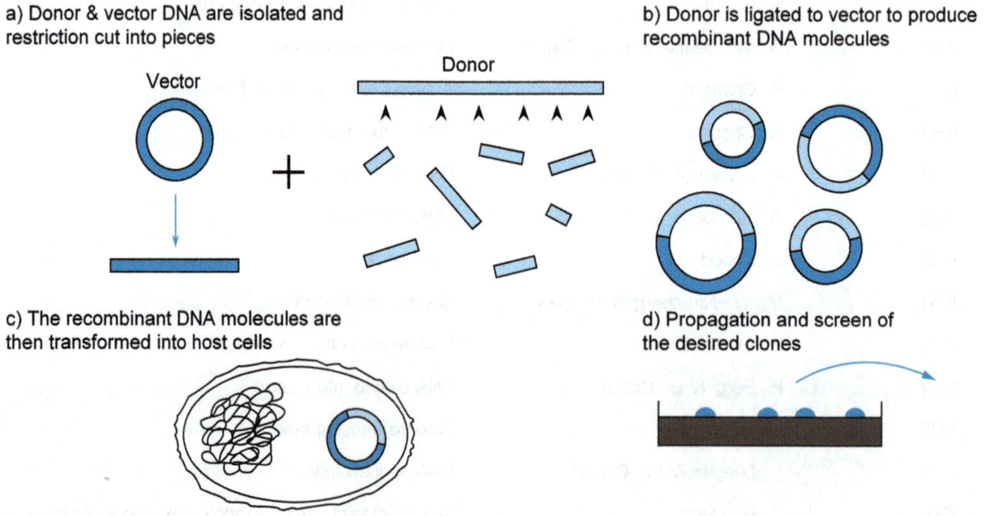

Fig. 1.1 Sketch map of gene cloning

The application of this new technology spread very quickly, and a sense of urgency and excitement prevails in the genetics research community. Now scientists could alter DNA sequence and join different DNA molecules from any organisms together to form a different organism. This was dampened somewhat by the realization that the new technology could give rise to potentially harmful organisms exhibiting undesirable characteristics. It is to the credit of the biological community that measures were adopted to regulate the use of gene manipulation and that progress in contentious areas was limited until more information became available regarding the possible consequences of the inadvertent release of organisms containing recombinant DNA. However, within 40 years, the gene cloning has entered a booming period. More and more **genetically modified organisms (GMOs)**, for example the GMO crop plants, have been produced particularly for commercial benefits; the debate has been re-opened about the safety of these organisms and the consequences of releasing GMOs into the **fragile environment**.

1.3 Significance and Alertness

Theoretically, genetic engineering has advanced the understanding of many aspects of genetic function and genomic organization in prokaryotes. It also helps distinguish between **exons**, the coding section, and **introns**, the non coding parts within the eukaryotic genes. It provides powerful means for searching in detail on the genomes of any living organism, if there is a need.

Practically, through gene cloning techniques, enzymes, vitamins, antibodies, vaccines, even synthesizing **human insulin**, **human growth hormone**, **alpha interferon**, **hepatitis B vaccine**, and other medically useful substances can be produced simply from bacterial cells. Genetic engineering also enables us to perform gene therapy for curing the crippling hereditary diseases like **haemophilia**, **phenylketonuria** etc. Based on this technique, some of the hereditary diseases can possibly be corrected by replacing "bad" genes with "normal" ones.

A number of valuable genes such as ***Nif* gene** (**Nitrogen fixation gene**) were finally introduced into other bacteria cells. The plants originally without the function of **nitrogen fixation** may be genetically adjusted to fix nitrogen in order to improve the **soil fertility** and crop production. The **transgenic agro-plants and livestock** are now really the highlight in the area of life science and technology.

In 1980, the "new" microorganisms created by recombinant DNA research were deemed patentable, and in 1986 the **U. S. Department of Agriculture** approved the sale of the first living genetically altered organism—a virus, used as a **pseudo rabies vaccine**, in this viral genome a single gene has been destroyed on purpose. Since then thousands of patents involving genetically altered microorganisms and plants have been released. Nevertheless, special concern has been focused on such achievements for fear that it might result in the introduction of unfavorable and possibly **dangerous traits** into microorganisms, plants and animals that were previously free of "**negative**" **traits**—e. g. , resistance to antibiotics, production of toxins, or a tendency to cause diseases.

Summary

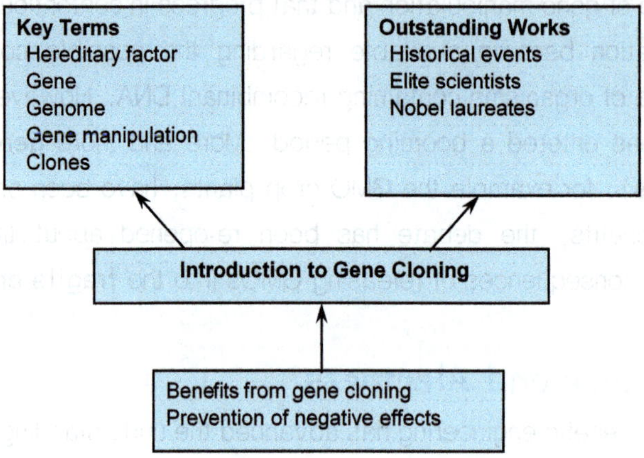
Key Terms
Hereditary factor
Gene
Genome
Gene manipulation
Clones
Outstanding Works
Historical events
Elite scientists
Nobel laureates
Introduction to Gene Cloning
Benefits from gene cloning
Prevention of negative effects

Chapter 2 Gene and Genome

正如前一章所述，基因组代表着生命个体的全部遗传信息。基因组的类型可有多种不同的表述，通常情况下分为真核生物基因组、原核生物基因组，还有病毒类生物基因组。真核生物基因组除了含有核基因组(指染色体 DNA)，还可能含有线粒体、叶绿体，质粒，甚至微体 DNA。真核生物基因组拥有的 DNA 体量可多达近百亿碱基对。原核生物基因组也可分为两类，真细菌(eubacteria)基因组和古细菌(archaebacteria)基因组，它们的基因组多数拥有几千万碱基对。病毒类生物的基因组比较小，甚至小到只含有三两个基因，但是病毒基因组的构成却很复杂，以 RNA 为基因组的病毒往往具有非常独特的生命周期。

“基因”这个专门术语特指基因组中的某段遗传单元，从分子生物学的角度考虑，基因是指一段连续的核酸序列，其具有编码一段肽链或转录一段 RNA 分子的特定功能。不同基因的集合体构成生命所需要的所有遗传信息。由于基因和基因组是生命的信息源泉，本章予以较为翔实的介绍。

2.1 Discovery of Hereditary Factor

2.1.1 Perception of hereditary factor

The conception of "hereditary factor" was initially raised by G. Mendel, who took eight years to breed the pea plants and later put forward his famous conclusion. Mendel lived in a town called Brno, which is today part of the Czech Republic. Starting in 1856, he bred pea plants in his monastery garden, to figure out how the traits of pea plants were inherited. Peas are flowering plants, or angiosperms. Their flowers, like those of most angiosperms, contain both male and female sex organs, so a pea plant can be self-pollinated or cross-pollinated. Mendel pollinated his pea plants using a paintbrush. He first dusted the paintbrush with pea pollen from one flower, and then transferred it to another one. Mendel got the first "true-breeding", or purebred plants that exhibited particular characteristics like different flower colors over multiple generations. Next, he crossbred plants with different traits and counted the numbers of progeny that exhibited the same characteristics as one of the parents. Finally, he crossbred the progeny and counted the ratio of one characteristic to another in this third generation. The work called for considerable patience. Among his remarkable findings were the **law of segregation** (which includes the notion of dominant and recessive genes) and the **law of independent assortment** (which says that an organism's individual traits are passed on independently of one another). Mendel published his work and the conception in the *Experiments in Plant*

Hybridization in 1865, but his theories were not widely embraced by other scientists until the 1900s.

In 1909, Danish botanist W. Johannsen coined "Gene" according to the "gen" in Danish and German, replacing the term of "hereditary factor" to describe the fundamentally physical and functional units of heredity. At that time, Johannsen had not yet noticed the physical existence of genetic materials, but used the term to make a universal unit for description on genetic traits of individual organism.

Almost during the same period, American geneticist T. H. Morgen had made great contribution to the genetics theory. Morgen used **fruit fly** to do the genetic study. His assistant had found a mutant with eye's color different form the normal red-eyed group, the mutant was bred with the normal one. The fruit flies of the next generation were all red-eyed, then the individuals of the second generation were cross bred, they found the mutants again from the third generation and they were all female. By examining chromosomes of fruit flies, Morgen located the hereditary trait of red eye onto the X-chromosome. This was the first materially identified inheritance unit and was named "Rosy" gene. The famous "**Chromosomal Theory of Inheritance**" established by Morgen was finally accepted by the academic world. Morgen was one of the Nobel Laureates in 1933.

2.1.2 DNA was proved as genetic material

The current idea that DNA is the genetic material began with an well known observation in 1928 by the British microbiologist F. Griffith, who was studying ***Streptococcus pneumoniae*** which is responsible for human pneumonia. The virulence of this bacterium depends on its **polysaccharide capsule**, which causes production of **smooth-edged colony** (**S**). The capsule protects the germ from the defense system of the human body. When Griffith mixed **a rough-edged** (**R**) **mutant** with the cellular extraction of heat-killed **wild type** S cells, and injected the mixture into mice, mice succumbed to pneumonia and died within a couple of days. The bacteria isolated from the dead mice were only S. When the mixture was grown in a culture, again only the living S cells were produced. It was concluded by Griffith that there must exist a **transforming factor** which genetically alters R cells into S.

The next remarkable work was published 15 years later by O. Avery and his colleagues. who demonstrated that the transforming factor was DNA. Avery used DNase, RNase, and protease individually to eliminate DNA, RNA and protein molecules in the transformation samples, and he found that the sample from DNase treatment resulted in no transformational effect. At that time, the enzymes used by Avery were not extremely pure, therefore other scientists who supported the genes-as-protein theory were suspicious on Avery's discovery.

An elegant confirmation of the theory of genes-as-DNA came from Blendor Experiment with *E. coli* phage T2 infecting the host cells in 1951. There are only two kinds of macromolecules in T2 particles: DNA and protein, and the confirmation methodology became simpler if one could identify the material being injected into the host cells. A. Hershey and M. Chase in Cold Spring Harbor Lab. used P^{32} to label the phage DNA and S^{35} to label the phage protein respectively, for DNA is the only P containing substance and protein is the only S containing material in T2. They believed that the radioactive substance being injected into host cells must be the T2 genetic material. The problem that remained then was how to tear off the attached phage particles after they inject the labeled genetic material into their host cells, and to separate the empty phages from their host cells. This work had a nickname called **Blendor Experiment**, for a common kitchen blendor was used as a major piece of the lab apparatus. The shearing force produced by blendor cut off the tails of empty phage particles, and separated them from their host cells. In the sample of S^{35} labeling, after centrifugation, it was found that 80% of S^{35} was in the supernatant which contained phage particles, and only 20% of S^{35} was in the pellet which was cell sediment. The remained 20% of S^{35} was proved some years later to consist mostly of T2 tail fragments that adhered tightly to the cell envelop. The result of S^{35} labeling remaining outside the host cells indicated that proteins might not be the genetic material of phage T2. A very different result came from the P^{32} labeled experiment, 70% of P^{32} labeling were associated with the cell pellet after blending. The 30% others were shown some years later to be the defective phages which could not inject their DNA into the host cells. From the Blendor Experiment, DNA was finally proved to be the genetic material. The related scientists were awarded the Nobel Prize.

Avery, Chase and Hershey conformed that DNA was the genetic material, but when it comes to living organisms as a whole, the question over generalizability of this conclusion remained, because some of viruses were found to contain no DNA, but only RNA and proteins. So far, apart from very few viruses, genes-as-DNA theory well applies to cellular level organisms and most viral species. The properties required of the genetic material are as follows: 1) Ability to carry genetic information and to transmit it to next generation; 2) Ability to pass its information to the progeny with only minimal mistakes; 3) Reliable physical and chemical stability, not easily be broken under physiological conditions; 4) Capability of mutation without major deviation from their parental information. Let's see how DNA meets the criteria of genetic material:

(1) DNA molecules in chromosomes are double-stranded, which could be as long as 10^8 base pairs. Huge amounts of the genetic information can be arranged in this giant macromolecule. DNA sequence transmits its information and made all kinds of RNA molecules and proteins being synthesized following its sequence.

(2) The replication of double-stranded DNA in most organisms is a strict process assisted with many enzymes and bioactive factors. Some of enzymes—the DNA polymerase I in *E. coli*, for instance, have proofreading functions which correct minor errors during DNA replication.

(3) The double helix structure and highly folded chromosomal configuration of DNA enhance its physical stability. Chemically, the sugar-phosphate backbond of DNA is extremely stable; Compared with RNA molecules, lack of 2′-OH in **deoxyribose** of DNA makes it tougher under higher pH conditions (the mechanism is called the β-elimination effect); The hydrogen-bonding of the two strands places all of the stacked bases into a gigantic cylindrical shaft, which prevents hydrophilic agents from entering the stacked structure unless the DNA helix opens up.

(4) There are two main mechanisms of DNA mutation: 1) Chemical alteration of the bases that result in a different hydrogen bonding or eliminate the pairing, which causes bases to change in the DNA sequence, hence the mutation passed onto a newly replicated daughter DNA molecule; 2) replication error by which a base might lost, or an incorrect base or an extra base is accidentally inserted into the daughter molecules. If an organism is ever to improve its **hereditary traits**—that is, to evolve—its DNA sequence must be capable of **base mutation** but only in a proper way.

Other forms of genetic materials

Only dsDNA (double-stranded DNA) genomes exist in cellular organisms, and probably more than 99% of the viruses are also possessing dsDNA genomes. However, there are a few species of viruses that use other forms of genetic materials, which are sometimes very useful to genetic engineers.

ssDNA genome

Studies of *E. coli* phage ϕX174 genome have been important for many reasons: it was the first organism to be discovered containing single stranded DNA (ssDNA) as its genetic material. Its circular ssDNA genome is enclosed in an **icosahedral** particle composed of three coat proteins and one internal protein. The DNA replication occurs in several steps: 1) Conversion of the parental ssDNA molecule. The virus uses their ssDNA as template to synthesize the (− strand), forming a covalently closed, double-stranded molecule called replication Form I (RF I). This occurs before transcription begins and thus depends on the host enzymes exclusively; 2) Replication starts from the origin, then many copies of RF I are produced; 3) Synthesis of (+ strand) DNA by rolling circle style for **encapsidation**. So, ϕX174 is also the first virus to be replicated *in vitro* to yield biologically active genomic DNA. The study of its replication mode led to a new hypothesis of rolling-circle replication, and for the first time its genomic DNA is packed *in vitro* into their capsid.

Filamentous ssDNA phages are also worth mentioning. There are three *E. coli* filamentous phages M13, fd, and f1, each contains circular ssDNA. Its life cycle shows many similarities to that of ϕX174, but the mode of phage DNA penetration and the release of the progeny particles are quite different. For example, as a lytic *E. coli* phage, M13 neither kills the host cells nor causes lysis by invading the host. By attaching to the tip of F pilus in an intact form, which is a unique way of infection by those phages. The release of the M13 progeny from the host is also mild; the cells are still alive and even keep grow slowly. For a typical **lytic phage**, **plaques** are the result of bacteria lysis, but for those filamentous phages, the slow growth of the infected host is sufficient to produce a less turbid plaque region against the host bacterial lawn.

Based upon the above phenomenon, the filamentous phages with ssDNA genome attract great attention from cloning researchers, for: 1) Either the dsDNA or ssDNA molecular forms can be easily obtained from the different periods of their life cycle; 2) Unlike icosahedral phages, a filamentous shape means there is redundant space for adding extra DNA sequence; 3) The phage genome can be isolated without contamination of the host chromosome under the non-lytic condition; 4) In the M13 genome, there is a spacer of 507 base with no function of coding, which can be modified by ***in vitro* mutagenesis** to make it an ideal insertion region for cloning. The modified M13 series have later become the most important vectors used especially for Sanger' DNA sequencing.

dsRNA genomes

When proteins are manufactured, the gene is first copied into RNA as an intermediate product. In other cases, the RNA molecules are the actual functional products. For example, some special RNAs known as **ribozymes** are capable of catalysis, and some of the microRNA also has a regulatory role. The DNA sequences from which such RNAs are transcribed are known as RNA genes. RNA is clearly not as suitable a genetic material as DNA, but nonetheless RNA genome survives in some viral species even though it lacks apparently beneficial features of chemical stability and there exists redundancy in the double helix structure. But as compensation for its chemical instability, the RNA viruses store their genes within a very inert viral envelope, hence its genome is protected by a nearly insoluble shell and isolated from the host cellular environment. Meanwhile, RNA viruses may synthesize their proteins as soon as they infect host cells and proceed with duplication without delay, and the life cycle of those viruses are very short, which means the RNA genome is rarely exposed to the host **cytoplasm**.

So far only a few samples of dsRNA virus have been found, for example, the phage *F6* in the bacterium ***Pseudomonas phaseolica*** differs structurally from most of the phages in that their capsid is enclosed in an envelope containing both protein and lipid, and carries genomically distinct genes. The genome is also called the **segmented genome**. A similar phenomenon has also been observed in a few cases of botanic RNA virus. Because the synthesis of progeny RNA genome is odious, a longer genomic sequence would cause tangling problem.

The ssRNA genomes seem more prevalent, which exist in some bacterial phages and many different kinds of eukaryotic viruses. The earlier known ssRNA phages are: f2, MS-2, and Qβ, each of which is a tailless icosahedron. Their life cycles are very simple, and have nothing to do with the DNA synthesis. The phage genome encodes RNA replicase to synthesize " − strand" from the genomic strand and then makes the " + strand" to produce progeny genome RNA. The ssRNA genome itself also acts as a primitive transcript for the viral peptide translation.

On the other hand, **RNA retroviruses**, such as HIV, require the **reverse transcription** of their genomes from RNA into DNA before their proteins can be synthesized. While RNA is not rare as genetic material in viruses, but RNA inheritance, particular in mammals, is indeed very rarely observed. In 2006, French researchers came across a puzzling example of RNA-mediated inheritance in mice. Those individuals with a loss-of-function mutation in the gene kit have white tails.

2.1.3 Chromosomal DNA and episomal DNA

Chromosomes

The total complement of genes in an organism may be stored on one or more chromosomes; the region of the chromosome at which a particular gene is located is called **gene locus**. A chromosome consists of a single, long DNA helix, on which from a few to as many as thousands of genes are encoded. Prokaryotes—bacteria and archaea—typically store their genomes on a single large circular chromosome, sometimes supplemented by additional small circles of DNA called plasmids which usually encode only a few genes and are easily transferable between individuals (see the part episomal DNA). Eukaryotic chromosomes are packed within the nucleus in a complex form of DNA wrapped with storage proteins called **histones**. The manner in which DNA is associates with the **histone**, as well as chemical modifications of the histone itself, are regulatory mechanisms governing whether a particular region of DNA is accessible for gene expression. The ends of eukaryotic chromosomes are capped by long stretches of repetitive sequences called **telomeres**, which do not code for any gene product but are present to prevent degradation of coding and regulatory regions during DNA replication. The length of the telomeres tends to decrease each time as the genome being replicated in preparation for cell division; the loss of telomeres has been proposed as an explanation for **cellular senescence**, or the loss of the ability to make cell divided.

Whereas the chromosomes of prokaryotes are relatively gene-dense, those of eukaryotes often contain so-called "**junk DNA**", or large regions of DNA that serve no obvious function. Simple single-celled eukaryotes have relatively small amounts of such DNA, whereas the genomes of complex multicellular organisms, including humans, contain an absolute majority of DNA without identified functions. However, it now appears that although protein-coding DNA makes up less than 2% of the human genome, about 80% of the bases in the genome might have some other functions, so the term "junk DNA" may be a misnomer.

Episomal DNA

Episomal DNA usually means the extrachromosomal DNA, which exists in the form of plasmid as free molecules in cytoplasm, or exists within some cellular organs such as mitochondrial DNA and chloroplast DNA.

With one exception only, the double stranded RNA plasmid in yeast **cells**, most plasmids are supercoiled circular DNA, such as those existing in most bacteria and in some species of eukaryotes. Under normal circumstances a plasmid is dispensable to its host, however, many plasmids carry genes that may be essential to host in certain environments. For example, the genes for antibiotic resistance are usually encoded on bacterial plasmids, which help its host to survive in the presence of an antibiotic. Some plasmids carry one or more genes, such as the *leu* gene in F′ plasmid, originally acquired from the host chromosome; and the **transferrable plasmids** can be passed between individual cells, even those of different species. Plasmids are often detected by the genes that they carry. Although some simple eukaryotes also possess plasmids with small numbers of genes, the majority of eukaryotic genes are stored on multiple linear chromosomes.

Plasmids normally use most of the replication machinery of their hosts, however each type of plasmids has its own regulatory function to control the copies of replication and the time required for synthesis. The high-copy-number (10～100, even higher) plasmids are usually welcomed by genetic engineers for it can be easily isolated. Many types of plasmids are found in *E. coli*, but three main types—the F, R, and Col plasmids—have been extensively studied, and made uses in molecular biological researches and in exploitation of cloning vectors:

(1) F, the sex plasmid, is able to transfer its copy to a host cell lacking the plasmid, and is also able in some cases to transfer chromosomal genes;

(2) R, the drug resistance plasmid, also has the ability to transfer the resistance to cells which lacks R;

(3) Col, the plasmid coding for **colicinogenic factor**, is able to synthesize **colicins** which is capable of killing closely related bacterial strains which lack Col plasmid. The

plasmids carrying other types of resistance genes have also been found in also in Gram positive bacteria.

The presence of plasmid DNA in a single bacterial colony can be detected by an elegant electrophoresis. A single colony is taken and grown for enlarging cell numbers, then make cells lysed and subjected to the separation from chromosomal DNA through the agarose gel electrophoresis. The host chromosome DNA is very large and can not penetrate the gel matrix, but the smaller plasmid, if present, will form a band at a position in the gel characteristic of its molecular weight. By this method, the molecular weights of plasmid DNA examined range from about 10^6 for the smallest one to slightly more than 10^8 for the largest one. Dimers or polymers of plasmids are frequently found in smaller plasmid containing bacteria, the forms are usually generated by genetic recombination. And smaller plasmids are usually non-transferrable, for at least 19 genes are needed for transfer in the case of F plasmid.

Both mitochondria and chloroplast contain DNA in the form of a circular DNA molecule, presumably derived from the ancestral bacterial chromosome. Over evolutionary time, these organelle genomes lost many genes that were unnecessary for life as an organelle inside a host cell. In addition, many genes that are still necessary have been transferred to the chromosomes in the nucleus. Consequently, the mitochondria of animals and eukaryotic microbes, have very little DNA left. For example, human mitochondrial DNA has only 13 protein-encoding genes, together with genes for several rRNA and tRNA molecules. However, about 400 different proteins are present in mitochondria. The genes for most of these reside in the nucleus and these polypeptides are imported into the organelle after synthesis on the ribosomes of the eukaryotic cytoplasm.

The chloroplasts of higher plants contain much more DNA than mitochondria, approximately enough for a hundred genes, but this is still much less than their bacterial ancestors. Chloroplast DNA also encodes for the proteins and some functions related to chloroplasts.

As the episomal DNA, only plasmid molecules are thoroughly researched and exploited to make very useful vectors for cloning.

2.2 Basic Knowledge of Gene

2.2.1 Different patterns of gene

Overlapped genes in viral genomes

The virologists had found in many viral species that their genomes seemed too small to carry enough sequences for encoding its protein molecules. The overlapped gene was first found in phage ϕX174 genome, when the molecule weights of phage ϕX174 proteins were added up, the figure (Fig. 2.1) indicated that the number of amino acids contained in these proteins far exceeds the coding capacity of this very small genome. Analysis of ϕX174 genome revealed, however, that it has efficiently evolved to solve the problem by repeated use of the sequences in the genes that are translated in different reading frames, clearly showing how the genes are overlapped. Gene A involved in DNA synthesis contains gene A* which regulates host DNA synthesis; and within the gene A* sequence there is a small region of gene B responsible for capsid formation. Gene E encoding the function on host autolysis is shielded within gene D which encodes synthesis of another protein component of capsid. This ingenious arrangement allows the D to express in assemblage of capsid while E is hidden, expressed only when the D finishes expression.

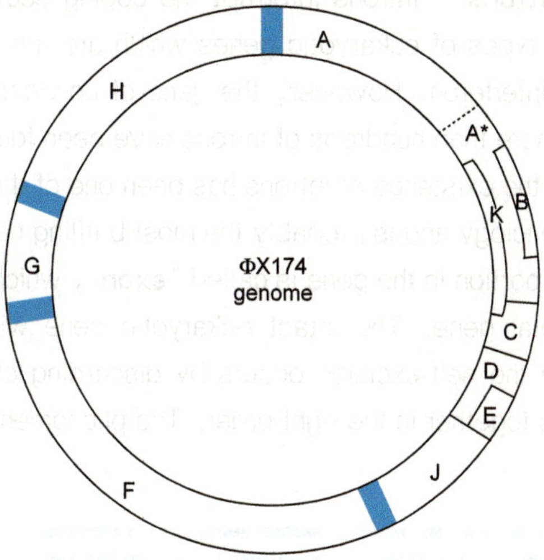

Fig. 2.1 Overlapped genes in ϕX174 genome

The pattern of **overlapped genes** explains the phenomenon of overloaded coding capacity in viruses, which offers as many genes as possible to the smallest life space, but leaves a big puzzle about how the viruses prevent the high risk of mutation, since a mere mutant base is enough to cause several overlapped genes to lose their functions. So far, none of the cellular organisms has been found to have overlapped genes, and only the slightly overlapped **polycistrons** in prokaryotes have been reported. During the long evolution course, the evolving cellular organisms have not inherited the pattern of overlapped gene, which may indicate that it's not worth taking so high a risk to reduce genomic sizes.

Intact and high-dense genes in prokaryotes

Various genetic techniques allow scientists to estimate the number of genes in an organism, by compared to the DNA content of one cell, in order to estimate the fraction of coding sequence. For instance, it is estimated that *E. coli* contains around 1800 genes, if a typical protein contains 500 amino acids, 2000bp (including the upstream and downstream sequences flanking the coding part) coding for a protein on average, at least 3.6×10^6 base pairs for 1800 proteins. An *E.coli* chromosome contains 8×10^6bp, so roughly half of the genome would encode proteins. In the known prokaryotic species, the genes are arranged in a high-dense form, neither overlapped nor interrupted.

Intervening gene

All eukaryotic genes are monocistronic genes. Usually the length of eukaryotic genes is much longer than prokaryotic ones, in the way of the presence of **intervening sequences**, called "**introns**". Introns interrupt the coding sections into many unlinked parts. There are a few types of eukaryotic genes which are not intervening genes, such as, the gene coding α-interferon. However, the gene of collagen contains 52 introns, in some extreme cases, more than hundreds of introns have been found in a single eukaryotic gene. The discovery of the existence of introns has been one of the major benefits from the recombination DNA technology and is probably the most baffling new phenomenon revealed in 1970s. The encoding portion in the gene is called "**exon**", which accounts for only about 10% to 50% of the total gene. The intact eukaryotic gene will be transcribed into a primary transcript, then the self-excision occurs by discarding of non-coding sections to link mRNA exon portions together in the right order. The phenomenon is named self-splicing (Fig. 2. 2).

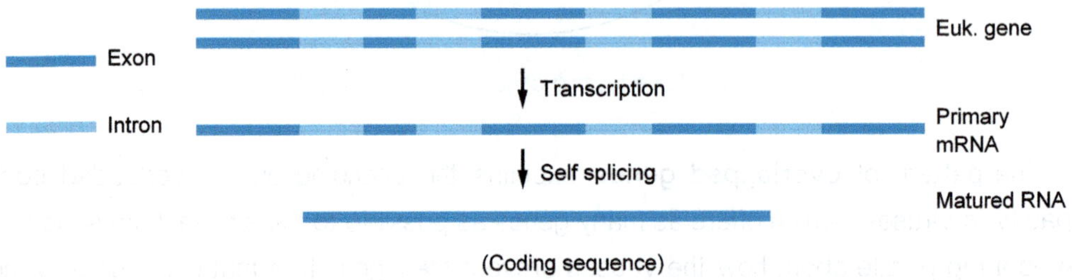

Fig. 2.2 Self-splicing

The overall biological significance of splicing remains to be studied in the long run. For it seems easier to synthesize mRNA directly without such highly risky splicing process. Many reports have suggested that the form of intervening genes serves a fundamental role in the regulation of gene expression, even cell differentiation in animal. The different modes of splicing on the same primary transcript might produce different matured mRNA molecules, and offer flexible encoding information for eukaryotes to handle with their complicated life cycle. The so called **alterative splicing** is believed to offer great opportunities for the organisms to accept evolutional chances.

2.2.2 How genes encode peptide

Although some of the genes encode rRNA, tRNA and other small molecular RNAs, most of genes encode peptide sequences with the intermediation of mRNA. The concept of relating protein peptide structure with gene function could be traced back to early 1900s, when A. Garrod figured out that one the human diseases called **alkaptonuria** was caused by genetically lacking of certain enzyme to catalyze metabolic reaction in patients' body. There were G. W. Beadle and E. L. Tatum who, for the first time, put forward the "**one gene-one enzyme hypothesis**". In this hypothesis, every metabolic reaction would be catalyzed by a particular enzyme which is the encoding product of a specific gene. Once a gene mutated, the related enzyme might lose its function because of the structural change. The "one gene-one enzyme hypothesis" didn't explain how a gene directs the peptide synthesis. Until 1957, V. M. Ingram in Cambridge had compared the sick hemoglobin from the patient who suffered **sickle cell anemia** to the normal one. He found that the 6th amino acid in hemoglobin peptide was no longer the Glu, but Val. Only one base pair different in the gene had caused the peptide sequence to change in one site, which had altered the hemoglobin structure and subsequently its function. The "one gene-one enzyme hypothesis" had been proved and revised as the "one gene-one peptide chain hypothesis".

In 1958, F. Crick put forward the "**Central Dogma**" **theory** to illustrate the relationships among DNA, RNA and protein, after he integrated large amount of data involved in genetic information flows. According to this "Central Dogma", the genetic information flow runs only from DNA to RNA then to peptide, not the other way around. In 1970, H. M. Temin had found that some of viruses have RNA genome, use RNA as template to synthesize complementary DNA strand, and then use DNA as a second template to make new viral genome. Based on this discovery, F. Crick has offered the "Central Dogma" a new revised form to illustrate the revetro-transcription from RNA to DNA.

The coding system revealing is almost a math deduction game, for there are altogether 20 different amino acids, 2-base coding for 1 amino acid would not meet the requirement, 3-base coding pattern sounded reasonable. Every amino acid is coded by different **triple codons** which were finally proved by genetic experiment; some of the amino acids might have more than one set of triple codons (Fig. 2.3). Fortunately, all the organisms on the earth use the same system of triple codon, except the slightly different coding triplet that exists in mitochondrial and chloroplast DNA. Obviously, the universal genetic codon system offers us a strong background to perform the gene cloning in a trans-species manner.

First position (5'end)	Second position U	C	A	G	Third position (3' end)
U	Phe	Ser	Tyr	Cys	U
	Phe	Ser	Tyr	Cys	C
	Leu	Ser	Stop	Stop	A
	Leu	Ser	Stop	Trp	G
C	Leu	Pro	His	Arg	U
	Leu	Pro	His	Arg	C
	Leu	Pro	Gln	Arg	A
	Leu	Pro	Gln	Arg	G
A	Ile	Thr	Asn	Ser	U
	Ile	Thr	Asn	Ser	C
	Ile	Thr	Lys	Arg	A
	Met	Thr	Lys	Arg	G
G	Val	Ala	Asp	Gly	U
	Val	Ala	Asp	Gly	C
	Val	Ala	Glu	Gly	A
	Val	Ala	Glu	Gly	G

Fig. 2.3 Amino acids are coded by different triple codons

2.2.3 Arrangement of genes in genome

How genes exist in genome

Coding and non-coding sections According to classic genetic perception, chromosomal DNA molecules are considered as kinds of a linear rope stringing the genes together, between the individual genes there are non-genetic materials connecting them one by one; The safer exchange of genetic segments could only occur between the intervals of genes but not within the genes. Genes were not only regarded as a genetic unit, but also an exchanging unit and a mutable unit. In 1955, S. Benzer had invented a method called intragenic mapping which helps scientists to understand that the smallest units of exchange and mutation would be an individual DNA base pair.

Alone with the development of gene cloning and DNA sequencing technology since mid-1970s, scientists finally stepped onto a new platform which allow the gene and genome to be dissected in a very defined way. As matter of fact, in all the eukaryotic, prokaryotic and viral genomes, there are always two major sections that could be identified: the transcriptable section usually called gene, and the portions beyond genes normally dressed "**spacers**". In spacer portion, the sequence playing no role for coding peptide does not mean it has no functions on the chromosome, several parts of spacer regions have been revealed as the back-bond to support the interaction of DNA region with the **scarford** RNA and protein molecules.

Whereas the chromosomes of prokaryotes are relatively **dense-gene,** in eukaryotes, it seems that the more highly evolved organisms comprise a smaller proportion of gene sequences. Genetic data for the **fruit fly *Drosophila*** suggest that it possesses about twice as many genes as *E. coli*, however, *Drosophila* genomic DNA is 20 times as much, therefore, only about 5% of the DNA consists of coding sequences. For years, we've known that only 1.5% of the human genome actually contains instructions for making proteins, the molecular - workhorses of our cells. But new data has shown that the rest of the genome (non coding majority) is still rife with "functional elements". That is, it's doing something. It contains docking sites where proteins can stick and switch genes on or off. Or it is read and "transcribed" into molecules of RNA. Or it controls nearby genes are transcribed (promoters; more than 70,000 of these). Or it affects how DNA is folded and packaged.

Mono- or poly- cistron The more genomic sequences were revealed, the more complicated we've learned about gene organization. Defining of a gene is done by carefully searching the sequence around the target area, since the sequences around the upstream and downstream parts of a gene segment carry many essential signals for gene expression, which must be included into a gene' territory. Many experts have described a gene as **a transcriptional** and **translational unit,** which contains the signal regions as well as the coding section for peptide expression. The definition should be revised after the concept of "polycistron" had been established by French scientists F. Jocob & J. Monod, who published the famous ***lac operon*** theory in 1960, revealing that 3 genes involved in lactose metabolism in *E. coli* cells being transcribed in one polycistronic unit. Later, varieties of prokaryotic polycistrons had been found, the results indicated in prokaryotic cells the genes coding for a group of enzymes related in a cascaded metabolic reaction, might be arranged in a polycistron form, and shared only one mRNA molecule, but being translated from this giant mRNA according the contained several translational signals, into a set of different peptides. Quite different in eukaryotic cells, so far, polycistrons are not found in those organisms; in the case of monocistron, the original definition of a gene needs no revision.

Gene copies In terms of the numbers of gene copy, it quite depends on the function of gene coding products. For most of the ordinal protein genes, one or two gene copies existing in their haploid genome would be enough for the protein synthesis, because the protein molecule might be rarely used or needed in a very small amount. According to human genomic studies, around 70% of the human genes belong to single-copy gene, and many of genes are not expressed all the time. Some of the genes, in contrast, exist in many copies, for example, the toad rRNA gene clusters might exist in more than 10,000 copies during the reproduction period. The synthesis of some fundamental facilities in the eukaryotic cells, such as histones, rRNA, tRNA molecules, and r-protein units are of vitally importance to life, and one can imagine that they should be guaranteed with sufficient and prompt supplies. The form of multi-copy-gene meets the special requirement.

Gene family In prokaryotes usually closely related genes are organized in operons and are transcribed in a polycistronic mRNA. The phenomenon is not the same in eukaryotes, but many related eukaryotic genes are functionally grouped as a set of genes called a gene family. The genes of histone members as well as rRNA members are grouped into gene family. A simple **multigene family** is one in which a few genes are arranged in a tandem array. The typical example is the set of genes for 5.8S, 18S and 28S rRNA of **toad Xenopus**, which comprises a single transcription unit containing all three rRNA sequences. The matured 5.8 S, 18S and 28S rRNA molecules are cleaved from the transcript by processing enzymes. The copy numbers of this three-gene family ranges from a few hundred to more than ten thousand per **diploid** cell of various types.

Activity of members of the set is usually but not always correlated. Recombinant DNA technology have been applied to clone the gene families and on occasion It has been found that all genes in a family can be cloned on a single plasmid, which indicates that such a family comprises a **gene cluster**. This recovery is useful in determining the order of the genes. Cloning results have also shown that many genes whose products are related but not coordinately regulated are also clustered. The term "gene family" has been expanded to include any gene cluster of functionally related genes.

Jumping gene (Transposon) The phenomenon of **jumping gene** was first reported in 1951, by Dr. McClintock in Cold Spring Harbor Laboratory after she studied maize genetics profoundly, and she was nominated as one of the Nobel Laureates in 1983. This process is usually detected as the appearance of a linear segment of DNA, a jumping gene or called later a **transposable element**, already contained in a genome at an original site of the same cell, it is usually called **transposon**. The term is a misnomer, because when transposition occurs, one copy of the transposon remains at the original site and a second copy appears at a new site. The transposable elements physically discovered in 1967 were the **IS elements** which caused **polar mutation** in *E. coli gal* and *lac* operons. Since then, many kinds of transposons had been discovered from prokaryotic, eukaryotic and viral genomes.

Recent studies revealed that transposons comprise a major component of eukaryotic genes, yet it remains controversial as to whether they are merely **genetic parasites** or instead some significant contributions to the eukaryotic genetic functions and their evolution. In plants, thousands of transposons were recently found to contain duplicated chromosomal fragments, the process was termed as "**transduplication**", although transduplication had been once considered as a potential rich source of novel segments in genome, finally the sequencing results revealed that they all appeared to be **pseudogenes** in rice genome.

An important observation on the base sequence of transposons is the presence of inverted repeat sequences at the termini of the elements, which plays critical role in the transposition mechanism.

In spite of eukaryotic transposable elements, from early studies transposons had been divided into several classes: 1) IS element and the composite structures containing them, 2) Tn3 class, 3) phage Mu class, and 4) the others. The Tn3 family is most important to the genetic engineers, for their structures are similar, which carry three genes, one encoding **β-lactamase** (conferring resistance to **ampicillin**) and two others for transposition process. The Tn3 family consists of rather large elements (about 5000bp) containing inverted repeats of 38bp, and none of them are flanked by IS-like elements. Tn3 family had been studied and cloned successfully; its ampicillin resistant gene is now cloned into many prokaryotic vectors.

Pseudogenes **Pseudogenes** are genomic DNA sequences similar to normal genes but non-functional, and they are regarded as defunct relatives of functional genes. It has been found in some cases that pseudogenes might result from the lost essential signals for gene expression, and in the other cases, from the mutation that cause the genes to become silent. Pseudogenes are sometimes referred as **gene remains**, since its genetic function might be replaced by the other members of the gene family, it remains in the genome and avoids death fatal. Since the sequences between pseudogene and its relatives are somehow similar, the phenomenon disturbs the genomic blotting, for the probe may hardly distinguish the slightly sequence difference.

2.2.4 Special sequences in eukaryotic genome

Satellite DNA

The percentage of G + C in different genomes varies widely throughout the living world, the extremes are 23% for *Clostridium* genus and 76% for *Sarcina* and almost 50% in *E. coli*. A convenient method to measure the base composition is by sedimentation of genomic DNA in buoyant CsCl solution. When some of the tested DNA had been broken randomly into about 1000 fragments, the sediment bands that resulted were very narrow and far from the major band, this indicated that the mean base composition of those fragments in the narrow band does not vary very much. This experiment was done for examining crab DNA, the observation that came out was very surprising, two discrete bands were observed, the main band that account for 70% of the total DNA was 52% G+C and the smaller band was 3% G+C, which was termed **satellite DNA**. Satellite DNA bands have been observed in many different organisms and may comprise 1～30 percent of their total genomic sequence. A critical characteristic of all satellite bands is that they contain repeated bases in genomic segment of long length. For example, in a cow a

1400bp sequence is repeated again and again to form satellite DNA band, and in a crab a single sequence of "AT" is repeated in 10^6 copies. Some of satellite DNA was suggested to have important role in eukaryotic mitosis.

Scattered repeated sequence

The **scattered highly repeated sequences** are so popular in all eukaryotes except unicellular organisms; however, these sequences do not show up as satellite DNA, for their base composition may not be different from the mean base composition of genome, for example, in human chromosome a famous *Alu* sequence (has *Alu* restriction site) is repeated in nearly 500,000 copies and scattered in many chromosomes, the number of the repeated copies is so great that one could figure out how those sequences may account for 20% of total genomic DNA. People had divided the repeated sequences into 3 classes: slightly repetitive (1 to 10 copies), middle repetitive (10 to several hundred copies) and highly repeated (several hundred to million copies). Some examples of the slightly repeated sequences are the histone genes and RNA genes. The middle repetitive sequences are generally not coding sequences but are thought to have regulatory roles in turning on and off the activity of many genes, although the significance of many repetitive sequences are still beyond explanation.

2.3 Sequence Signals Involved with Gene Expression

2.3.1 About replication origin

Replication origin

All known dsDNA molecules initiate a round of replication at a unique base sequence, called the **replication origin**, or **ori**. With A:T riched sequence, the **origin** is a frequently opening region, but specific to each organism, though there are some examples in which two organisms have the same ori.

θ and σ replication modes DNA replication can occur in two topological forms: θ and σ replication. When the replicating circle is schematically like the Greek letter θ (theta), this mode of replication is usually called θ replication. In a replicating DNA molecule, the region in which parental strands are separating and new strands are being synthesized is called a replication fork. θ mode can have either one or two **replication forks**, as shown in (Fig. 2.4).

The replication of some viral circular DNA genomes adopts another topological mode schematically like the Greek letter σ, which is also called **rolling-circle replication**. In this process, replication starts with a single-stranded cleavage in a double-stranded circle. This cleavage produces two chemically distinct ends: a 3′end and a 5′end. The

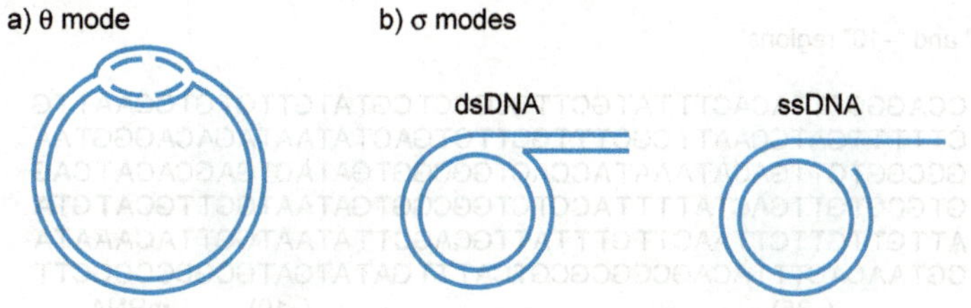

Fig. 2.4 θ and σ replication

DNA is synthesized by the addition of successive dNTPs to its 3′end with simultaneous displacement of the 5′end from the circle. As replication proceeds around the circle, the 5′ end "rolls out" as a tail of increasing length. If only the leading strand is being elongated, the replication will result in the formation of ssDNA genomic concatemer.

Termination of replication

There are different termination patterns: 1) For circular DNA molecules, in a unidirectional replication, one run of DNA synthesis terminates at origin, in a bidirectional replication, either there is a defined termination sequence, or two growing points collide and termination occurs wherever the collision point happens to be. 2) For linear DNA molecule, two lagging strands at two free ends might not finish completely the replication, such as telomeres in human chromosome, which causes serious problem even involving aging phenomenon. 3) Termination has another topological problem when circular dsDNA replicates semi conservatively, and the result is a pair of circles that are linked as in a chain, called catenane. Bacteria use **topoisomerase** to perform the separation of two catenated molecules. 4) Many viral DNA genomes replicate in the form of "rolling circle", producing multimeric branch from rolling circle first, then be cut at the adjunction by so called "**terminase**" into the monomeric genome, such as in the case of λ phage.

2.3.2 Transcriptional signals

Transcription is carried out by RNA polymerase. In prokaryotic cells there is only one kind of RNA polymerase, but more than one in eukaryotic cells. The *E. coli* RNA polymerase consists of five subunits, two α units, and one each of types β, β′, and σ. The term **core enzyme** is used to describe the σ-free unit, for σ subunit dissociate from the **holo-enzyme** easily and only plays the role of leading the holo-enzyme to lie on the precise region to initiate transcription.

Promotor

The first step in transcription is binding RNA polymerase to the upstream region of a gene. The particular binding sites are called "**promotor**" which are specific sequences of

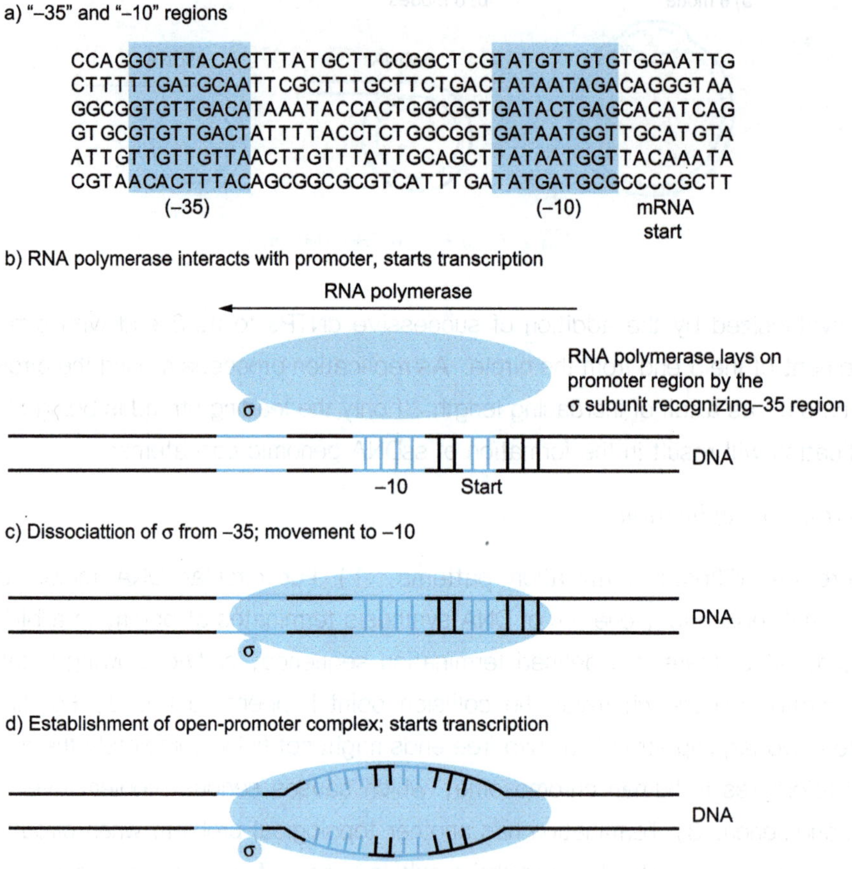

Fig 2.5 Transcription signal and mechanism

more than 100 bases at which several interactions might occur. Promoters exist at the upstream of " – region". Polymerase must recognize a promoter, attach in a proper configuration, open the dsDNA region to gain access to the bases to be copied, and then initiate peptide synthesis. There are some sequence similarities among prokaryotic promoters(**Fig 2.5 a**), so called " – 35" region is rich in T and A, and " – 10"region rich in T, A, and G, which is also called "Pribnow box". " – 35" box is the region for σ subunit recognition, and " – 10" is the region for RNA polymerase core enzyme to open the dsDNA and start transcription. The transcription efficiency is quite depending on how the base sequence is arranged in " – 35" and " – 10". For cross-specie cloning works, a strong native promoter or an engineered promoter must be assembled into the vector to promote a successful transcription of the desired clone.

Termination signal of prokaryotic transcription

For a long time there was a puzzle of how the RNA polymerase disassociates the template and finishes the RNA synthesis. There must be some specific sequences existing somewhere around the downstream part of the genes. Sequencing information from those related regions has revealed at least: 1) there is an inverted-repeat bases arrangement which is capable of intrastrand base pairing, to form a "stem-and-loop" configuration in the RNA transcript (Fig. 2.6); 2) the stem part is rich in G : C pair and the loop part is rich in base U; 3) a third region is a sequence of A : T pair in DNA which yields a sequence of 6~8 uracils at 3′end of prokaryotic RNA molecules.

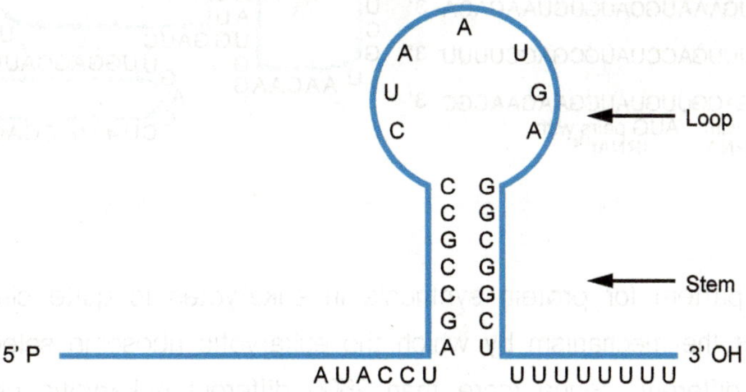

Fig. 2.6 "Stem and loop" structure for terminating transcription

In some of bacterial species, stem-loop *structure* might be the only mechanism of RNA synthesis termination and does not have an absolute requirement for Rho factor, but others might need both mechanisms, for instance in the case of λ phage. Rho-dependent termination is very important in regulatory processes, because the activity of termination can be switched off by regulatory proteins called antiterminators.

2.3.3 Signals for translational initiation

The sequence AUG in mRNA serves both as an initiation codon and Met codon. Since Met occurs within peptide chains there must be some signal that identifies a particular AUG triplets as start codon. In 1975, an attractive hypothesis was put forward by Australia scholars J. Shine and L. Dalgarno to explain that the prokaryotic translation is initiated at the **purine** rich region (5~7 bases) at the upstream (3~11 bases distance) to the initiation codon in mRNA and the complementary region of **pyrimidine** (5~7 bases) near 3′end of 16S rRNA. Later, this hypothesis has been proved by abundant sequencing results, and the purine rich region in the initiator region of mRNA was named Shine-Dalgarno site (S.D. site)(Fig. 2.7). Even prokaryotic phages have S.D. site in their mRNA. Different bacterial and phage species have been examined; their length and strength of base-paired regions were not the same, that's why different mRNA molecules are varied in frequency of translational initiation.

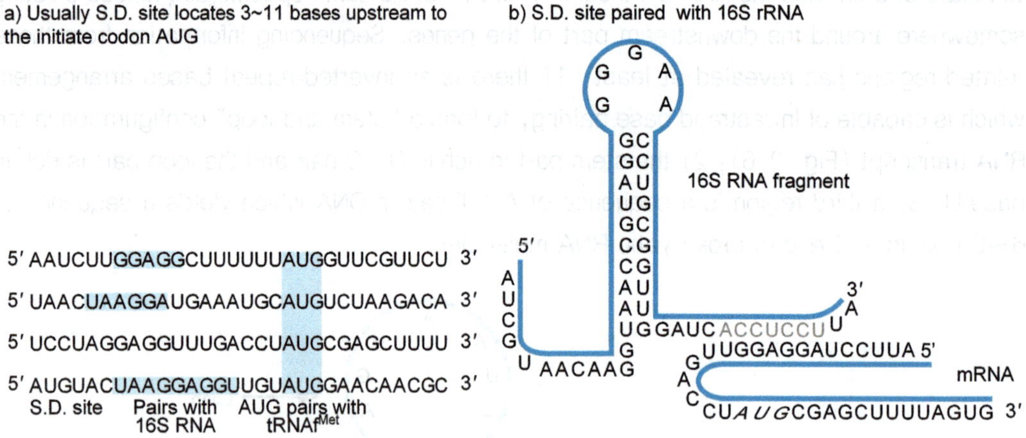

Fig. 2.7 Mechanism of S. D. site as initiation region for translation

The basic pattern for protein synthesis in eukaryotes is quite similar to that for prokaryotes, but the mechanism by which the eukaryotic ribosome selects the initiation codon is quite different. Since more than 2500 different eukaryotic genes have been sequenced and compared with their mRNA 5′ end sequences, the bioinformatics data shows: about 24% of the mRNA contains TATA box at the upstream (24～36 bases) to the initiation codon, the location of TATA seems relatively concentrated in a narrow section; around 47% contains GCGC box at the upstream (23～128 bases) and CAAT box (51～159 bases) to the first codon in mRNA. There is no evidence that similar base-pairing happens between those special regions and their rRNA molecules, the scanning hypothesis or the idea is that a ribosomal subunit attaches to the mRNA at or near its 5′ end and then drifts long the mRNA in the 5′ to 3′ direction until it encounters a AUG codon.

Now, it's easy to insert a short S. D. site sequence into a bacterial cloning vector to promote the expression of a foreign gene in prokaryotic host. Sometimes, if needed, translational initiation region can be altered through *in vitro* mutagenesis to enhance or attenuate the frequency of translation for a particular gene.

2.4 About Genomic and Metagenomic Studies

Genomic project

Since F. Sanger & W. Gilbert published their innovative results of DNA sequencing techniques, huge amount of genes have been sequenced and recorded in gene banks, and the genomic sequences of many organisms have been worked out in detail. It's all known now that a single copy of human genome comprises of more than 3 billion base pairs. Recent years, so called "post genomics study" has become active, varieties of software have been developed which help blast up as many information from the "Gen-bank" as possible, and dig out secondary information about life's secrets.

Metagenomic study

In 1998, Handelsman gave his idea on study of the genomes of total microbes found in nature, and he first raised the conception of "**metagenome**". "Meta-" might means "overreach", "above", "beyond", and "higher level". Since then, many scholars switch their focuses from the study of individual genome to higher disciplinary survey, sequencing, blast, and establishment of meta-genomic database. Metagenomic study might help human being to understand broadly the mass life phenomenon, reveal the unknown ecological situation, and recover the rest species that have not been screened out from the nature. Novel strains and their special products are expected to have totally new functions which might be valuable to human society.

Summary

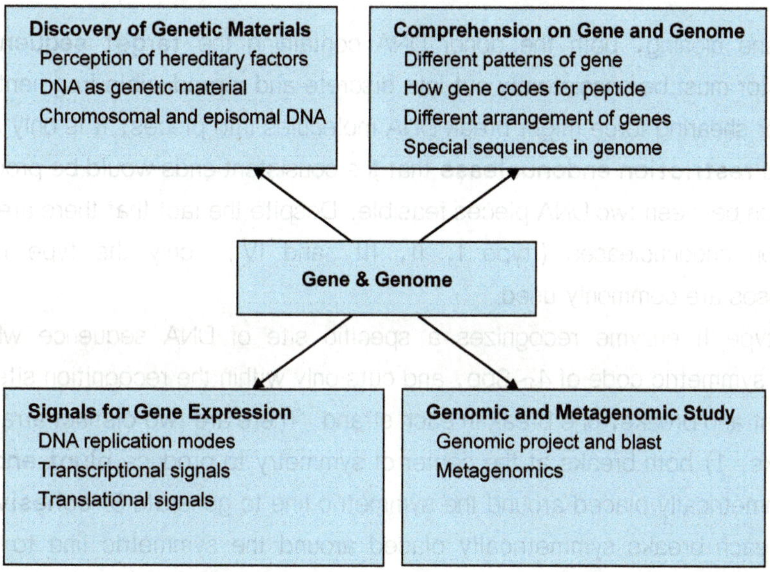

Chapter 3　Basic Techniques for Experiments with DNA

从某种角度出发，进行基因工程有点像进行一台外科手术，虽然手术台上的操作本身是至关重要的环节，但在手术之前，医生需要对手术对象进行详细的检测，对病灶进行预处理，还要有的放矢地为之准备各种外科器械、药物试剂，以及专门的医疗设备。在开展基因克隆的主要步骤之前，人们也往往需要选择各种适合的酶分子工具和核酸分子工具，甚至各种标记物；需要预处理 DNA 分子；扩增 DNA 的拷贝数；分离、提纯目的 DNA 和载体 DNA 分子；鉴别 DNA 片段大小以及序列信息；甚至需要修饰目的 DNA 片段。以下的章节有助于我们了解各种常规的 DNA 处理方法，这些基本方法与技术的建立大大促进了基因工程的发展。

3.1　Enzymatic Tools for Gene Manipulation

3.1.1　DNA restriction endonucleases

For gene cloning, both the donor DNA containing the **target sequence** and the cloning vector must be consistently cut into discrete and reproducible fragments. Although ultrasonic or shearing force might break DNA molecules into pieces, it is only after cutting by bacterial **restriction endonuclease** that the consistent ends would be produced which make ligation between two DNA pieces feasible. Despite the fact that there are many kinds of restriction endonucleases (type Ⅰ, Ⅱ, Ⅲ, and Ⅳ), only the type Ⅱ restriction endonucleases are commonly used.

Each type Ⅱ enzyme recognizes a specific site of DNA sequence which usually consists of symmetric code of 4～8bp, and cuts only within the recognition site to produce two single strand breaks, one break in each strand. There are two distinct arrangements of these breaks: 1) both breaks at the center of symmetry to produce **blunt ends**, 2) each breaks symmetrically placed around the symmetric line to generate 5′ **cohesive (sticky) ends**, 3) each breaks symmetrically placed around the symmetric line to generate 3′ cohesive (sticky) ends. An example of each type is shown in (Fig. 3.1).

Hundreds of different restriction endonucleases are commercially available today. Restriction enzymes are named according to the bacterial producer from which they were isolated. The generic and specific names of the organism in which the enzyme is found are used to provide the first part of the designation, which comprises the first letter of the generic name and the first two letters of the specie name. Thus, a restriction enzyme from a strain of *Escherichia coli* is termed *Eco*, the one from ***Bacillus amyloliquefaciens*** is termed *Bam*, and so on. Further descriptors may be added, depending on the symbol of

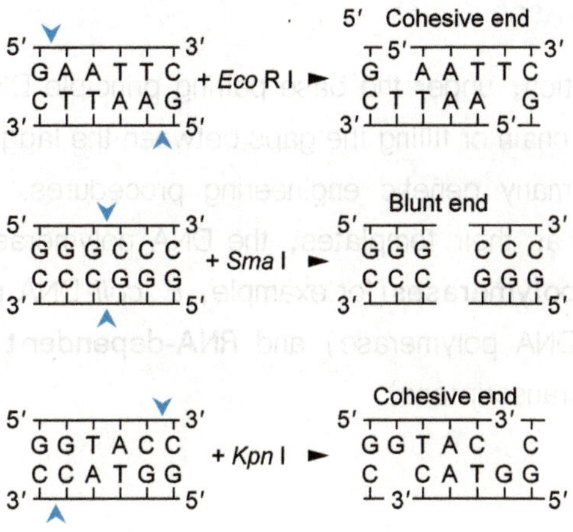

Fig. 3.1 Restriction enzyme cleavage

bacterial strain involved and on the presence or absence of extra-chromosomal elements, and Roman figures are used when more than one enzyme could be isolated from the same microbial strain, for example, *Eco*RI and *Eco*RV are different restriction enzymes, but from the same *E. coli* strain.

In practical terms, the restriction enzyme is incubated (usually 37℃) with DNA in a compatible buffer solution and let cut of the DNA into fragments of a defined size with consistent ends. The individual cutting condition is usually described by the commercial specification.

3.1.2 DNA ligases

DNA ligases are enzymes with synthetic function, which attach two pieces of DNA covalently to one another. Two different fragments can be paired by virtue of complementary cohesive ends, and one fragment can be circularized since its two termini are cohesive. The ligases are also able to ligate blunt ends but with lower efficiency, that reminds the manipulator to avoid the blunt ends ligation as possible.

T4 DNA ligase is the most commonly used enzyme, which is purified from *E. coli* cells infected with **bacteriophage T4**. The reaction occurs between the 3′ hydroxyl group of one DNA strand and the phosphate group of the partner strand. ATP provides the energy for the reaction. DNA ligation can occur efficiently only between compatible ends of DNA fragments, and this specificity gives the investigator some control over how different DNA fragments ligate together in a reaction mix.

The most common use of ligation is to incorporate a piece of donor DNA into a cloning vector. T4 DNA ligase works best at 37℃, but is sometimes used at much lower temperatures (4～15℃) to prevent **thermal denaturation** of the short base-paired regions that hold the cohesive ends of DNA molecules together.

3.1.3 DNA ploymerases

During DNA replication, under the base pairing principle DNA polymerases play the roles of elongating DNA chain or filling the gaps between the lagging strands, which would be frequently used in many genetic engineering procedures. Accounting for different nucleotides being used as their templates, the DNA polymerases can be classified as **DNA-dependent DNA polymerases** (for example, *E. coli* DNA polymerase I, T4 and T7 DNA polymerase, *Taq* DNA polymerase) and **RNA-dependent DNA polymerases** (for example, viral reverse transcriptase).

DNA polymerase I

E. coli DNA polymerase I consists only of one **single peptide** but folds into two **domains** to form a rather strange 3D structure. **Polymerase function** and 3′→5′ **exonuclease activity** exist in its large domain, the former activity catalyzes the deoxynucleotide polymerization by adding tri-phosphate nucleotides to the 3′ end of a primer and elongates the 2th DNA strand, and the latter activity catalyzes the clipping of the mismatched bases from 3′ direction to ensure only the right pairing happens and this is so called the "proof reading ability". In the small domain, the existing 5′→3′ **exonuclease activity** clips the unwanted bases from 5′ direction to help clean up the mis-pairing.

The 5′→3′ exonuclease activity at a single-stranded break (a nick) can occur simultaneously with polymerization. That is when a 5′-P nucleotide is removed, and the DNA synthesis for replacement can be made by the polymerizing activity. The nick moves along the direction of DNA synthesis, and the phenomenon is called **Nick Translation**.

Klenow fragment

The two peptide domains of DNA polymerase I can be cleaved by **proteolysis** with **trypsin** or **subtilisin**. Both proteases cleave at the amino acid 324 (**Fig. 3.2**). The

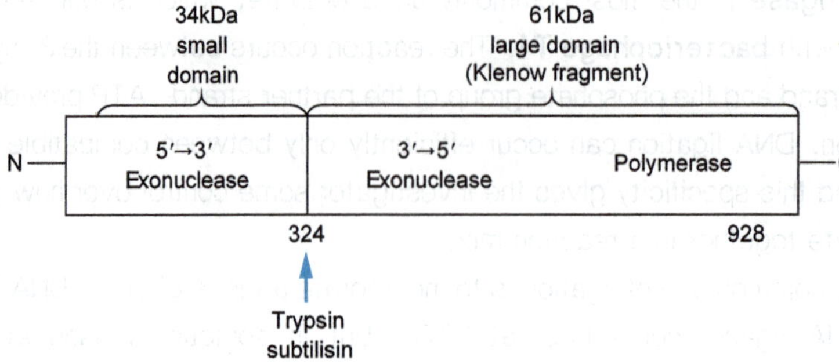

Fig. 3.2 DNA polymerase I

produced large domain known as the **Klenow fragment** contains only the DNA polymerase activity and 3′→5′ exonuclease activity, but lacks the 5′→3′ exonuclease activity. Klenow fragment are now prepared by the expression of the large domain sequence, and proved to be very useful for preparing the dsDNA from ssDNA template, adding radiolabels to the growing chain during primed synthesis and playing key roles in DNA sequencing based on Sanger-Coulson (dideoxy) method.

Taq DNA polymerase I

The activity of *Taq* DNA polymerase I is utilized in the **polymerase chain reaction** (PCR) for *in vitro* selective amplification of specific regions of a DNA molecule. *Taq* DNA polymerase I had been isolated from an extremely heat-tolerated bacteria named ***Thermus acquaticus***, which shows higher polymerase activity, contains no 3′→5′ exonuclease activity, and is more resistant to thermal denaturation than other kinds of DNA polymerases.

Bacteriophage T4 and T7 DNA polymerase

E. coli, infected by bacteriophage T4 or T7 produces viral DNA polymerases known as bacteriophage T4 or T7 DNA polymerase. T4 polymerase possesses a very active single-stranded 3′→5′ exonuclease (with activity of many folds stronger than that of the Klenow fragment), but lacks a 5′→3′ exonuclease activity. This enzyme is frequently used to fill 5′ protruding ends with labeled or unlabeled dNTPs or to generate blunt ends from DNA with 3′ overhangs.

T7 polymerase has a very high 3′→5′ exonuclease activity in addition to its polymerase activity. The enzyme used today is the one under ***in vitro*** mutagenesis; it has lower 3′→5′ exonuclease activity, higher processing activity, and faster polymerization rate, which make it a suitable tool used in Sanger' DNA sequencing.

Sequenase

Sequenase sounds very commercialized, and is implied as an engineered polymerase used only for Sanger's sequencing for its high fidelity. One would not allow any misreading of genomic information for it is possible that only one base mistake could ruin the whole cloning result.

Reverse transcriptase

Reverse transcriptase (RTase) exists in some of the virus particles which make ssRNA as their genome. In the infective cycle of those reverse-transcriptional viruses, the molecular biological life cycle is quite different. The viral ssRNA genome is usually attached by the reverse transcriptase, and both being injected into their host cells. Almost immediately upon entering the host cells, the RNA-dependent DNA polymerase activity starts the 1st DNA strand synthesis by pairing with the sequence of RNA template, then the enzyme turns around, clips off the RNA strand by their nuclease activity, and synthesizes the 2nd DNA strand by using the 1st DNA strand as its alterative template. The host polymerase is believed to have function on the synthesis of 2th DNA strand. It's not surprising that the viral primitive double DNA strand contains a hair pin structure at one of the ends, which functions the second priming activity and would be cleaved by some cellular endonuclease, to allow the further integration of provirus into the host chromosomal DNA.

For most of the eukaryotic genes are interruptive and their introns would be as many as hundred pieces, directed isolation of eukaryotic genes would not get rid of the non-coding sections. Therefore isolation of the matured mRNA molecules after self splicing process and making the matured mRNA reverse-transcripted into DNA would be the only choice. AMV reverse transcriptase had been isolated from avian blood cancer sufferers; the enzyme plays the critical roles in the synthesis of cDNA and construction of cDNA libraries.

3.1.4 Nucleases

The enzymes so called **nucleases** degrade nucleic acids by breaking the phosphodiester bond that holds the nucleotides together. Restriction enzymes are good examples of endonucleases, which cut within a dsDNA strands at specific sites, but **deoxyribonuclease I** (**DNase I**) cuts either single-stranded or double-stranded DNA at essentially random sites. DNase I is used to destroy DNA molecule or, in extremely low concentration, to nick the double strands DNA with only one or few cuts.

Other nucleases, which degrade DNA from the termini of the molecule, are known as **exonucleases**. Many useful exonucleases, for example, **exonuclease Bal 31**, **exonuclease III**, **S1-nuclease** (**nuclease cuts only single strand DNA**) and RNase H are often used as tools in genetic engineering. These enzymes differ in their precise mode of action and provide the genetic engineer with a variety of strategies for processing DNA or RNA molecules. Nuclease Bal 31 is a complex enzyme, when Bal 31 is present at a high concentration its activity effectively shorten DNA molecules no matter in single or double strand form, from both termini. Exonuclease III digests from 3′ ends of dsDNA molecules, and produces both protruding 5′ termini. Nuclease S1 is a single-stranded polynucleotide cutting enzyme, which cuts either on RNA or DNA, and either on single free ends or the looped unpaired section of polynucleotide. RNase H, there are members of the RNase H family, which can be found in nearly all organisms, from archaea to bacteria and eukaryotes. RNase H's ribonuclease activity cleaves only the RNA strand in the DNA-RNA hybrid molecule, but not the dsDNA or dsRNA molecules.

3.1.5 Other enzymes for DNA modifying

Recombinant DNA technology requires many different enzymes with various activities. Some of these are listed in Tab. 3.1.

Tab. 3.1 Some of the enzymes used for recombinant DNA technology

Enzyme	Activity
Alkaline phosphatase	Remove 5′ phosphate groups of DNA molecules
Polynucleotide kinase	Catalyzes the transfer of gamma phosphate group from ATP to the free hydroxyl end of the 5′ DNA or RNA. In this way the enzyme can also be used to add the radioactive labeling to a desired DNA sequence
Terminal transferase	Transfers uses tri-phosphate nucleotides as subtracts and adds the free nucleotides onto the 3′ terminus of a DNA strand. The enzyme has been used to make protruding ends of DNA molecules and in most of the cases to create homologous tailings for further ligation in early cloning time
DNase I	Degrades double-stranded DNA by hydrolyzing internal phosphodiester linkages
E. coil exonuclease III	Sequentially removes nucleotides from 3′OH ends of DNA molecules, except from protruding 3′ OH termini
Klenow fragment	Proteolytic product of *E. coil* DNA polymerase I that has both polymerase and 3′ exonuclease activities and no 5′ exonuclease activity because fractionation of the digestion products removes the fragment with the 5′ exonuclease activity; a Klenow fragment with only DNA polymerase activity due to a mutation in the 3′ exonuclease sequence is also available
Mung bean nuclease	Single-stranded DNA and RNA endonuclease
Nuclease BAL 3l	Degrades both 3′ and 5′ ends of DNA without internal cleavages
Poly(A)polymerase	Adds AMP from ATP to the 3′ end of mRNA
Reverse transcriptase	Retroviral RNA-directed DNA polymerase
RNase H	Degrades the RNA strand of a DNA-RNA hybrid molecule
S1 nuclease	Degrades single-stranded DNA
T4 polynucleotide kinase	Catalyzes the transfer of the terminal (γ) phosphate from a nucleoside 5′ triphosphate to a 5′ hydroxyl group of a polynucletide
T4 DNA polymerase	DNA polymerase and 3′ exonuclease activities
T7 DNA polymerase	DNA polymerase and 3′ exonuclease activities
Taq DNA polymerase	Heat-stable DNA polymerase from *Thermus aquaticus*
E. coli RNA polymerase	Consisting of 5 subunits, is the only enzyme performing RNA synthesis in the cells. Discovery of *E. coli* RNA polymerase wrapping up a region of 40bp around the upstream part of prokaryotic gene, had revealed the specific A:T rich sequence in the Pribnow Box
Phage T7 and T3 RNA polymerases	Catalyzing rapid transcription, are used for *in vitro* transcriptional assays on the cloned genes
β-Agarase I	Digests agarose; is used to retrieve separated DNA molecules from agarose gels

3.2 Oligo-nucleotide Tools

In some cases it's useful and some time necessary to add short DNA fragment with designed sequence to assist the gene manipulation. Alone with the advanced technique of DNA synthesis, one can easily prepare all kinds of **oligo-nucleotide tools** to assist the gene cloning.

3.2.1 Homologous oligonucleotide

At the very beginning of the cloning age, few DNA restriction endonucleases were available and the cutting ends between donor and vector might not be paired. Professor Jackson had invented a so called **homologous tailing technique**, by which donor and vector molecules had been terminally transferred with oligo-G and oligo-C respectively, and then the ligation had started as oligo-G paired with oligo-C.

Oligo-T are used for pairing with poly(A) tail of eukaryotic mRNA molecules, and it plays as the man-made primer to assist the revetro-transcriptation in cDNA synthesis.

3.2.2 Linkers and adaptors

Some of DNA restriction endo-nucleases produce blunt ends which would not be easily ligated compared to sticky ends ligation. This is possible if a designed short segment, called linker(Fig.3.3a), of 8~12 base pairs containing a cohesive restriction site is linked to the both ends of the blunt cut molecules, and at a later time, the linker sequences in between the blunt ends create new restriction site in the target DNA, that would facilitate cloning further treatment. For the type II restriction enzymes offer symmetric cuttings, there is no need to synthesize the 2nd oligo-nucleotide segment, and self pairing will happen when the single strand oligo-nucleotide molecules are treated with annealing steps. After gradually heating the linker molecules to 94℃ for a while, then cooling down to room temperature the paired linker molecules are ready for use.

For construction of circular DNA molecule form two segments by linkage with the same linker, there are two linkage directions, but only one would be correct. In this case, two linkers with different cutting sites will be used for the conjunction work to ensure the right ligating result.

Even more, people had constructed polylinker sequences which contain a dozen of restriction sites, and inserted those segments into some of the useful vectors, for example, M13 mp vectors. The vectors with polylinker sequences have greatly facilitated the genetic engineers.

Another choice is preparing **Adaptors**(Fig.3.3b), which was invented by Professor

Wu in Connell University, with the blunt site at one end, and sticky end at the other side. In order to prevent the self ligation between two sticky ends, the 5′ P group can be removed by alkaline phosphatase, and the resulted 5′-OH would not be ligated to the 3′-OH at the other end. At the moment, those adaptor segments can only be ligated to the blunt ends of target DNA. After phosphorylation to the 5′-OH of the protruding sticky ends by polynucleotide kinase, the normalized sticky ends are ready for the next ligation.

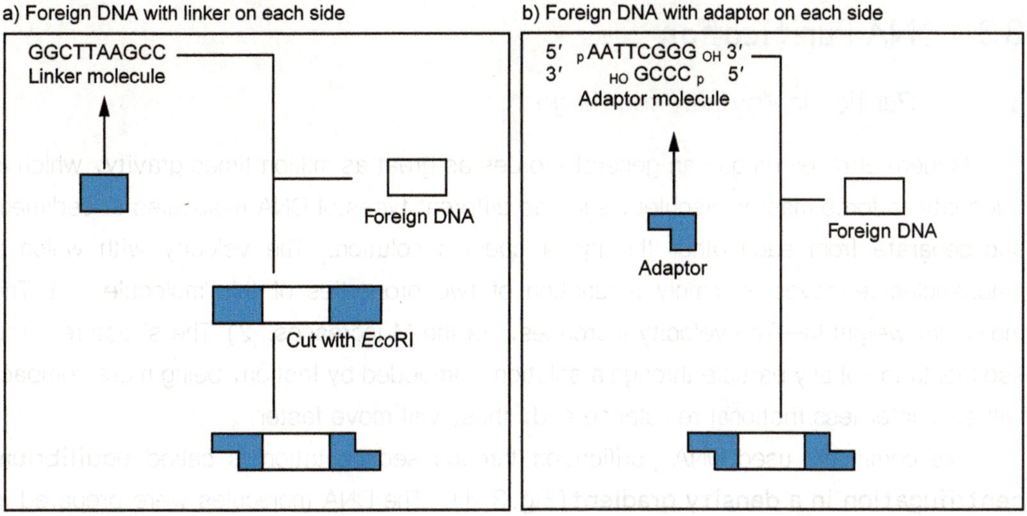

Fig. 3.3 The use of linker and adaptor for construction of DNA fragments with cohesive termini

3.2.3 DNA primer

Doesn't like the naturally occurred primer, the synthesized priming segments are no longer the RNA molecules which would be considered as less stable and less convenient to be synthesized. DNA priming molecules are normally synthesized according to real cases, and used mostly for PCR operation and fishing of target sequence. Based upon the accumulative genomic information and varieties of well programmed software, one could design the sequence of those priming molecules rationally according to the blasting results and make the DNA **primer** being exactly matched the target sequence. In most of the cases, those oligo-nucleotide tools are synthesized commercially following the defined sequence information on claim, and usually worked very well.

3.2.4 Uncompleted matched oligo-nucleotide

Sometimes, a DNA primer designed on purpose with **uncompleted matched oligonucleotide** can be used to initiate ***in vitro* mutagenesis**. This technique now becomes the most powerful method for substitution of one or few bases at a specific site in DNA molecules. The length of uncompleted matched oligonucleotide is normally 9～19 bases, which are matched to the single strand of target sequence except the middle part (1 or 2 bases). Using proof-reading activity weakened DNA polymerase I to protect uncomplete matched oligonucleotide, after several runs of DNA replication, the positive clones undergoing the **site directed mutagenesis** are screened.

The oligonucleotide tools had been further exploited into a new technique called "**PCR site directed mutagenesis**".

3.3 DNA Purification

3.3.1 Purification by ultracentrifuge

Modern ultracentrifuge can generate forces as great as million times gravity, which is adequate to force macromolecules, such as different types of DNA molecules to sediment and separate from each other through a specific solution. The velocity with which a macromolecule moves is mainly a function of two properties of this molecule: 1) The molecular weight M—The velocity increases, as the M increases. 2) The structure—The sedimentation of any particle through a solution is impeded by friction, being more compact will encounter less frictional resistance and, thus, will move faster.

The commonly used DNA purification through sedimentation is called **equilibrium centrifugation in a density gradient** (Fig. 3.4). The DNA molecules were prepared in the form of RNA & protein free samples. In order to separate the covalently closed circular DNA (cccDNA) from linear chromosomal fragments, CsCl is an ideal medium, for its solution can be adjusted to a density approximately equal to that of DNA molecules. Under powerful centrifugal force, Cs^+ and Cl^- ions sediment somewhat to achieve an equilibrium concentration distribution in which there exists a nearly linear concentration gradient. The supercoiled cccDNA molecule, like plasmid, would sediment faster than the linear DNA in this system. A chemical called **ethidium bromide** (**EB**), which can easily be inserted into DNA double strands and emit red fluorescent ray under UV light, is also added to the DNA sample. Linear DNA molecules would absorb more EB for they have open ends and can release the tension freely, but the cccDNA absorbs much less. By EB insertion, the diversity of the compactness of these two kinds of DNA molecules was enlarged.

After several hours centrifugation, cccDNA is well separated from linear chromosomal DNA. Further purification procedure would be very simple: Punching the tube at the top to let air flow in the tiny hole, then punch the bottom to collect the red bands. cccDNA must be in the lower band, for it moves faster. The salt and EB can be later removed by solvent extraction, the purified DNA samples are ready for any kinds of manipulation.

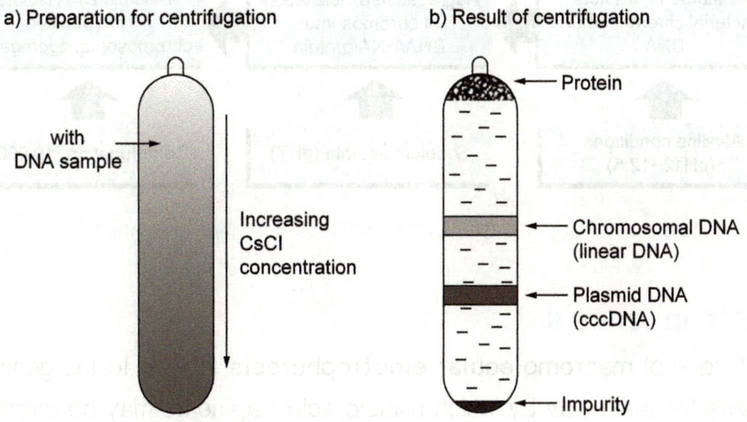

Fig. 3.4 Caesium chloride density gradient centrifugation

3.3.2 DNA purification of mini-sample

Based upon the topological diversity of cccDNA and linear DNA, some other simplified methods can be used for small scale plasmid purification(Fig.3.5).

(1) Alkaline denaturation of linear DNA. The extracted DNA sample is heated in an alkaline buffer of pH 12.0～12.5, alkali denatures the linear DNA but not the cccDNA molecule. The denatured DNA strand will interact with each other to form a loosely metric structure which can be spinned down with the other impurities.

(2) Plasmid purification through **lysozyme-heat treatment**. This is almost the simplest method for purifying *E. coli* cccDNA. Put 1.5mL *E. coli* culture into *Eppendorf*, spin down the cells, wash the cell suspension and repeat for 4times. Suspend the cells in 0.15mL Tris-EDTA buffer containing lysozyme, and put the eppendorf in boiling water for 40 seconds. Collect the lysate by centrifugation and purify the cccDNA from the lysate through electrophresis.

(3) Chromosomal DNA purification. Purification of chromosomal DNA involves gentle lysation of the cells by **lysozyme** and EDTA without heating. It's necessary to add RNase first to get rid of RNA pieces and later protease to digest proteins including RNase. all the peptide residual is further denatured and extracted away by phenol to ensure the chromosomal DNA is keeping off RNA and protein,and disperse well in the buffer. In the final step, cold ethanol of high purity is slowly added to the system, which causes chromosomal DNA precipitation, and being collected by centrifugation. In a more tricky technique, a clean glass bar is put into the DNA solution to guide ethanol draining in slowly, the dehydrated DNA molecule will be twined up around the spinning glass bar.

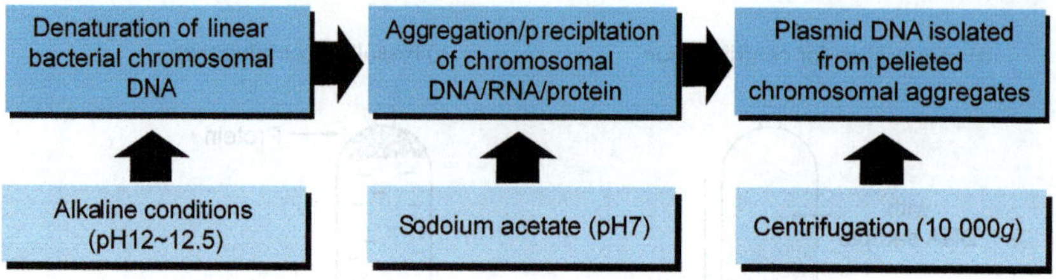

Fig. 3.5 Alkaline denaturation procedure for plasmid purification

3.4 Electrophoresis

The technique of **macromolecular electrophoresis** is vital to the genetic engineer, as it represents the main way by which nucleic acid fragments may be directly visualized for picking up or measuring of length. The method relies on the fact that nucleic acids are **polyanionic** at neutral pH; that is, they carry multiple negative charges from phosphate groups on the phosphodiester backbone of the nucleic acid strands. This means that the molecules will migrate towards the positive electrode when placed in an electric field. As the negative charges are distributed evenly along the DNA molecule, the charge/base ratio is constant; thus, mobility depends on the length of DNA fragment. The technique is carried out using a gel matrix, which separates the linear nucleic acid molecules according to their lengths.

3.4.1 Slab electrophoresis

In a routinely slab gel electrophoresis, usually a thin slab of an **agarose** or polyacrylamide gel is prepared containing small slots (also called **Wells**) into which samples are placed. The electric field applied to the gel pushes the negative charged DNA molecules penetrating and passing through the gel **matrix**. The migration rate increases as the DNA molecular weight, M, decreases. For unknown reasons the distance moved, D, depends logarithmically on M, obeying the simple equation: $D = a - b \log M$, in which a and b are experimentally determined constants that depend on gel matrix, electric voltage, buffer, temperature. Thus, the molecular weight of DNA samples can be measured by referring to the M and D of DNA molecular weight standard running in the slab.

Agarose electrophoresis

The property of gel matrix has important effects on the efficiency of DNA separation, which is critically dependent on the porosity of the matrix. Two gel types are commonly used: agarose and polyacrylamide. Agarose is extracted from seaweed and can be purchased as a dry powder, which can be easily melted in buffer at an appropriate concentration, normally in the range 0.3%~2.0% (m/V). On cooling, the agarose coagulates to form the gel. **Agarose gel electrophoresis** is usually run for separation of longer DNA fragments in a submerged way in the horizontal apparatus.

Polyacrylamide electrophoresis

Polyacrylamide-based gel electrophoresis (**PAGE**) is sometimes used to separate small pieces of nucleic acid molecules; in the applications such as DNA sequencing, the pore size required is much smaller than that achieved with agarose. PAGE electrophoresis is carried out by placing the nucleic acid samples in the well of gel, and applying an electric potential difference across it until a marker dye (usually **bromophenol blue**, added to the sample prior to loading) reaches the end of the gel. The band of running nucleic acids in the gel are usually visualized by being stained with the intercalating dye such as ethidium bromide, and then spotted under UV light. Nucleic acids show up as fluorescent orange bands, which can be photographed to provide a experimental record. The data can be used to estimate the sizes of unknown fragments by construction of a calibration curve using a set of molecular length standards of known sizes, as migration is inversely proportional to the log10 of the number of base pairs. This is particularly useful in DNA length measuring, and more often cooperated with the technique of restriction mapping.

In addition to its use in the analysis of nucleic acids, PAGE is performed extensively for analysis of proteins. The methodology is different from that used for nucleic acids, but the basic principles are similar. One common technique is protein SDS-PAGE, in which the detergent SDS (**sodium dodecyl sulphate**) is added to break disulfide bonds of proteins into the near-random coil form, and cover the protein molecules with evenly negative charges. In this way, the inherent charge of the protein is masked, and the charge/mass ratio becomes constant. Thus, proteins can be separated according to the peptide length in a similar way to DNA fragment separation (Fig. 3.6).

a) Apparatus

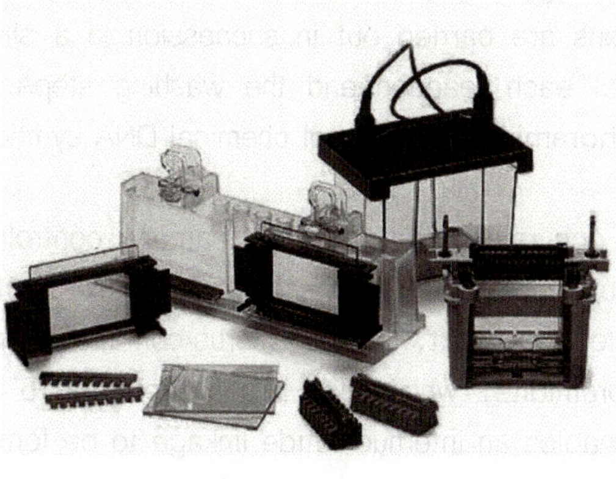

b) Protein SDS electropheresis

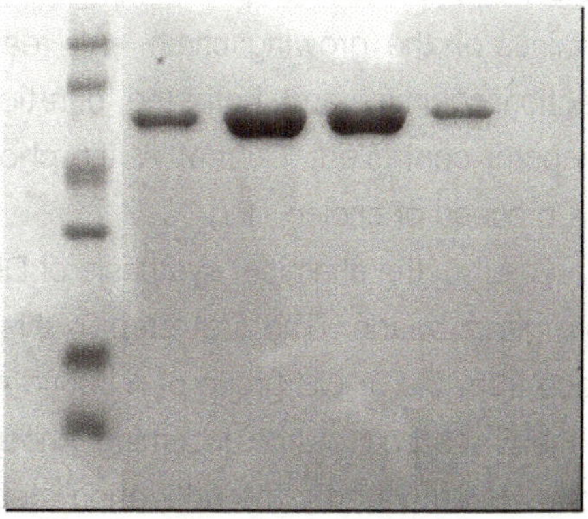

Fig. 3.6 SDS-PAGE

3.4.2 Capillary electrophoresis

In the early 1980s, a milestone for **column based electrophoresis** was set which introduced **capillary zone electrophoresis** (**CZE**) with on-column optical detection. The inner-diameter of CZE capillaries had being scaled down to 80μm, but the voltages applied had been as high as 30kV without overheating problem. Very small amount of sample is used and the separation time could be shortened to less than 30min. In the mid 1980s so called **fused silica capillary column** quickly arouse interest from biochemists and genetic engineers, for it results in ultra-high speed separation of either small or macromolecules, provides a solution to some formidable tasks including DNA sequencing and proteomic studies.

3.5 *In Vitro* Synthesis of DNA

***In vitro* DNA** synthesis can be performed by chemically linking nucleotides or by enzymatic polymerization. In reactions done with chemical reagents the DNA is single-stranded and is short because the process is not very efficient. Chemical synthesis of DNA is primarily used for making short primers or **oligonucleotides**.

3.5.1 Chemical synthesis of DNA

Making DNA chemically rather than biologically was one of the first new technologies to be applied by the biotechnology industry. The ability to make short synthetic stretches of DNA is crucial to using DNA replication in laboratory techniques. DNA polymerase cannot synthesize DNA without a free 3′-OH end to elongate. Therefore, to use DNA polymerase *in vitro*, researcher must supply a short primer.

Chemical DNA synthesis does not follow the biological direction of DNA synthesis; rather, in the chemical process, each incoming nucleotide is coupled to the 5′-hydroxyl terminus of the growing chain. All reactions are carried out in succession in a single reaction column, and both the duration of each reaction and the washing steps are computer controlled. Currently, the **phosphoramidite** method of chemical DNA synthesis is a process of choice (Fig. 3.7).

Briefly, the chemical synthesis of DNA occurs on a solid surface, namely, controlled-pore glass beads. The first step in this multistep process is the removal of a protecting group from the 5′-OH group of a bound nucleotide. Next, the phosphorus-containing group at the 3′-carbon of the incoming phosphoramidite, which also has a protected 5′-OH group, is activated. The activation step enables an internucleotide linkage to be formed between the 3′ end of an incoming nucleotide and the available 5′-OH end of a bound nucleotide. The 5′-OH ends of bound nucleotides that do not react with an activated phosphoramidite are capped to prevent them from participating in ensuing cycles. The

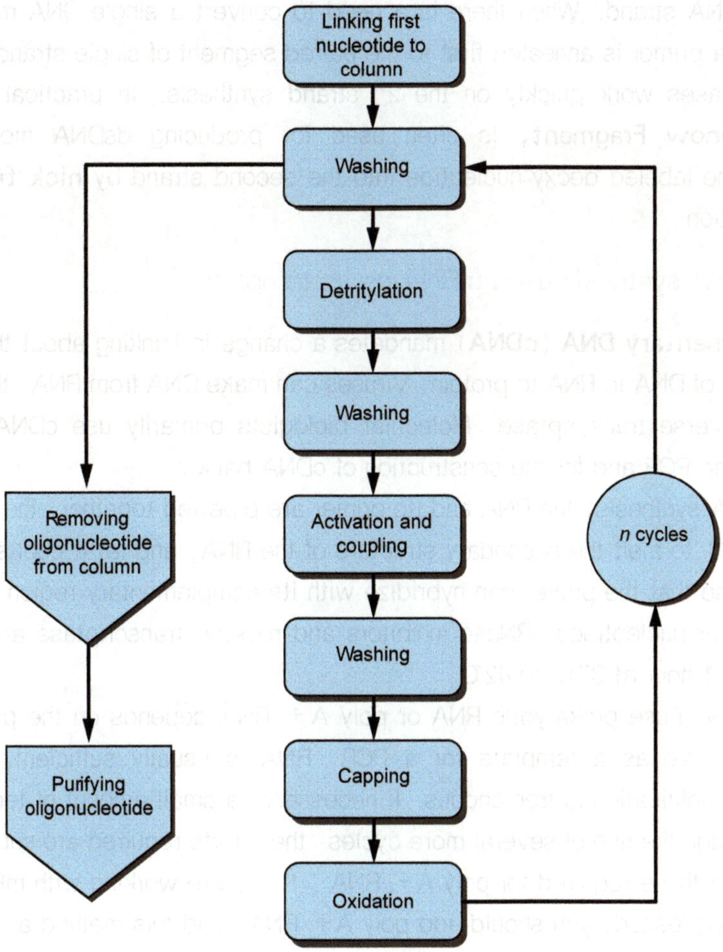

Fig. 3.7 Flowchart for the chemical synthesis of DNA oligonucleotides. After *n* time runs of coupling reactions (cycles), a single-stranded piece of DNA with *n* + 1 nucleotides is produced

internucleotide linkage, which is a phosphate trimester bond, is converted to a more stable pentavalent phosphate trimester bond. The cycles are repeated with successive phosphoramidites and continue until the complete oligonucleotide sequence is formed. After the last cycle, the amino-protecting groups are removed from the guanine, cytosine, and adenine residues. The oligonucleotides are cleaved from the solid supports and purified on the basis of size. Finally, a phosphate group is added either enzymatically or chemically to the 5′ end of each oligonucleotide to produce a bona fide DNA strand.

3.5.2 The second DNA strand synthesized by DNA polymerase

As mentioned in 3.1, DNA polymerases are able to copy the template to synthesize the second DNA strand. When there is a need to convert a single DNA molecule into a double one, a primer is annealed first to the paired segment of single stranded DNA, then DNA polymerases work quickly on the 2th strand synthesis. In practical process, the enzyme, **Klenow Fragment**, is often used for producing dsDNA molecule, or to incorporate the labeled deoxy-nucleotide into the second strand by **nick translation** for further detection.

3.5.3 cDNA synthesis used mRNA as its template

Complementary DNA (**cDNA**) mandates a change in thinking about the direction of the sequence of DNA to RNA to protein. Viruses can make DNA from RNA; the responsible enzyme is reverse transcriptase. Molecular biologists primarily use cDNA as an initial material for the PCR and for the construction of cDNA banks.

For cDNA synthesis, the RNA and its primer are pipetted together, the preparation is heated to 70℃ to melt the secondary structure of the RNA, and let it cool slowly to room temperature so that the primer can hybridize with its complementary region in RNA. After that, the buffer nucleotides, RNase inhibitors and reverse transcriptase are added, and incubated for 1 hour at 37℃ to 42℃.

Whether you use prokaryotic RNA or poly A+ RNA depends on the purpose. If the cDNA is to serve as a template for a PCR, RNA is usually sufficient, because the subsequent amplification is tremendous. If necessary, a small amount of template can be adjusted through the use of several more cycles; the efforts required are substantially less extensive than those required for poly A+ RNA. If you are working with mRNA, which is only slightly expressed, you should add poly A+ RNA, and this method also works if the cDNA is to be used as a hybridization probe.

Advantage: You can synthesize complete cDNA, beginning at the poly A+ tail and ending at the 5′ end of the mRNA.

Disadvantage: The reverse transcriptases used to synthesize cDNA have an average length of 1 to 2kb, whereas mRNA can easily be 10kb long. The protein coding portion of interest is usually in the vicinity of the 5′ end. As a consequence, it is frequently impossible to demonstrate long mRNA, or they are poorly demonstrated. In this case you can take advantage of a **random hexamer primer** (random hexamers), which hybridizes somewhere along the mRNA so that all mRNA segments are represented in the cDNA and the non-mRNA structures. If you are interested in a specific mRNA, you can also make use of a specific primer, although you must frequently optimize the conditions first, or the yields will be smaller than that observed from either of the two other types of primer.

3.6 DNA Sequencing

The ability to determine the sequence of bases in DNA is a central part of modern molecular biology and provides what might be considered the ultimate structural information. Rapid methods for sequence analysis were developed in the late 1970s, and two technique were used in laboratories worldwide. In recent years the sequencing techniques have been revolutionized by automation, which has improved the efficiency of sequencing to the point where **Genome Project** is possible.

3.6.1 Maxam-Gilbert (chemical degradation) sequencing

This method was invented by Prof. Gilbert with the aid of technician Maxam. Based on the precise treatment of chemical degradation on DNA strands, a purified fragment of DNA is required as the starting material. This need not be cloned into a plasmid or viral DNA vector, the technique is applicable to any DNA fragment. The DNA is isotope-labeled with P^{32} at the 5′ ends (or 3′ ends) of each strand, and the strands are internally cut, separated, and purified to give a population of labeled strands for the next chemical reactions. The next step involves dividing the sample into 4 groups, 4 chemical treatments specific for 4 different bases in the DNA strands are trickily designed. Since the chemical modification conditions are so finely chosen that, on average, only one typical reaction will be introduced randomly into each copy of the DNA molecule. The modified bases are then removed from their sugar groups and the strands cleaved at these positions using the chemicals such as piperidine. The theory is that, four groups of different reaction on the DNA sample will produce a set of nested fragments. In practice, the 4 groups of nested fragments can be separated in electrophoresis gel to form a readable sequence (Fig. 3.8b).

3.6.2 Sanger-Coulson (dideoxy or enzymatic) sequencing

Although the end result is similar to that attained by the chemical degradation method, the principle of Sanger-Coulson procedure is totally different from that of Maxam-Gilbert. In this case, sample of the DNA to be sequenced is treated by Klenow fragment of DNA polymerase I. The template for this reaction should be single-stranded DNA, and a primer must be used to provide the 3′ terminus for DNA polymerase to begin the synthesis of the 2nd strand. The production of nested fragments is achieved by the additional 4 groups of ddNTPs (dideoxy-NTP) in each reaction. ddNTPs lack a hydroxyl group at the 3′ position of deoxyribose, so they play the role as terminator of DNA chain elongation. Along with four normal dNTPs and the enzyme, four ddNTPs (carrying A, or G, or T, or C) are included in a series of four reactions, 4 groups of dideoxy form will be incorporated into the growing DNA chain infrequently and randomly. Each reaction, therefore, produces a series of fragments terminating at a specific nucleotide, and the four reactions together provide a set of nested fragments. The DNA chain is labeled by including a isotope-labeled dNTP in the reaction mixture. This is usually [α-S^{35}]dATP (O in the phosphate has been changed into S), which enables the bands on the electrophoresis gel to become sharper, more

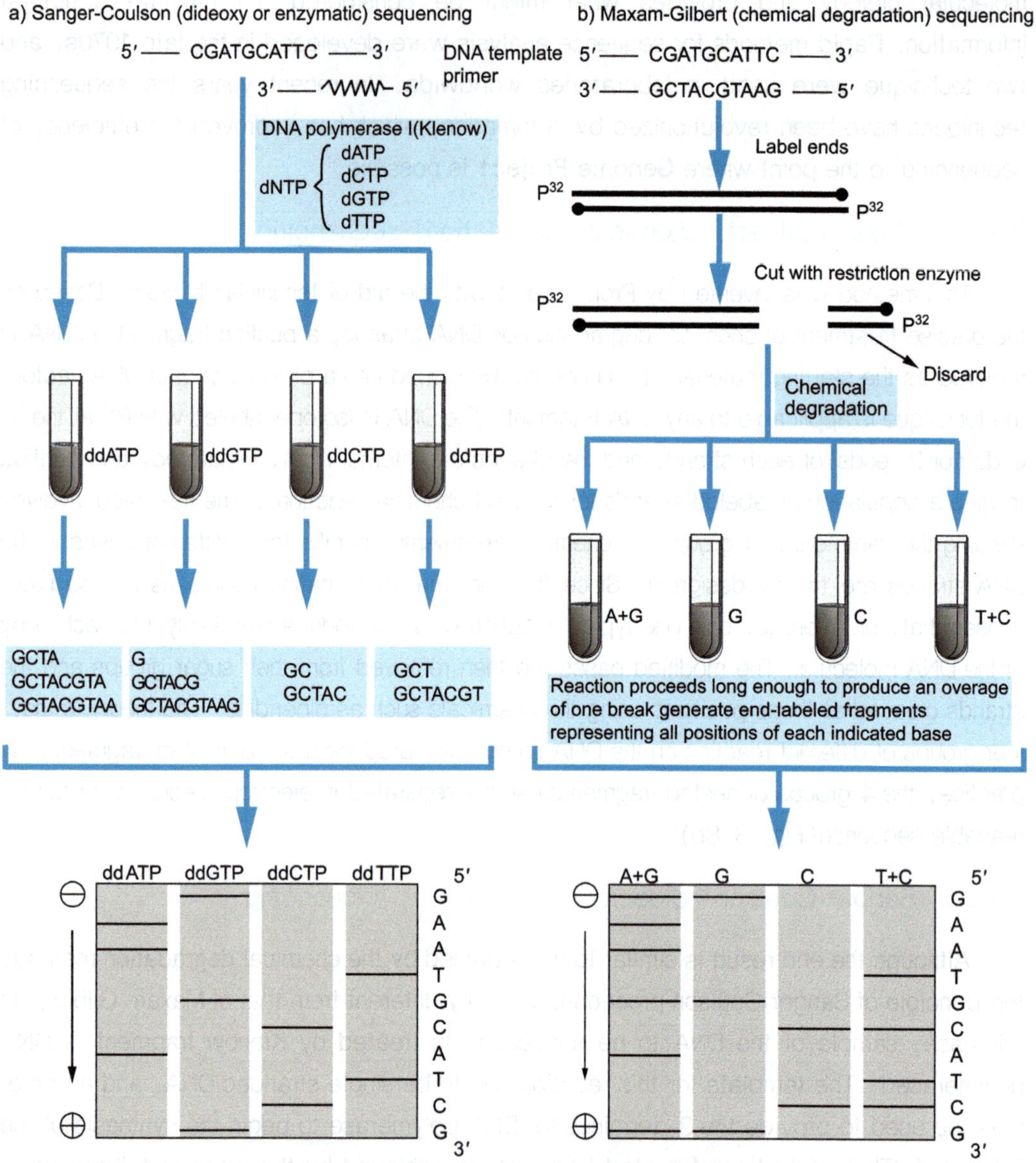

Fig. 3.8 Comparison of two methods for DNA sequences

bands can be read from a single gel than using of P^{32}-labeled dNTPs(Fig. 3.8a).

The generation of nested fragments for dideoxy sequencing is more complicated than for chemical degradation sequencing, and usually involves sub-cloning into different vectors. Many plasmid vectors are now available, and some types can be used directly for Sanger's DNA sequencing. Another method is to clone the DNA into a vector such as the bacteriophage M13, which produces single-stranded DNA after infection. This provides a suitable template for running Sanger' Dideoxy-sequencing.

3.6.3 Development of DNA sequencing

One of the major technologic progresses that enabled sequencing to move from single-gel lab-based systems up to large-scale "production line" sequencing was the automation on many sections of the process. Whereas a good lab scientist or technician could sequence maybe a few hundred bases per day, this was not going to solve the problem of determining genome sequences as opposed to a huge pool of gene sequences. Improving the technology by orders of magnitude was required. This was achieved by improvements in sample preparation and handling, with robotic processing enabling high-volume throughput. In a similar way, the automation of the sequencing reactions and linear continuous capillary electrophoresis techniques have enabled the scale-up of the sequence-determination.

Some of the pioneer DNA sequencing sectors have developed novel separation, detection, and imaging techniques for real-time monitoring in capillary electrophoresis. These techniques are applied to substantially increase the speed, throughput, reliability, and sensitivity in DNA sequencing with highly multiplexed capillary arrays. By gross estimation that it should be possible to eventually achieve a raw sequencing rate of 40 million bases per day in one instrument based on the standard Sanger protocol. The substantial increase in sequencing rate is a result of several technical advances. 1) The use of commercial linear polymers for sieving allows replaceable yet reproducible matrices to be prepared that have lower viscosity (thus faster migration rates) compared to polyacrylamide. 2) The use of a charge-injection device camera allows random data acquisition to decrease data storage and data transfer time. 3) The use of distinct excitation wavelengths and cut-off emission filters allows maximum light throughput for efficient excitation and sensitive detection employing the standard 4-dye coding. 4) The use of index matching and 1∶1 imaging reduces stray light without sacrificing the convenience of on-column detection.

Continuing efforts include further optimization of the separation matrix, development of new column conditioning protocols, refinement of the excitation/emission optics, design of a pressure injection system for 96-well titer plates, validation of a new 2-color base-calling scheme, simplification of software to allow essentially real-time data processing, implementation of voltage programming to shorten the total run times, and scale up of the technology to allow parallel sequencing in up to 1000 capillaries.

In addition to the challenges of sequence determination, a parallel challenge lies in the need to develop sufficient computing power to deal with the vast amounts of data generated by the newly improved technologies employed in genome sequencing. In fact, it could be argued that the most critical part of the whole process is the data analysis side of things—without the ability to interrogate sequence data, the sequence remains essentially silent (Fig. 3.9).

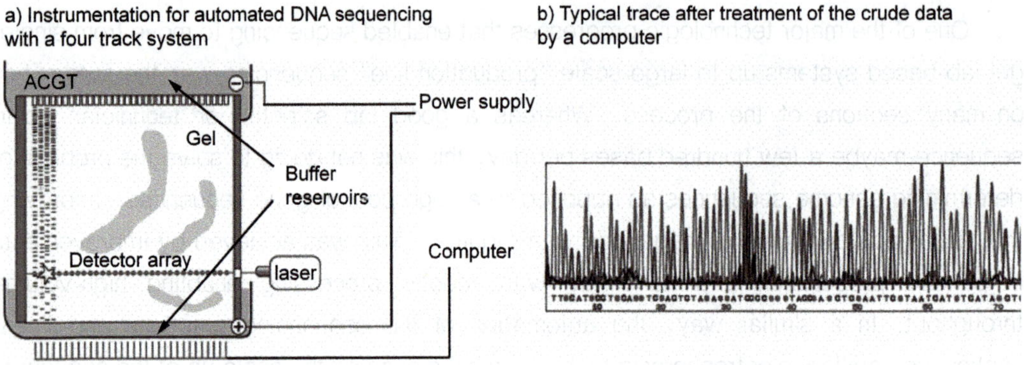

Fig. 3.9 DNA sequence read out form an automatic sequencing system

3.7 Macromolecular Hybridization and Blotting

3.7.1 Southern blotting

Southern blotting, The technique for detection of specific DNA molecules in a mixture was invented by E. M. Southern of Oxford University in 1975. This is probably the most widely used method in genetic engineering. In this protocol, the DNA fragments, generated by restriction digestion are subjected to agarose gel electrophoresis, and then transferred from the gel to a porous nitrocellulose or nylon membrane by capillary function of absorbent paper to soak over night the solution from the gel to the membrane. For DNA, the specific separated molecules are detected in membrane by molecular hybridization with labeled nucleic acid probes. The so called "Southern blotting" procedure can be completed in one day in a very inexpensive way, and only simple laboratory equipments are needed.

Although other methods such as **vacuum blotting** and **electroblotting** have been devised, The original method, so called **capillary blotting**, is still used extensively because it is simple and efficient. When the fragments have been transferred from the gel and bound to the membrane, it becomes a replica of the gel. The membrane can then be hybridized with a radioactive probe, in a similar way the DNA in colony or plaque can be transferred to membrane, and blotted. As with all hybridization, the key is the availability of a suitable probe. After hybridization and washing, the membrane is exposed to X-ray film and an autoradiograph prepared, which provides important information for afterward design of gene cloning(Fig. 3.10).

More recently, the probes need to be labelled either with radioactive isotopes (P^{32}) or with **chemiluminescence** in which alkaline phosphatase or **horseradish peroxidase**

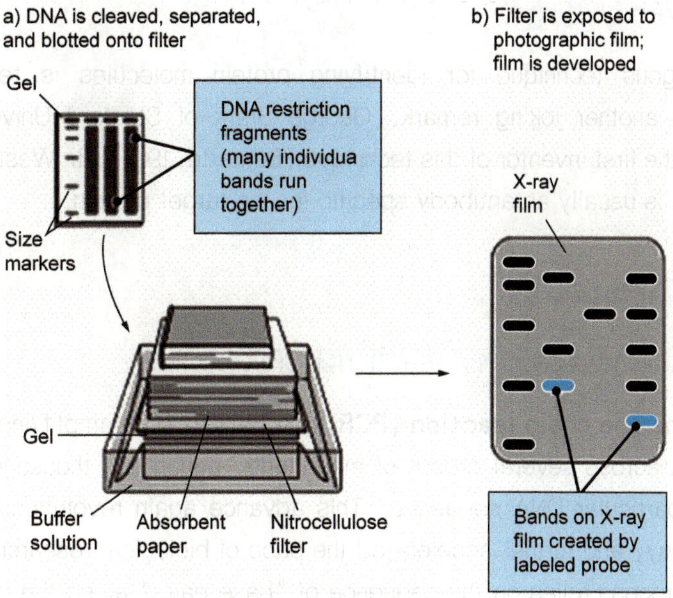

Fig. 3.10 Southern blotting

breakdown chemiluminescent substrates produce a detectable emission of light. The chemiluminescent labelling can occur in two ways: either the probe is attached to the enzyme, or the probe is labelled with a **ligand** (e.g. biotin) for which the antibody (e.g. **avidin** or **streptavidin**) is attached to the enzyme. X-ray film can detect both the radioactive and chemiluminescent signals, and many researchers prefer the chemiluminescent signals because they are faster, more sensitive, and avoid of radioactive contamination.

It's worth to mention that Southern Blotting had been used in several key studies, for example, the original proposal for gene mapping of human genome was based on **Restriction Fragment Length Polymorphisms** (**RFLP**) detected by DNA blotting; Introns were first seen from the result of rabbit genomic DNA hybridized with the probe of rabbit β-globin gene; The first human **DNA fingerprints** were produced by hybridizing restriction digests of human DNA with the **mini-satellite probes**.

3.7.2 Northern blotting

Inspired by the technique of Southern Blotting, in 1977, J. C. Alwine etc. of Stanford University have developed a similar protocol called "**Northern blotting**" in order to identify specific RNA , so called "Northern" is actually a joking remark corresponding to the term of "Southern". The identification of particular RNA following separation by gel electrophoresis, blotting and probe hybridization is similar to the procedure of "Southern blotting", but with more strict aseptic manipulation.

3.7.3 Western blotting

The analogous technique for identifying protein molecules is termed **Western blotting**, also another joking remark. George Stark of Stanford University has been believed to be the first inventor of this technique (Burnette, 1981). In Western blotting, the probe of choice is usually an antibody specific for the target protein.

3.8 PCR Techniques

3.8.1 The polymerase chain reaction (PCR)

The polymerase chain reaction (**PCR**) is a technique for amplifying a few copies of a piece of DNA across several orders of magnitude, generating thousands to millions of copies of the particular DNA sequence. This advance again revolutionized recombinant DNA methodology, and further accelerated the pace of biological research.

The PCR process relies on the sequence of "base pairs" along the length of the two strands that make the complete DNA molecule. In DNA there are four deoxy-nucleotides derived from the four bases: adenine (A), thymine (T), guanine (G) and cytosine (C). The strands or polymers that comprise the DNA molecule are held to each other by hydrogen bonds between the base pairs. In this arrangement, A only pairs with T while G only pairs to C, and this unique system folds the entire molecule into the now well-recognized double-helix structure.

PCR procedure involves three steps: DNA **denaturation**, **annealing** and then **extension** by DNA polymerase (Fig. 3. 11). In Step 1, the double-stranded DNA is heated (95～98℃) and separates into two complementary single strands. In Step 2, the working temperature is lowered down to 60℃, while the synthetic oligonucleotide primers, the short sequences of nucleotides with usually about 20 nucleotide base pairs long, are added and bind to the single strands in places where the strand's DNA complements their own. In Step 3 (37℃), the primers are extended by DNA polymerase in the presence of all four **deoxynucleoside triphosphates**, resulting in the synthesis of new DNA strands that are complementary to the template strands. The completion of the three steps comprises a PCR cycle, and the real power of it is that, with 25～30 cycles, this experimental synthesis leads to **massive amplification** of DNA which can then be used for analytical or cloning purposes. A major recent advance has been the development of automated thermal cycles (PCR machines), which allows the entire PCR to be performed automatically in a few hours.

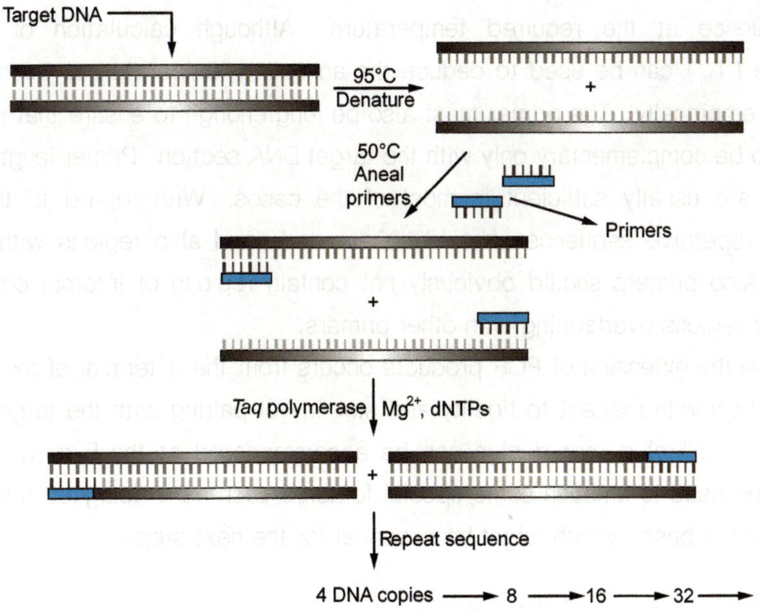

Fig. 3. 11 The polymerase chain reaction

The double-stranded DNA is heated and separates into two single strands. The synthetic oligonucleotide primers then bind to their complementary sequence and are extended in the direction of the arrows, giving a new strand of DNA identical to the template's original partner

3.8.2 Design of PCR primers

The synthesis of an oligonucleotide primer is not difficult, since many commercial sources provide such kind of service. But the specific sequence of the primer has to be determined by scientists. To design a primer, there are several aspects that have to be considered. Perhaps the most obvious aspect is how to define the sequence of the primer, and where the sequence information comes from?

It may be derived from the amino acid sequence of the coding peptide, in which case the genetic code degeneracy has to be considered. In synthesis of primer there are two approaches, by incorporating a mixture of bases at the wobble position, a mixed primer can be made, with the "correct" sequence represented as a small proportion of the mixture. Alternatively, the base inosine (which pairs equally well with any of the other bases) can be incorporated as the third base in degenerate codons.

If the primer sequence is taken from an already-determined DNA sequence, it may be the gene with the same function but from different organism, or may be a cloned DNA that has been sequenced during previous experimental work.

Regardless of the source of sequence information of the primers, there are some general considerations that should be addressed. The primer length is important; it should be long enough to ensure that the stable hybridization happen between the primer and the target sequence at the required temperature. Although calculation of the melting temperature (T_m) can be used to deduce the annealing temperatures, this is often best determined empirically. The primer must also be long enough to ensure that it is a unique sequence to be complementary only with the target DNA section. Primer lengths of 20~30 nucleotides are usually sufficient in most of the cases. With regard to the annealing specificity, repetitive sequences should be avoided, and also regions with single-base sequence. Also primers should obviously not contain regions of internal complementary sequence or regions overlapping with other primers.

Because the extension of PCR products occurs from the 3′termini of the primer, this region is critical with respect to fidelity and stability of pairing with the target sequence. Some "looseness" of primer design can be accommodated at the 5′end, and this can sometimes be used to ambush some special features such as a designed restriction sites or a mismatched base, which might be essential for the next steps.

3.8.3 More exotic PCR techniques

Reverse transcriptase PCR (RT-PCR)

There are many variants of the extended PCR techniques, and one of the most important protocol is **reverse transcriptase PCR (RT-PCR)** for amplification of cDNA derived from mRNA molecules. Using mRNA as the template is a common variant of the basic PCR, and it can be useful in determining low levels of gene expression by analyzing the PCR product of cDNA prepared from the mRNA transcript.

Nested PCR

Nested PCR is designed for overcoming some of the problems associated with a large number of PCR cycles, which usually lead to error prone synthesis. The technique essentially increases both the sensitivity and the fidelity of the basic PCR protocol. It involves using two sets of primers.

Inverse PCR (IPCR)

Often a stretch of DNA sequence is known, but the desired target sequence lies outside of this region. **Inverse PCR (IPCR)** involves isolating a fragment, with restriction site in it, which contains the known sequence plus the flanking unknown sections. By circularizing the fragment, and then cutting inside the known sequence, the fragment is effectively inverted.

Many other variants of the PCR protocol will not be discussed in detail, but be listed in Tab. 3.2.

Tab. 3.2 Some variants of the basic PCR process

Acronym	Technique name	Application
RT-PCR	Reverse transcriptase PCR	PCR from mRNA templates
IPCR	Inverse PCR	PCR of sequences lying outside primer binding sites
RAPD-PCR	Random amplified polymorphic DNA PCR	Genomic fingerprinting under low-stringency conditions
AP-PCR	Arbitrarily primed PCR	As RAPD-PCR
TAIL-PCR	Thermal asymmetric interlaced PCR	PCR using alternating high/low primer binding stringency with arbitrary and sequence-specific primers
mrPCR	Multiplex restriction site PCR	PCR using primers with restriction site recognition sequences at their 3′ ends
AFLP	Amplified fragment length polymorphism	Genome analysis by PCR of restriction digests of genomic DNA
CAPS	Cleaved amplified polymorphic sequence analysis	Genome analysis for single nucleotide polymorphisms (SNPs) by PCR and restriction enzyme digestion
GAWTS	Gene amplification with transcript sequencing	PCR coupled with transcript synthesis and sequencing using reverse transcriptase
RAWTS	RNA amplification with transcript sequencing	Variant of GAWTS using mRNA templates
RACE	Rapid amplification of cDNA ends	Isolation of cDNAs from low-abundance mRNAs

Summary

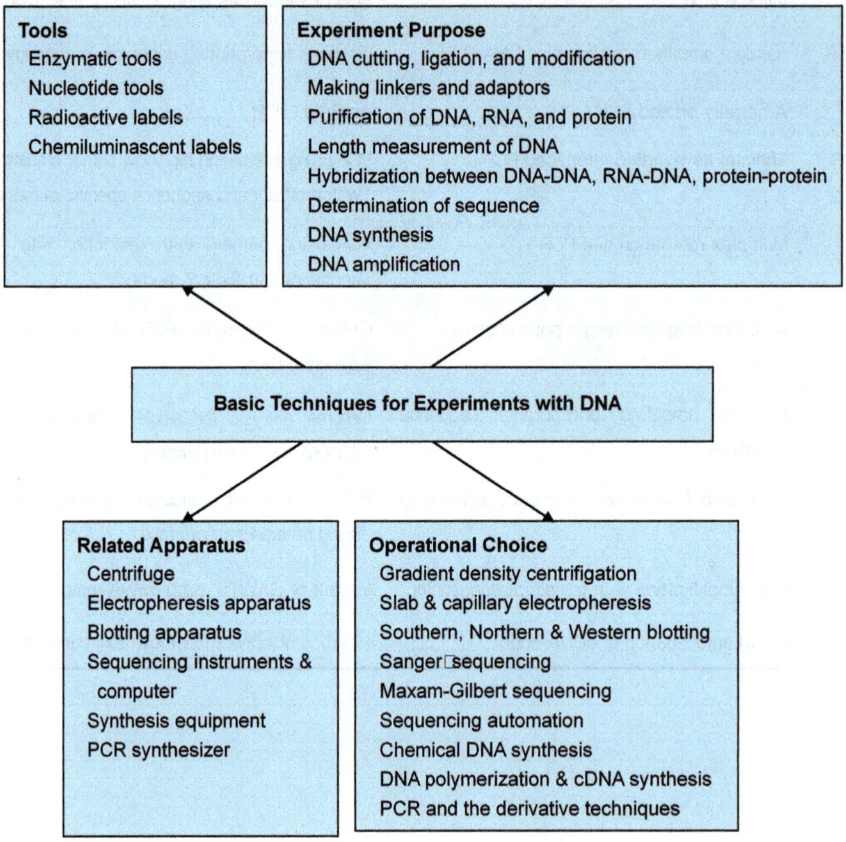
Tools
Enzymatic tools
Nucleotide tools
Radioactive labels
Chemiluminascent labels
Experiment Purpose
DNA cutting, ligation, and modification
Making linkers and adaptors
Purification of DNA, RNA, and protein
Length measurement of DNA
Hybridization between DNA-DNA, RNA-DNA, protein-protein
Determination of sequence
DNA synthesis
DNA amplification
Basic Techniques for Experiments with DNA
Related Apparatus
Centrifuge
Electropheresis apparatus
Blotting apparatus
Sequencing instruments & computer
Synthesis equipment
PCR synthesizer
Operational Choice
Gradient density centrifigation
Slab & capillary electropheresis
Southern, Northern & Western blotting
Sanger sequencing
Maxam-Gilbert sequencing
Sequencing automation
Chemical DNA synthesis
DNA polymerization & cDNA synthesis
PCR and the derivative techniques

Chapter 4 Cloning Procedure

基因克隆技术历经 40 年的发展，衍生出诸多精妙绝伦的技术，但是万变不离其宗，基因克隆的基本步骤离不开以下的若干环节，如获取克隆所需的供体 DNA 片段；纯化、改造、重构适合的载体分子；合理选择接受重组 DNA 分子的宿主细胞品系；将重组分子导入到受体细胞中；筛选、鉴别目的克隆子；必要时还要对克隆子进行再次修饰改造等。本章还介绍了与基因克隆有关的若干衍生技术。

4.1 Basic Procedure of Cloning

Genetic engineering involves manipulation of the genetic material towards a desired purpose in a directed and pre-determined way. This is alternately called **recombinant DNA technology** or **gene cloning**. Strictly 'to engineer' means to design, construct and manipulate according to a set of plans. Through forty years' development, varieties of novel techniques have been derived from the early genetic cloning methods, but the basic steps would be similar: 1) Isolation of the gene (or other piece of DNA) to be cloned; 2) Insertion of the gene into another piece of DNA called a vector, which will allow it to be taken up by bacteria and replicated within them as the cells grow and divide; 3) Transfer of the recombinant vectors into bacterial cells, either by transformation or by infection using viruses; 4) Selection and screening of the cells which contain the desired recombinant vectors; 5) Sub-cloning if necessary to get better clones.

All these are made possible by some basic methodologies which form the essential ingredients of genetic engineering. Although cloning technology has become more sophisticated in recent years, the basic premises on which the procedures are based have not changed substantially, and routes such as those indicated are still used in many applications (Fig. 4. 1).

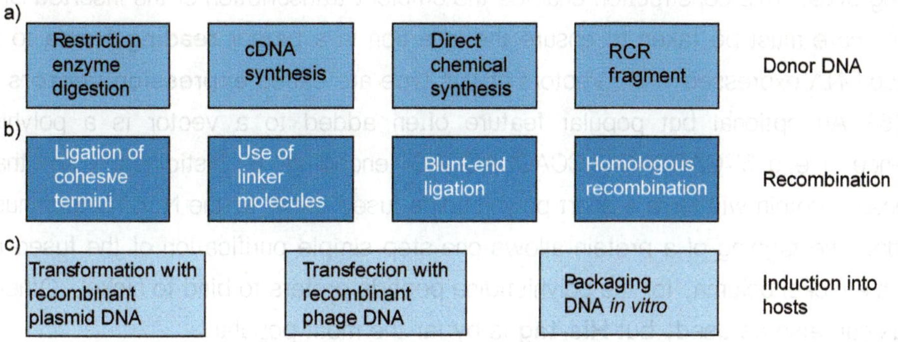

Fig. 4. 1 Frame of basic cloning strategy

In short, gene cloning or genetic engineering is essentially the insertion of a specific piece of "foreign" DNA into a host cell, in such a way that the inserted DNA is replicated and handed on to daughter cells during cell division.

4.2 Selection of Cloning Vectors

The introduction of a foreign DNA into a host cell in many cases requires the use of a vector. Vectors are DNA molecules used to transfer a gene into a host (microbial, plant, animal) cell, and to provide control elements for replication and expression. The vector to be used is determined by the type of host cells and the objectives of the cloning experiment.

4.2.1 Plasmid vectors

For cloning use, plasmids preferably contain a number of structural elements.

(1) **Replication origin**: The origins in plasmids might have different capacities on control of plasmid replication. It's necessary to select the plasmids with very strong origin which guarantees the inserted section has as many copies as possible.

(2) **Cloning sites**: No matter whether the naturally existing plasmids contain restriction sites or not, usually the potential vectors are constructed artificially in the recognition sequences for a number of restriction cutting. This cluster of cutting sites, referred to as multiple cloning sites (MCS), serves to facilitate the convenient insertion of a foreign DNA.

(3) **Selectable markers**: These are usually antibiotic resistance genes, such as ampicillin resistance (Amp^r) and tetracycline resistance (Tet^r). For plant s and animals cloning, some of the fluorescent markers have been constructed, by which the desired clones can be easily spotted out in dark. Non-transformed cells (not picking up the plasmid vector) will not survive in a growth medium containing the antibiotic, or will not illuminate the fluorescent light .

(4) Some plasmid vectors also contain a **promoter region** upstream of the **multiple cloning site**. This construction enables the efficient transcription of the inserted DNA. Of course, care must be taken to ensure the insertion in a proper **reading frame** to let the inserted DNA expressed well. Vectors of this type are called **expression vectors**.

(5) An optional but popular feature often added to a vector is a polyhistidine sequence (e.g. 5′-CACCACCACCACCACCAC encoding 6 histidines) so that the expressed protein will have a short polyhistidine fused either at the N- or C-terminus of its peptide. The tagging of a protein allows one-step simple purification of the fused protein through a nickel column, for the polyhistidine peptide prefers to bind to nickel. Other types of tags can also be used, but **His-tag** is by far the most popular.

Plasmids can be grouped into high copy or low copy plasmids, depending on the number of molecules of the plasmid found in a bacterial cell. Low copy refers to 1～25 plasmids per cell. High copy refers to 100 copies and more per cell. High copy plasmid vectors are the choice if a high yield of the recombinant DNA is desired. On the other hand, it may not be desirable to use high copy plasmid vectors if the gene product is expected to cause adverse effects to the host cells. **Copy number** depends on the origin of replication, which determines whether the plasmid is under relaxed or stringent control. Some time the size of the plasmid and its associated insert may affect he copy number, the pUC plasmid and its derivatives can reach very **high copy numbers** within the host cell.

The following description of the pUC series plasmid vectors, taken as an example, illustrates how a plasmid vector would construct and operate. The pUC series contain a number of mechanisms that have become standards in many other vector systems(Fig. 4.2).

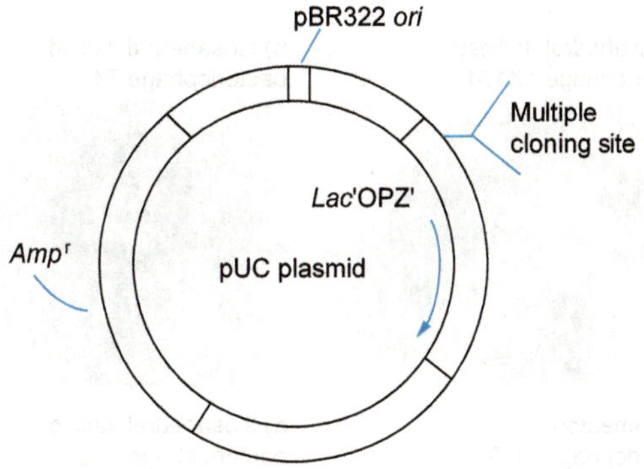

Fig. 4.2 Structural organization of a pUC plasmid vector

(1) The origin of replication from the pBR322 plasmid (one of the early *E. coli* cloning vectors that gives rise to many more recently developed vectors).

(2) The **β-lactamase** gene (**ampicillin resistance**, Amp^r), from a transposon segment playing the role of selectable marker. In transformation step, the cells are plated on a medium containing ampicillin. Majority of the *E. coli* cells are non-transformants, which do not survive for the lack of the ampicillin resistance gene. Only transformants (those that have picked up the plasmid) are ampicillin resistant, and will show up as white colonies.

(3) The partial section of ***lac* operon** in pUC. It makes use of a gene called *lacZ*, which encodes the enzyme **β-galactosidase**. The normal substrate for this enzyme is lactose, which it breaks down into glucose and galactose. However, this enzyme also break down a compound called **5-bromo-chloro-3-indolyl-β-D-galactosidase** (IPTG) or **X-gal** for short. The useful thing about X-gal is that when it is broken down it produces a blue color. This means that clones with a functional *lacZ* gene can easily be identified because they are blue in color. The plasmid only contains a part of the *lacZ* gene that

encodes a fragment of the enzyme, and this partial gene is called *lacZ′*. Recombinant clones can be identified because they do not express β-galactosidase, and hence they will be white on plates containing X-gal and IPTG. Colonies containing recircularized vector will be blue.

(4) **A multiple cloning site (MCS)**. The multiple cloning site in PUC is in the middle of *lacZ′* gene so cloning into it will result in insertional inactivation of the gene and the resulting clones will be unable to break down X-gal. The clones with insertion will be white, and are easily distinguishable from those containing the recircularized vector, which will be blue.

4.2.2 Bacteriophage vectors

Another type of vector is derived from bacteriophage that includes three structural categories (Fig. 4.3): **icosahedral tailess**, **icosahedral head with tail** and **filamentous**.

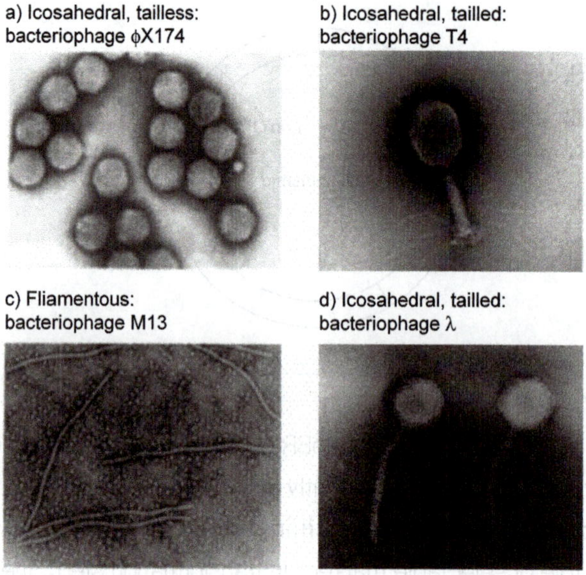

Fig. 4.3 Phages samples

The λ vectors and its derivatives are used mostly for the construction of cDNA or genomic libraries. M13 vectors were used primarily to obtain ssDNA for DNA sequencing before the development of more recent sequencing methods.

Phage λ vectors

Bacteriophages λ carry nearly 50 genes in a linear 48.5kb DNA, which are ligated in a circle after injection into *E. coli* host. It may either be duplicated to form many progeny particles (lytic phase) or the λ genome may be integrated into host genome by site-specific recombination (lysogenic phase). When λ packing DNA there are size limits, only the DNA between 75%～105% of its normal length can be packed. Taking account of λ essential segment, foreign DNA fragment, if the DNA is less than 23kb, it can be cloned into the specific constructed λgt vector series. Others called replacement vectors have been well developed for conveniently accepting foreign DNA; usually, they contains:

(1) **Cos site** which consists of 12bp cohesive ends, plays a unique role for transition of λ DNA from linear to circular form or vise versa, cos site can be recognized by λ ***in vitro*** packaging system.

(2) **A set of restriction sites** in between the nonessential segment of λ can be easily cut off and replaced by the foreign DNA with the same sticky ends.

(3) **Markers** such as "Blue & White" mechanism has been employed and the related section have been constructed into some λ vectors, which help direct identification of the desired clones.

In the *in vitro* packaging system, the λ packing components are grouped into two parts, one for head package, another for tail assembly, and the cloned λDNA molecules are packed into the head first, then fully assembled into a phage particles. The infected *E. coli* cells are spread on agar plate to let some plagues show up. The plagues with marker identity would be the objects for the clone identification.

M13 vectors

The genome of phage M13 consists only of ssDNA, but after it infects host *E. coli*, the M13 DNA duplicates as dsDNA circle (replicative form, **RF**) within the host cells. The special capability of M13 DNA make it also a very useful vector for a number of applications, such as, DNA sequencing and site-directed mutagenesis.

Except of the naturally existing M13 genes, some of the critical sections have been incorporated into M13 vector, which includes:

(1) A *lac Z′* system (with Blue & White marker).

(2) A multiple cloning site constructed in the *lacZ′* region (Fig. 4.4).

M13 vectors need special host strains, It is essential for M13 infection to the *E. coli* host strain contains the **F′ episome** (specialized plasmids containing an F factor that encodes sex pili in the male *E. coli* cells). The genes for proline biosynthesis are removed from the *E. coli* genome and inserted into the F′ episome, so that the F′ episome is retained when the strain is grown on minimal media (deficient in proline). The F′ episome is also inserted with a mutated *lacZ* gene (lacking the α-peptide sequence), referred to as *lacZ*△M15, *which is used for* α-complementation, and a mutated *lac I* gene, referred to as *lacI*^{q}, which results in the overproduction of the *lac* repressor. Therefore, the genotype of the M13 host strain is *lacI*^{q}*Z*△M15.

The M13 phage DNA is not infectious, but *E. coli* cells can pick up both ss and RF forms with $CaCl_2$ treatment in the same way as plasmids. The M13mp vectors are constructed in pairs (e. g. M13mp18 and M13mp19) with opposite orientation of the recognition sites in the multiple cloning site regions (Fig. 4.4). This enables cloning of a DNA fragment, in both orientation, to be sequenced in both directions.

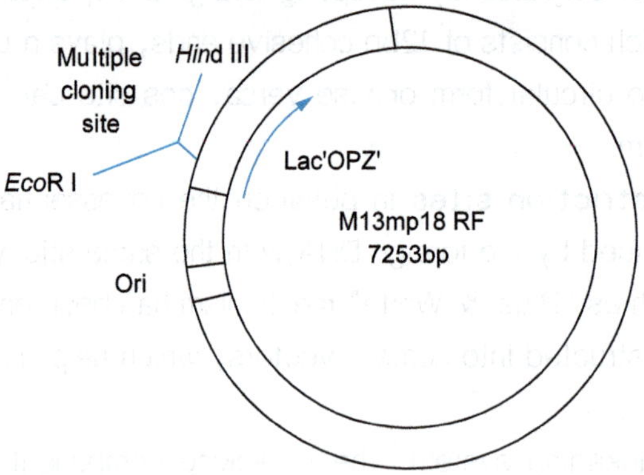

Fig. 4.4 M13 vector showing multiple cloning sites in two orientations

4.2.3 Cosmid

Cosmids are plasmids containing a bacteriophage λ ***cos* site**. The hybrid structure enables insertion of large DNA fragments. λ *cos* site is required for recognition of λ genomic unit in phage packaging (Fig. 4.5).

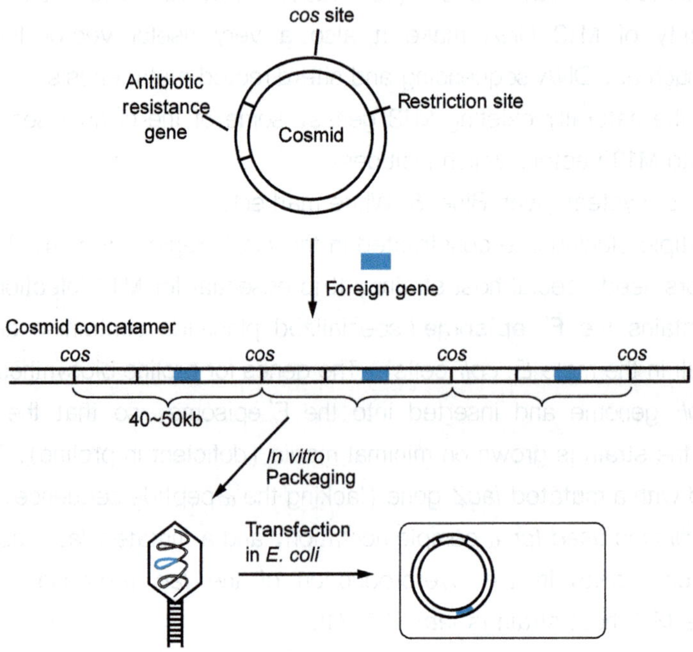

Fig. 4.5 Cosmid replicates like a plasmid and is packaged like phage λ DNA

In normal life cycle, λ DNA molecules produced after replication are joined by the cohesive ends (*cos* site) to form a concatamer (long chains of many genomic units). In the steps that follow, the concatamer is enzymatically cleaved, and each individual λ DNA molecule is packaged into phage particles. Therefore, by incorporating a *cos* site into a plasmid, the resulting hybrid vector can be used for *in vitro* packaging. The size of a foreign DNA that can be inserted into the cosmid, and packaged into phage particles, is restricted to a range of 35～45kb, assuming the cosmid size is ～5kb. As a general rule, the total length of cosmid plus the DNA insert should be 75%～105% of the size of phage λDNA for efficient packaging. The phage particles are then used to infect *E.coli*. The recombinant cosmid self-circularized inside the *E.coli* cell, and replicates like a plasmid.

4.2.4 Other vectors

Artificial chromosomes vectors

The development of vectors for cloning very large pieces of DNA was essential to enable large genome sequencing projects to proceed at a reasonable rate, although genomes such as *S. cerevisiae* have been sequenced mainly by using cosmid vectors to construct the genomic libraries required. However, even insertion sizes of 40～50kb are too small to cope with projects such as the Human Genome Project. The development of **yeast artificial chromosomes** (**YACs**) has enabled DNA fragments in the megabase range to be cloned, although there have been some problems of insert instability. YACs are the most sophisticated yeast vectors and represent the largest-capacity vectors available. They have **centromeric** and **telomeric** regions, and the recombinant DNA is therefore maintained essentially like a yeast chromosome.

A further development of artificial chromosome technology came with the construction of **bacterial artificial chromosomes** (**BACs**). These are based on the F plasmid, which is much larger than the standard plasmid cloning vectors and, therefore, offers the potential of cloning larger fragments. BACs can accept inserts of around 300kb, and the instability problems of YACs can be avoided by using the bacterial version. Much of the sequencing of the human genome has been accomplished using a library of BAC recombinants. Vectors based on the phage P1 have also been developed, both as phage vectors and also as P1-based artificial chromosomes.

Vectors for plant cells

Ti (tumor-inducing) plasmid is widely utilized to introduce DNA into plant cells. The Ti plasmid is isolated from Agrobacterium tumefaciens, a soil bacterium that infects plants, causing the formation of tumor tissue (see Section 8.2.1).

Vectors for mammalian cells

Genetically engineered animal cell lines are useful for the production of human therapeutic proteins, and also provides a convenient system for studying gene regulation and control in eukaryotic cellular processes. There are, in general, two types of methods for transferring DNA into mammalian cells: 1) mediated by virus infection, or 2) transfected with mammalian expression vectors.

Viral-mediated transfer provides a convenient and efficient means of introducing eukaryotic genes into mammalian cells. This method involves the use of a number of viruses, such as simian virus 40 (SV40), bovine papilloma virus (BPV), Epstein-Barr virus (EBV), and retrovirus (see Section 9.2.1). Baculovirus is also included, although insect cells are used as the host in this system. It is not necessary, however, to use a viral vector to express foreign genes in animal cells, particularly if transient expression (several days to weeks) is desired. Mammalian expression vectors for this purpose are derived from plasmid DNA carrying regulatory sequence from viruses.

4.3 Preparation of Donor DNA

Gene cloning requires, as an initial step, isolation of a specific gene encoding for the protein of interest. Locating and selecting a single gene among thousands of genes in a genome is not a simple task.

4.3.1 Donor DNA from genomic library

For prokaryotes, identification of a particular gene is usually made by first constructing a **genomic library**. The total genomic DNA is isolated, purified, and digested with a restriction enzyme. The short DNA fragments are then cloned into a suitable vector, for example, bacteriophage λ vectors or phagemids. The resulting recombinant λDNAs are assembled into phage particles by *in vitro* packaging, with all the required phage proteins provided by λ phage mutants that cannot replicate. Now, we have a library of all genomic DNA in short fragments (usually average ~15kb) cloned into bacteriophage λ vectors packaged into viable phage particles.

The phage particles are used to infect *E.coli* cells. Phage transfection results in clear plaques on a bacterial lawn. Each plaque corresponds to a single phage infection. The next step is to screen the plaques for the clone(s) containing the gene of interest. A commonly used technique is DNA hybridization as described. A radiolabeled short DNA probe complementary to the gene sequence is used to identify the particular recombinant clone. This technique obviously implies some prior knowledge of a short segment of the gene sequence for constructing oligonucleotide labels. The sequence (usually 18 to 22 bases) can be deduced from 1) sequences from comparative species, 2) the known sequence of the protein encoded by the gene, or 3) N-terminal or peptide sequencing of the protein.

An alternative is immunological method, based on the detection of the translation product of the gene. This screening method requires that the λ vector used is an expression vector (such as λgt11,) the insertion of the gene is in frame, and the availability of antibodies raised against the protein. A third alternative method used for screening libraries involves the use of PCR to amplify the gene of interest. This requires some knowledge of short sequences at the 5′ and 3′end to synthesis primers for the PCR reaction. After the PCR amplification, the genomic DNA or cDNA can be used directly for cloning.

4.3.2 Donor DNA from cDNA synthesis

Construction of genomic library for gene isolation is usually applied to prokaryotes. For fungi, plants and animals, the identification of an eukaryotic gene from genomic library is not suitable for at least two reasons: 1) The large genome size in plants and animals requires screening of an astronomical numbers of clones for the gene of interest; 2) Eukaryotic genes contain introns, the non-coding regions cause complications in subsequent expression. The use of a **cDNA library** can circumvent these problems. The construction of a cDNA library begins with the isolation of total RNAs from a specific cell type that produces the protein of interest, the isolation of mRNAs from the total RNAs, followed by the conversion of the mRNA molecules to complementary DNA (cDNA) strands.

The isolation of mRNA makes use of the fact that eukaryotic mRNA has poly(A) tail at the 3′ end, that can hybridizes with oligonucleotides of poly(T) immobilized on a column matrix. When total RNA extract is passing through such affinity column, mRNA will be retained, while the bulk of the cellular materials drains off. The isolated mRNA sample is then treated with reverse transcriptase to synthesize complementary DNA strands (cDNA) using the mRNAs as templates, to form RNA: cDNA hybrids. The RNA strands in the hybrids are nicked with **ribonuclease**, and nick-translated by DNA polymerase I to replace the RNA strand with a new (second) cDNA strand, to form dsDNA molecules.

The resulting cDNAs are inserted into a suitable vector (plasmids or bacteriophage λ vectors), and introduced into *E. coli*. The cDNA library is thus screened for the clones carrying the gene of interest by DNA hybridization or immunological detection method. The gene of interest can also be isolated by PCR amplification. The identified clone(s) is then cultured and the recombinant DNA is purified, using the cDNA mixture directly without cloning. The gene sequence is excised from the vector, followed by determination of the complete nucleotide sequence. The isolated gene sequence must be determined for at least two reasons: 1) To confirm that the identified sequence is the gene of interest; 2) For proper construction of the regulatory region and in frame ligation for gene expression. A full-length gene may not be identified in the first trial. The gene isolated may be truncated (missing segments at the 5′ end, for example). In such cases, the partial gene can be used as a probe to "fish" out longer length cDNA clones.

4.4 Joining of Donor and Vector by Ligation

4.4.1 Ligation after restriction cutting

After restriction cutting of the donor piece and the vector DNA, the ligation between these two DNA molecules is usually carried out by T4 DNA ligase at 37℃. Normally the blunt ends ligation needs higher concentration of the enzyme and longer time, that's why the cohesive cutting and ligation would be the right option unless there is only one choice for blunt end ligation. When the donor DNA was cut by two different restriction enzymes at each ends, the direction of the inserted piece will be decided by the cutting sites in the vector we choose, it would be better to orient the upstream side of the foreign gene right behind the expression signals pre-arranged in vectors.

4.4.2 Other recombination methods

Recently, different techniques have been developed to provide some alterative methods for DNA recombination (Tab. 4.1). Some of the methods don't need the restriction cutting as in classic cloning. If both donor and vector have homologous regions, the cohesive ends can be created by other enzymes (for example, **T4 DNA polymerase**), and sometimes, the integration vectors help the foreign DNA be integrated into the host chromosome.

Tab. 4.1 Different techniques for DNA recombination (Alterative cloning)

Method	Feature	Condition
Classic cloning	Restriction cutting, ligation	*in vitro* manipulation
UPS	Univector plasmid-fusion system	With plasmid-phage fusion vector
Gateway	Gateway cloning system	Based on λ integration
RecJoin	Enhance fused protein expression	With fusion vector
MAGIC	Mating assist- integrated cloning	With mating plasmids
SLIC	*in vitro* homologous recombination	T4 DNA polymerase creating protrude ends
In fusion PCR	PCR with homologous single ends	With two protruding ends More than 15 base
Co-transformation	With tags at N,C ends of peptide	Have overlapped section between 2 pieces

4.5 Selection of Host Cells

The type of host cell used for a particular application will depend mainly on the purpose of the cloning procedure. If the aim is to isolate a gene for structural analysis, the requirements may call for a simple system that is easy to use. If the aim is to express the genetic information in a higher eukaryote such as a plant, a more specific system will be required. These two aims are not necessarily mutually exclusive; often a simple primary host is used to isolate a sequence first, then introduce it into a more complex system for expression.

4.5.1 Commonly used prokaryotic host cell lines

The main types of host cell are shown in Tab.4.2. An ideal host cell should be easy to handle and propagate, should be available as a wide variety of genetically defined strains, and should accept a range of vectors. The bacterium *Escherichia coli* fulfills these requirements and is employed in many cloning protocols. *E. coli* has been studied in great detail, and many different strains were developed by microbial geneticists after they investigated the genetic mechanisms of this prokaryotic organism. Such studies provided the essential background information on which genetic engineering is based. *E. coli* can be considered as one of the simplest host. Much of the gene cloning that is carried out routinely in laboratories involves the use of *E. coli* hosts, with many genetically different strains available for specific applications. In addition to *E. coli*, other prokaryotes may be used as hosts for gene cloning experiments, with examples including the species of *Bacillus* for better secretion of proteins, *Pseudomonas* for degradation of contaminant, *and Streptomyces* for antibiotic related metabolites.

Tab.4.2 Types of host used for genetic engineering

Major group	Prokaryotic/eukaryotic	Type	Examples
Bacteria	Prokaryotic	Gram−	*Escherichia coli*
		Gram+	*Bacillus subtilis*
			Streptomyces spp.
Yeast & Fungi	Eukaryotic	Microbial	*Saccharomyces cerevisiae*
		Filamentous fungi	*Aspergillus ndulans*
Plants	Eukaryotic	Protoplasts	Various types
		Intact cells	Various types
		Whole organism	Various types
Animals	Eukaryotic	Insect cells	*Drosophila melanogaster*
		Mammalian cells	Various types
		Oocytes	Various types
		Whole organism	Various types

One disadvantage of using prokaryotes to clone eukaryotic genes is the expression problem. prokaryote. Prokaryotic cells lack the membrane-bound nucleus (and other organelles) as found in eukaryotic cells. Certain eukaryotic genes may not function in *E. coli* as they would in their normal host and the environment which can hamper their isolation by selection mechanisms that depend on gene expression. Also, if the production of a eukaryotic protein is the desired outcome of a cloning experiment, it may not be easy to ensure that a prokaryotic host produces a fully functional protein.

4.5.2 Some of the eukaryotic host cell lines

The yeast ***Saccharomyces cerevisiae*** is one of the most commonly used eukaryotic microbes in genetic engineering. The genetic background and their plasmids have been intensively studied compared with other eukaryotic microbes, and many vectors specifically developed for yeast hosts are available now. Nevertheless *Saccharomyces cerevisiae* has been used in brewery and bakery for several thousand years, it's commonly regarded as the most safe hosts among the eukaryotic species. Recently, more and more human hormones and physiologic factors are cloned into *Saccharomyces cerevisiae*, and made into the valuable pharmaceuticals. Plant and animal cells may also be used as hosts for gene manipulation experiments. Some aspects of genetic engineering in plant and animal cells are discussed in the Part II of this book.

4.6 Introduction of Recombinant DNA into Host Cells

After insertion of a foreign DNA into a vector, the next step is to introduce the resulting recombinant DNA into a suitable host cell. The process of introducing DNA into living cells is called transformation. In a special case where the introduced DNA is a viral DNA, the process is called transfection. The choice of methods depends on the type of host systems in use, as well as the objective of cloning. Some of the procedures for transformation have been briefly mentioned in the discussion of vectors.

4.6.1 Transformation with competent cells

Pretreatment with $CaCl_2$ before transformation

In the case of using *E. coli* host, recombinant DNA can be easily introduced into the host if the cells have been pretreated with $CaCl_2$ or a combination with other salts. The treated cells are called competent cells which mean they are able to take up DNA readily. As in routine experiment, the competent host cells are incubated with recombinant DNA together in a buffer containing 100mmol/L calcium salt solution, under zero degree for 20 minutes, then suddenly being removed to 42℃ water bath (**heat shock** step). It's believed that this step helps recombinant DNA molecules penetrate the cell envelope. The heat shock treatment is only lasting for 40 seconds, then the host cells are replaced again into 0℃ ice bath. Right after transformation step, the host cells should be cultivated in rich broth to allow the recombinant DNA to be duplicated and the new traits expressed.

A common method for introducing a foreign DNA into mammalian cells involves coprecipitation of the DNA with calcium phosphate (mixing the purified DNA with buffers containing calcium chloride and sodium phosphate), and the mixture is presented to the cells in the cultured medium. The individual DNA usually integrates as multiply copies in the cell genome. A wide selection of competent cells featuring a full range of cloning efficiencies and genotypes is commercially available.

With other cell types, transformation generally requires additional treatments. Fungal, cells containing complicate cell walls in some cases need to be digested to produce protoplasts before DNA can be picked up.

4.6.2 Electroporation

Many cells, both prokaryotic and eukaryotic, can be induced to take up DNA by subjecting them to an electric shock, the term **electroporation** appeared. Cells are generally placed in a cuvette between two electrodes and a **potential difference** (i.e. shock) applied across the electrodes. The potential difference is usually generated by charging a capacitor and then allowing it to discharge across the electrodes. The parameters to be altered are the field strength of the shock and the rate of decay of the field. It should be varied according to the particular cells used. It seems likely that the electric shock disturbs the structure of the membrane, making it briefly permeable. The treatment usually leads to death of a fraction of the cells. Establishing optimal conditions for electroporation usually requires a balance between the numbers of cell death and the fraction of cells that are induced to take up DNA. That's why it's important to perform pre-running to establish the optimal parameters for electroporation.

4.6.3 Other methods

There are other transformation methods. The "**gene gun**" method can be applied to plant cells, animal cells, tissues, and organelles, yeast, bacteria and other microbes. Transformed plant cells can be regenerated into transgenic plants carrying the cloned DNA. However, animal cells cannot be regenerated into transgenic animals. For the production of transgenic animals, the DNA is injected into the **pronuclei** of the fertilized egg using a micropipette. For expression purposes, the gene of interest must be properly constructed with a promoter region and other control elements to direct tissue-specific production of the protein. The transformed zygote is implanted into a surrogate mother to give birth to transgenic offspring (detailed in Part II).

4.7 Selection, Screening and Sub-cloning

4.7.1 Directed selection of clones

Success in any cloning experiment depends on being able to identify the desired gene sequence among the many different recombinants that may be produced. There are two terms that require definition before we proceed: **selection** and **screening**. Selection is where some traits are expressed in a visible way for us to pick up the desired individual clone, or be resistant to the pressure has applied on cell growth, after the positive clones accepted recombinant DNA. The cells with the desired characteristics are therefore selected by their ability to survive under the pressure. This approach ranges in sophistication, from simple selection for the presence of a vector up to direct selection of cloned genes by complementation of defined mutations. A successful cloning work is very much dependent upon a well designed selection protocol.

4.7.2 Undirected screening methods

Screening is an undirected procedure by which a population of viable cells is subjected to some sort of analysis that enables the desired sequences to be identified. Because only a small proportion from the large number of transformed colonies or bacteriophage plaques being screened will contain the DNA sequence(s) of interest, screening requires methods that are highly sensitive and specific. There are various screening strategies developed, some of the techniques have been so important that the inventors were recognized as worldwide top scientists. In practice, both selection and screening methods may be required in a single cloning experiment, and may even be used at the same time if the overall procedure is carefully designed. A few examples of screening methods are introduced as follows:

Insertional inactivation

The presence of cloned DNA fragments can be detected if the insert interrupts the coding sequence of a gene. This approach is known as **insertional inactivation** and can be used with any suitable genetic system. Antibiotic resistance can be used as an insertional inactivation system if DNA fragments are cloned into a restriction site within an anitbiotic-resistance gene. The X-gal system can also be used as a screen for cloned sequences. If a DNA fragment is cloned into a functional β-galactosidase gene, any recombinants will be genotypically *lacZ′* and will therefore not produce β-galactosidase in the presence of IPTG and X-gal. Plaques containing such phage will therefore remain colourless. Non-recombinant phage will retain a functional *lacZ* gene and, therefore, give rise to blue plaques. This approach can also be used with the **α-complementation** system; in this case the insert DNA inactivates the *lacZ′* region in vectors such as the M13 phage and pUC plasmid series.

Blotting techniques

The power of nucleic acid hybridization lies in the fact that complementary sequences will bind to each other with a very high degree of fidelity. In practice this depends on the degree of homology between the hybridising sequences, and usually the aim is to use a probe that has been derived from the same source as the target DNA. However, under certain conditions, sequences that are not 100% homologous can be used to screen for a particular gene, as may be the case if a probe from one organism is used to detect clones prepared using DNA from a second organism. Such heterologous probes have been extremely useful in identifying many genes from different sources. There are three main types of DNA probe: 1) cDNA, 2) genomic DNA, and 3) oligonucleotides.

Blotting without separation of sequences can be a useful technique for determining the amount of a specific sequence in a sample—often this is used to measure transcript levels in gene expression studies. There are further variations on the blotting theme. If nucleic acid samples are not subjected to electrophoresis but are spotted onto the filters, hybridisation can be carried out as for Northern and Southern blots. This technique is known as dot blotting and is particularly useful in obtaining quantitative data in the study of gene expression. In some variants of the apparatus the nucleic acid is applied in a slot rather than as a dot. Not surprisingly, this is called slot blotting. The final technique is known as Western blotting, which involves the transfer of electrophoretically separated protein molecules to membranes. With SDS-PAGE, Western blotting can identify proteins specifically if the appropriate antibody is available. The membrane with the proteins fixed in position is probed with the antibody to detect the target protein in a similar way to immunological screening of plaque lifted from expression libraries.

PCR screening

The polymerase chain reaction (PCR) can be used as a method for screening clone library, although there are many cases in which the more traditional approach to library screening may be more efficient. One way of reducing the number of individual PCRs being performed is to carry out PCRs on pooled samples of clones. This is known as **combinatorial screening**.

Immunological screening

An alternative screening beyond nucleic acid probes is to identify the protein product of a cloned gene by immunological screening. The technique requires that the protein is expressed in recombinants, and is often used for screening cDNA expression libraries that have been constructed in vectors such as λ gt11. Instead of a nucleic acid probe, a specific antibody is used. Antibodies are produced by animals in response to challenge with an antigen, which is normally a purified protein. There are two main types of antibody preparation that can be used. The most common are polyclonal antibodies, which are usually raised in rabbits by injecting the antigen and removing a blood sample after the immune response has occurred. The immunoglobulin fraction of the serum is purified and used as the antibody preparation (antiserum). Polyclonal antiserum contain antibodies that recognize all the antigenic determinants of the antigen. A more specific antibody can be obtained by preparing monoclonal antibodies, which recognize a single antigenic determinant. However, this can be a disadvantage in some cases.

Monoclonal antibody production is a complicated technique in its own right, and good-quality polyclonal antisera are often sufficient for screening purposes.

Screening based on mRNA translation in vitro

In some cases the identity of a particular clone may require confirmation. This is particularly true when the plus/minus method of screening has been used, as the results of such a process are usually somewhat ambiguous and cannot give a definitive identification. If the desired sequence codes for a protein, and the protein has been characterized, it is possible to identify the protein product by two methods based on translation of mRNA *in vitro*. The two methods are known as **hybrid-arrest translation** (**HART**) and hybrid-release translation (HRT, sometimes called hybrid-select translation). Although these techniques are no longer widely used in gene analysis, they do illustrate how a particular problem can be approached by two different variations of a similar theme—a central part of good scientific method.

4.7.3 Sub-cloning if necessary

No one guarantees that only from one run of cloning work can the engineers get the ideal clones from the screening step. A successful result sometime includes the **sub-cloning** procedure which works precisely on the primary clones based on the more detailed sequence information.

Restriction mapping

Obtaining a **restriction map** for cloned fragments is usually essential before additional manipulations can be carried out. This is particularly important where phage or cosmid vectors have been used to clone relatively large pieces of DNA. If a restriction map is available, smaller fragments can be isolated and used for various procedures, including subcloning into other vectors, the preparation of probes for chromosome walking, and DNA sequencing. If a cloning protocol has involved the use of bacterial or yeast artificial chromosome vectors, with inserts in the size range of 300~500kb, then so-called long-range restriction mapping may be needed initially.

DNA sequencing for cloned gene

The development of rapid methods for DNA sequencing, as outlined in Section 3. 6, has meant that this task has now become routine practice in most laboratories where cloning is carried out. Further sequencing the cloned gene provides much useful information about coding sequences, control regions, and other features such as intervening sequences. Thus, full characterization of a gene will inevitably involve sequencing, and a suitable strategy must be devised to enable this to be achieved most efficiently. The complexity of a sequencing strategy depends on a number of factors, the main one being the length of the fragment that is to be sequenced. Most manual sequencing methods enable about 300~400 bases to be read from a lane of sequencing gel. If the DNA is only a few hundred base pairs long, it can probably be sequenced in a single step. However, it is more likely that the sequence will be several kb in length; thus, sequencing is more complex.

Modification and re-ligation

Performing of sub-cloning would be very dependable, in some cases, the primitive vector can be replaced by another one for the sakes of safety, higher copy number, efficiency, or easily operative; and the donor piece would be modified by sequence adding, or shortening, or replacement, for enhancing the expression of foreign gene to improve both the quality and quantity of the desired product. Ligation of the reconstructed donor and vector would be performed again as mentioned above. Hopefully the clones from the sub-cloning procedure would be over advantaged compared to the primitive clones.

4.8 Derivative Technology Based on Cloning

Along with the advance of gene cloning more derivative techniques have been invented, which greatly accelerate the progress of genetic engineering. In the following section, two of the innovative derivative technologies are briefly introduced.

4.8.1 *In vitro* mutagenesis

By insertion, deletion or base substitution, the DNA sequence would be precisely altered and thus the amino acid sequence of the related protein might be changed. it's believed that the artificial evolution of the treated organism might occur in a very short period through in vitro mutagenesis. By the way, from scientific points of view, the advanced methodology also provides the researchers a free kingdom for investigation of the relation among the sequence, structure, function and even the signal transmit.

Site-directed mutagenesis

One variant of this technique is called **oligonucleotide-directed** or **site-directed mutagenesis** and is elegantly simple in concept. The requirements are a single-stranded template containing the gene to be altered, and an oligonucleotide (usually 15～30 nucleotides in length) that is complementary to the region of interest. The oligonucleotide is synthesised with the desired mutation as part of the sequence.

The technique of site-directed mutagenesis should become more useful in rationally designing enzymes with new properties. It is dependent on detailed knowledge of structure and enzyme mechanism and does not require screening or selection because the effects of each substitution can be individually analyzed. Currently, the most powerful application for site-directed mutagenesis is its use in conjunction with other mutagenesis techniques. Selection of an enzyme containing several amino acid substitutions, site-directed mutagenesis can be used to individually test each of the substituted amino acids and thus yield information on mechanisms of catalysis.

Random mutagenesis and DNA shuffling

Random mutagenesis removes the need for predictive alterations to DNA and protein sequences and is potentially more powerful than site-directed methods.

The process generates a library of recombinants that encode the protein of interest, with random mutagenesis (often using techniques such as **error-prone PCR**) applied to generate the variants. Thus, a large number of different sequences can be generated, some of which will produce the desired effect in the protein. These can be selected by expressing the gene and analyzing the protein using a suitable assay system to select the desired variants. Additional rounds of mutagenesis and selection can be applied if necessary.

An extension to the above technique called **DNA shuffling**, which can be used to the mix pieces of DNA from variants that show desired characteristics. This mimics the effect of recombination that would occur *in vivo* and can be an effective way of "fast-tracking" the directed evolution technique.

Some successful examples of random mutagenesis are about improvement of enzyme properties. For example, Japanese scientists succeeded in creating thermostable and broader pH tolerant **cholesterol oxidases** by random mutagenesis. This engineered cholesterol oxidase exhibits 100% stability at 40℃ for at least 4 weeks, which has been used in clinical diagnosis to analyze cholesterol levels in various specimens. And the optimum pH was desirably changed in the case of another strain to encompass a broad range from acid to alkali.

Mutagenesis by targeted gene knockout

Targeted gene disruption (**gene knockout technology**) is used to inactivate a given gene(s). Compared with site-directed mutagenesis, gene knockout has the advantage of ensuring the insertion of the selectable marker gene and the mutation in the target gene. The limitations are that mutations are not easily obtained, so it can be difficult to dissect the function of essential genes. Despite theses drawbacks, gene knockout offers one of the most powerful means now available for studying gene function in a living animal. Such studies will accelerate efforts to translate newfound knowledge of the human and mouse genomes into better strategies for diagnosing, treating, and preventing human cardiovascular heart defects.

4.8.2 Antisense techniques based on RNA interference

The discovery in 1993 of a gene expressing small non-protein coding RNAs in a small worm, *Caenorhabditis elegans*, led to the discovery of a previously unknown class of endogenous, single-stranded 19～21 nucleotide molecules called **microRNAs** (**miRNAs**) that can regulate up to 30% of mammalian genes. Based on the discoverer, Craig Mells, was awarded in 2006 the Nobel Prize for Physiology and Medicine. These miRNAs have been shown to inhibit specific mRNA molecules and prevent translation of large numbers of messages at a time. The naturally occurring miRNA is usually composed with a short sequence complementary to a segment of mRNA, by which it blocks the translation from this mRNA. The potential ability of miRNA is to ignite the cell's own gene-silencing machinery, makes **RNA interference** phenomenon the most attractive for cloning technology and clinical research. Current practical applications with interference RNAs are almost exclusively related to technical invention, so called **Antisense technology**, which involving synthesis of oligo-nucleotide sequence with complementary pairing ability to the target mRNA and cloning into the related host. It has long been recognized that genes can be silenced by means of antisense technology. An antisense gene can also be constructed by reversing the nucleotide sequence of the gene. The sense and antisense RNA strands are antiparallel and, thus, form a double-stranded RNA molecule similar to double-stranded DNA. The formation of this double-stranded RNA molecule results in decreased translation of the sense RNA. Antisense technology was used in transgenic plants to control ripening—the Flavr Savr tomato. This technology has also been extensively practiced in transgenic organisms and in gene therapy.

Summary

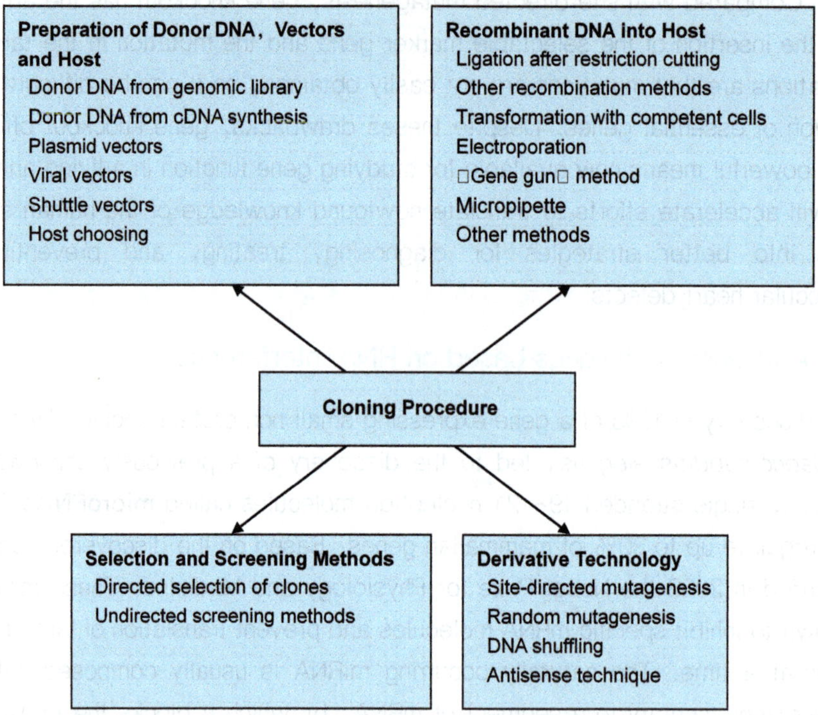
Preparation of Donor DNA, Vectors and Host
Donor DNA from genomic library
Donor DNA from cDNA synthesis
Plasmid vectors
Viral vectors
Shuttle vectors
Host choosing
Recombinant DNA into Host
Ligation after restriction cutting
Other recombination methods
Transformation with competent cells
Electroporation
□Gene gun□ method
Micropipette
Other methods
Cloning Procedure
Selection and Screening Methods
Directed selection of clones
Undirected screening methods
Derivative Technology
Site-directed mutagenesis
Random mutagenesis
DNA shuffling
Antisense technique

Chapter 5 Application and Impacts

作为20世纪70年代诞生的最具革命性意义的基因工程问世不久，就得到科学界广泛的重视，同时也得到工业界、医学界和农业工作者的高度关注。几年后，这项技术很快就被应用于生产人的激素和生理活性因子，原先只能从人体中提取的极其宝贵的医疗制剂，通过克隆技术得以批量生产。从此以后，基因工程技术在医药、农业、食品与发酵工业、化工、材料等领域得到不同程度的应用，今天基因克隆产品或转基因食品已经进入千家万户，直接影响着人类的生活与社会进步。不过，基因克隆技术应用在生物遗传、环境生态、伦理道德、国际贸易等方面，甚至可能成为生物武器的负面效应，也引起人们一定程度的忧虑和警觉。科学必须与正义联手才能造福于人类。

5.1 Initiation of Recombinant DNA Industry

The scientists who developed gene cloning were at the first glance to realize this technology had great commercial potential. Instead of dreaming about the varieties of application resulting from this new invention, they were really concerned about whether the discoveries would be prevented from being commercially produced.

In 1975, with the name Asilomar, the famous conference center on California coast, there was bursting out serious debate on the guidelines concerning the regulation of recombinant DNA techniques. Those troubled years that were dominated first by the regulation and then by the deregulation of cloning research and application, we now tend to associate with sociopolitical attitude on safety, but not the typical scientific approaches. Anyway, the early application cases can be traced back to 1977, with advent of the technology, the first cloned human protein—**somatostatin** was produced by the first genetic engineering company—Genentech. More scientists began to cooperate with industries to clone the sequences coding pharmaceutically useful proteins into *E. coli* or yeast cells, and cultivated those microbial hosts by fermentation for production of hormones and bioactive factors.

5.1.1 Early commercial purpose

The aim of producing cloned proteins in the late 1970's, was mainly focused on exploitation of new route to allow the human or animal proteins to be made from microorganisms, and to produce those valuable pharmaceuticals with low costs and high efficiency. Early cloning products were: **insulin**, the important hormone to regulate carbohydrate and fat metabolism in the body; **human growth hormone** (**HGH**), the key factor in growth control; somatostatin, the small circle peptide to control the release of growth hormone; **interferon**, the antiviral agent, which belongs to the large class of cytokines, with the ability to " interfere" in the viral replication in the host cells; **plasminogen activator**, the thrombolytic **agent** which causes **fibrinolysis** at the site of a blood clot; and **blood clotting factor VIII**, the blood coagulation factor for treatment of type **A haemophilia** patients.

Another prospect was to harness *E. coli* or yeast to produce **viral coat proteins** for use as safer vaccines, such as the engineered **surface antigenic** vaccine against **Foot-Mouth Disease**. The recombinant **yeast-derived vaccine** would be a better choice when the antiviral vaccines cannot be prepared from the animal cell culture system, for example, this technique has been applied in the manufacture of the vaccine against **hepatitis B virus**.

Beyond the pharmaceutical industry, there were hopes that the genetically engineered microbial hosts would carry out processes of importance to the petroleum, chemical and mining industries, and the new germ lines for agricultural plants and animals would be easily bred based on the **transgenic technique** (Part II).

"US Genentech Company", established in 1976, should be remembered for this is worldwide the first genetic engineering production sector. The company produced somatostatin in 1977, then insulin in 1978, and human growth hormone in 1979. More genetic engineering companies were set up at the end of 1970's, and they were pioneer industries which contributed not only the cloned products, but also the practical methodologies for cloning application, which allowed the desired genes to be well expressed in specific recipient hosts and the cloned protein molecules easily isolated and the original bio-functions well kept.

5.1.2 Early cases of industrial cloning

Human insulin produced from bacterial cells

Human insulin, the peptide secreted from pancreas into bloodstream, is one of the most important hormones. Mature insulin consists of two peptides, one of 21 amino acid (A chain) and the other of 30 amino acid (B chain), which are covalently held together by disulfide bridges. It sounded easier to clone A and B segments into *E. coli* host individually, isolate and purify peptides of A and B, and then mix both peptides to let disulfide bridges form. But actuarially this procedure had produced the active insulin molecule in a very low proportion, since there are 6 residues of Cys that exist in A and B chains, randomly bridging between any two Cys would produce nearly a hundred different linkage patterns, but only one of them would have insulin activity.

Then the people remember that insulin is an enzymatic cleavage product of its precursor **proinsulin** which has a surplus C chain (35 amino acid) in between the A and B chains. The structure of insulin is actually determined by proinsulin which directs the correct forming of those disulfide bridges.

In the second approach to producing human insulin, the proinsulin sequence was cloned into *E. coli* and let it be well expressed into the peptide consisting A, B and C sections. The peptide chain of proinsulin folded into its natural 3-D structure, all disulfide-bridges were linked properly as in the natural pattern, then the C chain of proinsulin was finally cleaved away by some specific proteases at the well designed adjunctions between A and C, also C and B, to yield pure human insulin with correct structure.

cDNA cloning to produce γ-immune interferon

No one knows why the DNA sequences in human genome coded for α- and β-interferon are not interrupted by introns, expect γ-**immune interferon**, although interrupted genes are dominated in human genome. For producing of γ-immune interferon, without stripping way the three introns in between the exons, the direct cloning of whole set of its interrupted DNA sequence would certainly be fruitless. To solve the problem, human spleen cells were exposed to a bacterial toxin in order to induce the γ-interferon mRNA synthesis. The mRNA collection was injected into frog' oocytes and let it be translated, the translational products were then assayed for the γ-interferon activity. In this case, the distinguish work was done with an early innovative technique, named cDNA cloning. The extraction of matured γ-interferon mRNA was used as a template for revetro-transcriptase to synthesize a complementary DNA strand, and then the 2nd DNA strand synthesized, paired to the 1st. the resulting cDNA molecule has been ligated to a shuttle vector with both sequences form SV40 and bacterial plasmids, then the recombinant DNA was introduced into *E. coli* host first and propagated, then the amplified recombinant cDNA was isolated and transformed to COS cells, in this way, the SV40 sequence allows the cDNA with γ-interferon coding section to be expressed. Since then, γ-interferon can be obtained in commercially amounts to meet the needs of treatment on viral infection or cancer.

The cDNA cloning techniques had been individually optimized by many cloning companies, to produce varieties of peptide which were originally coded by interrupted sequence in human genome.

Synthetic DNA sequence cloned into human growth hormone

Human growth hormone (HGH) is badly needed either for increasing the height of dwarf persons, or for curing the wounded skin or burns. The extraction of HGH from human brain would never become a commercial practice to meet the medical needs, since the amount of naturally existing HGH is far less than required.

The idea of HGH production by cloned *E. coli* had drown great attention from industries at the end of 1970's, since it's a single peptide that consists of only 191 amino acids and is not glycosylated. It seems to be an easy work for gene engineers to manipulate the HGH cloning, but actually the early work was involving a combination of an uncompleted cDNA molecule which was reverse-transcripted from **human pituitary cell** mRNA and a chemical synthesized short sequence. In an earlier case of cDNA synthesis, the upstream regions of cDNA were normally incomplete caused by S1 nuclease over digestion, which indicated the translational signals and the coding sequence upstream might be lost from the cDNA molecule. In order to provide the complete sequence, a DNA fragment coding from the 1st amino acid to the 24th was synthesized chemically, and the DNA sequence coding for peptide from 25th to 191th amino acid was obtained from cDNA technique. The linkage of these two segments was cleverly designed by arrangement of two homologous cohesive restriction sites in it. More jobs were subsequently done, for example, immediately in front of the 1st codon an initiate triplet of bases (ATG) was added, and all the **signal factors** involved with protein expression were properly arranged. Now production of the cloned GHG runs into a matured procedure and provides enough products to meet the requirements.

Cloning viral surface protein to produce safer HBV vaccine

In 1979, the complete genome from human hepatitis B virus (HBV), a partially single-stranded DNA molecule around 3200 bases, was cloned and sequenced. This allow the scientists to deduct the peptide sequences of two HBV proteins, the **core protein** associated with DNA existed in the center of HBV, and the major **surface antigenic protein** which is part of the membrane-like envelope of HBV. The core protein of hepatitis B has been made by cloning into *E. coli* and produced in commercial quantities, and has been sold to diagnostic labs in many countries for detection of HBV in blood of the suspect infectors.

The real goal of making HBV vaccine was to prevent this most prevalent virus spreading further to harm more people. Since the cloning to *E. coli* was much easier than the other organisms, the first attempt of getting high expression of HBV **major surface antigen** in this commonly used prokaryote was rather disappointing, even the powerful bacterial promoter had been linked ahead of the cloned sequence. Later the gene of major surface antigen was linked behind a yeast promoter, ligated with an autonomously replicating yeast plasmid, and introduced into *Sacchramyces* recipient cells. The cloned HBV surface protein molecules, so called "22nm particle", are in fact the empty virions, consisting only of surface antigen but no DNA and the associated core protein. Compared to the natural particle of HBV, it closely resembles the surface protein in structure and the immune-activity. No doubt, safer application of engineered vaccines would be a critical clue in the industrial cloning strategy.

5.2 General Aspects of Application

Not at all surprising, gene cloning technology has reached the state of arts after 40 years' research and exploitation. Aimed at different purposes, huge amount of industrial strains and agricultural organisms had been genetically engineered to offer them better hereditary traits and higher added value; human hormones and bio-active peptides had been produced form engineered microbial producers in safer, cheaper and reproducible way; altering the coding sequences of protein molecules had either improved or suppressed the protein expression in order to enhance or block the critical branch points in metabolic pathway, by which the quality of bio-production would be greatly improved; the work of combination of several sequence fragments from different gene resources had produced the so called **multi-antigenitic vaccines**; the method of ***in vitro* mutagenesis** had been performed on all kind of genes, precisely at any single base or a defined section, to change protein structure or enzymatic property for practical uses; based on **DNA shuffling techniques**, the genes from several strains carrying different sequences of **fructose-glucose isomerase** had been chopped with the same restriction enzyme, and the restricted pieces were mixed together for random ligation, finally the best clone was selected from this **man-made evolution** procedure; and even the randomly synthesized sequences had been massively cloned into *E. coli* cells to construct a peptide pool for further selecting the so called novel bio-medicines. The **immunodiagnostic agents** and DNA probes DNA for accurate identification on rare diseases can also be produced from specific techniques of gene cloning. According to the **British National Economic Development Office**, the worldwide total Sales of Bio-production has already exceeded £60 billion per annum by early this century.

It's also noticed that the cloning applications need huge amount of budget and teams of very intelligent scientists and engineers, and the bio-industries were also characterized as high investment and high risk sectors. Probably, that's why either the academic research or practical application had progressed slowly in most developing countries. Some of developing nations had successfully collaborated in the past with western biotechnology companies, but it was noted that during the biotechnology booming period (1990's), on the contrary, the percentage of arrangements implemented by US bio-companies with developing nations dropped from 20% to 3%.

The ability of developing countries to avail themselves of promises of this new technology, to a large extent, depends on their capacity to integrate the up-to-date biotechnology into their own system of innovation and application, in accordance with their own situation and need. In China, the cloning research started in the late 1970's, since then, varieties of bio-products had been exploited and many species of agricultural plants and animals had been cloned and put into daily application. Given that the global population has surpassed 7 billions at the end of October, 2011, and the proportion of the aged people in many nations are increasing, successful application of gene cloning might hold the answer to providing enough foods and varieties of pharmaceuticals , and in the meanwhile maintaining the ecological balance for human being's common benefits.

5.2.1 Gene cloning applied in medical fields

Cloned pharmaceuticals

Indeed, the majority of investment on "new" genetic engineering production over 30 years has been in the area of healthcare, especially in the exploitation of novel drugs. Many pharmaceuticals based on gene cloning are appearing almost daily in the marketplace, which include: human and animal hormones; small peptides with special regulational function; and bio-active factors. Although the pioneer application works started at the end of 1970's, but the gene engineered drugs were not broadly approved for marketing until well into 1980's. Tab. 5.1 lists the examples of early bio-pharmaceuticals.

Tab.5.1 Examples of early bio-pharmaceuticals

Products	Medical uses	Time of appearing
Insulin	Diabetes	1982
Somatostatin	Growth control	1983
HGH	Growth deficiency	1985
Interferon	Cancer, viral infection	1985
Tissue plasmogen activator	Cardiovascular disease	1987
Erythropoietin	Anaemia	1989
Interleukin-2	Cancer	1992
Coagulation factor VIII	Haemophilia	1992

Note: Data partially from Jonh E. Smith, 2004, Biotechnology, 148-152

Before 1990, there were not many bio-pharmaceuticals officially approved for marketing, probably due to the fact that the technique itself at that time was not that advanced, and the commercial investors had not foreseen the promising future. With the domestic progress of science & technology and the rapid growth of global economy, pharmaceutical markets were increasingly developed. Then hundreds of cloning drugs have been manufactured and put into clinic application, since the first batch of the bio-pharmaceuticals appeared not to have serious problems in the short term. Till the end of the last century, hundred of local Bio-industry Parks had been established which attracted investment from oversea investors. Some of China-made bio-pharmaceuticals based on the cloning technology are listed as follows: insulin, human growth hormone, α, β, and γ-interferon, interleukin, tumor necrosis factor, plasminogen, single cell colony stimulating factor, urokinase, epidermal growth factor, transforming growth factor, brain natriuretic peptide, atrial natriuretic peptide, superoxide dismutase, cytokines, anti-cytokines, toxins guide, and vaccines against HBV, flu viruses, and animal infection viruses. Rice, wheat, maize, soy bean, cotton, and rape etc. were the crops into which foreign genes have been introduced. Those transgenic plants are now grown into the staple crops. **Transgenic farm animals** are still kept under strict monitoring, although some of the mammalians had been cloned into their mammary gland with genes which coding human peptides, for production of valuable bio-pharmaceutical.

The current global bio-pharmaceutical market is still growing. A lot of patients benefited from those fantastic "cloned" drugs and diagnostic agents. We also notice that many related research results are now still kept in the labs, for the "cloning" products are not easy to pass the strict approval on new drugs, and the biomedicine market should be kept in a reasonable range.

Vaccines

Viral vaccines normally consist of viral particles that either have been inactivated or weakened so as to ensure that they can no longer cause infection. Using these traditional viral vaccines there is always a risk that one or more viral particles might survived after inactivation, then terrible accidents might occur which cause man-introduced contamination and further pestilence. Producing viral vaccines by cloning part of viral genome into *E. coli* or yeast would be much safer, and in principle would be less expensive than that of propagating virus in animal cells.

In the early 1980's, scientists started to work on the cloned vaccine against Food-and-Mouth Disease Virus (FMDV), in this case, the single stranded RNA genome of FMDV was isolated and reverse-transcripted into cDNA, which was later introduced into *E. coli* host and well expressed in quantity of several million molecules per cell. Usually, it had been observed that the immune-ability of cloned vaccines, consisting only of the major surface protein, might be weaker than that of inactive viral particles.

After the cDNA of FMDV had been sequenced, the peptide sequence can be deduced. Researchers noticed that the critical peptide segment can be produced from a synthesized DNA sequence to make new vaccines, and the positive antigenic results coming out from HBV and influenza virus peptides were very encouraging. After that, the manufacture of multi-antigenic vaccines had been exploited, by ligating together of several DNA sequences coding for different viral peptides.

Gene therapy

Gene therapy can be referred to as any medical treatment strategy that involves with the cloning of genes or genetic materials into human body for alleviation or elimination of diseases. The aim of gene therapy is to replace or repress defective genes with DNA sequence encoding the right genetic message, in order to provide positive genetic instruction for treatment of hereditary disorders.

In the majority of gene therapies, the clinical trails had used the cloned viruses or retroviruses to deliver recombinant DNA through their infective mechanisms. The viral vectors are able to reach human chromosome, some of them even integrate into the host DNA sequence. it's necessary to distinguish between **germ-cell gene therapy and somatic-cell gene therapy**. In the former, redresses occur at the individual's genetic make-up, which can be passed on to their offspring. Most of the ethics and sociologists had suggested that this type of therapy should not be permitted in any countries in the near future lest some unknown genetic traits are passed on and run out of control in eco-system. In the latter, only the somatic cells are receiving the cloned sequence which is not supposed to pass onto the recipient's offspring.

There are some successful cases of single-gene therapy, for example, injection of rectified genes into target tissue of bone marrow for **curing thalassaemia**; of liver for treatment of liver cancer; of muscle or brain to rectify **muscular dystrophy**; and of fibroblasts for curing of **haemophilia B**. Somatic-cell gene therapy against complex multi-factorial diseases, e. g. **Parkinson's symptoms** and cancer, must be a long way off, and dealing with the diseases involving with many genes which interact with environmental factors would be tougher for the doctors and genetic engineers.

In addition, the marketing of DNA probe for disease monitoring and **genomic fingerprinting** has expanded quickly to several billion dollars per year, which provides evidential data for gene therapy or for criminal identification.

5.2.2 Agricultural application

Agriculture is most important to human society, and is undergoing, in many parts of this world, a revolutionary restructuring to achieve higher productivity and better quality. There are increasingly desires to innovate the breeding techniques for cultivation of new germ lines with the specific traits related to higher production, added value, and easier biological control.

Plant breeding by cloning

The first GMC (genetically modified crop) tobacco was cultivated in 1983, and **delayed ripening tomato**, the first commercialized GMC emerged in 1984, in USA. Since then some other countries such as Canada, Argentina, and Brazil began to construct and cultivate GMC. There has been a remarkably successful commitment, as witnessed by the superior traits of the cloned crops, such as rice, wheat, corn and potato. The cloning of agricultural plants does not only allow rapid propagation and higher production, but also has some other important uses, for example, cloning of antimicrobial or antiviral sequences into crops to eliminate pathogen or viruses, and cloning of **anti-herbicide** or anti-pesticides genes to control weeds and insects. Transformation of special sequences into plants had also raised the productivity and quality of **the secondary metabolites** of valuable plants, and stimulated their commercial application. The proportion of transgenic crops increased dramatically in US recent years, for example, the outputs of transgenic soy beans and maizes have now reached 93% and 90% respectively of their total in 2011, according to the up to date news issued by White House Office of Science & Technology of USA.

Targets of forest genetic improvement are clear: apical dominance, shorter cycle time, higher wood quality, resistant to diseases and pests, better outward appearance and smaller land base needed. The genes related to those traits had yet to be identified and cloned into some of the forestry species.

Application of transgenic animal

Animal agriculture benefits human being by providing meat, milk and egg. The form of poultry, pig, cattle, sheep and fish represents a major aspect of animal products worldwide. The first recorded cloned animal "**Super Mouse**" had received the growth hormone gene, and its progeny became twice as bigger as than the parents. The most unusual and commercially feasible application, so called "**Molecular Ranch**", appeared at the end of 1980', by cloning of human genes into certain lactating animals such as cow, sheep, rabbit and pig, valuable proteins would be secreted along with milk from the mammal bioreactors, and later be extracted and finally used pharmaceutically. Particular relevance involving with animal cloning would be the transgenic breeding for germ lines of disease free, rapid growth, and super nutritional quality. Yunnan researchers have revealed a series of recent results in which the very unique **Yunnan minipig** from the inbreeding of 25 generations had been successfully integrated and expressed with ***Porcine endogenous retrovirus*** in 2010, then the human genes can be introduced into those minipig bodies to make them the ideal species for **human organ transplantation**. Some of fluorescent genes had been cloned into mice, pigs, and even dogs, and showed very exiting result with displaying function, for example, in 2009, the Korean scientists have successfully cloned four transgenic beagles that glow red under ultraviolet light, and the transgenic beagles were named "**Ruppy Puppy**".

5.2.3 Approaches for environmental application

Application of environmental microbial metagenomics

The advanced cloning and sequencing techniques enable scientists to collect and analyse the genomic information from environmental microbial communities. In 1998, Professor Handelsman has put out the technical term "**Metagenomics**". By the subsequent studies, massive genomic information from uncultured and unknown environmental microflora were obtained, for example, novel antibiotics were found from metagenomic study on **streptomyces**, and novel cellulases were screened from soil microflora existing in paper milling environment. Also the **human commensal flora**, discharged along with manure into soil and water, has been metagenomically studied, which helps human beings understand some of our own health problems. Many genes coding specific enzymes with chemical conversion abilities have been found, which added originally unknown **biocatalysts** to our list.

Bioremediation

Bioremediation provides a technique for cleaning up pollution by enhancing biodegradation that occurs in nature. Along with the development of economy, environmental pollution becomes more serious. Oil spills, fertilizer remains, chemicals effluents, and domestic waste discharges caused soil, water and air to be heavily polluted. The cloned microbials with combinatory biodegradation ability were constructed, such as in the case of isolating genes with petroleum hydrocarbon degradation functions from *Pseudomonas*, and putting them into oceanic microorganisms to treat oil spills on sea surface. In other cases, the genes involving with the degradation of aromatic compounds were cloned into soil bacteria for treatment of chemical pollution in soil. But indeed, it's not a proper approach to do the cleaning up after pollution happens, and more important works must be done to prevent the pollution.

Pollution control

Using enzymes rather than chlorine for pulp and paper manufacture is an intellectual choice for controlling pollution from pulp waste water discharges, the strains carrying cellulase, **hemicellulase**, and **ligninase** were reconstructed to treat pulp preparation in order to eliminated the chemical treatment, and in the meanwhile to improve the paper quality. Pretty the same purpose, some of the cloned strains carrying **amylase**, **pectinase**, **catalase** and **protease**, yet tolerant to higher pH, were used to treat cotton or wool fibers to reduce the pollution from the **chemical finishing process** in textile industries. Recovery of toxic heavy metals from the **mine tailings** by microorganisms helps reduce pollution and in the meanwhile saves the mine resource, some of the novel strains have been genetically improved to tolerate extreme pH and temperature condition and used in the **bioleaching** process.

Making full use of all the bio-resources sounds comprehensive and reasonable. The bio-recycling practice becomes more important when people notice that not much space is left to them and also not many resources remain as endless supply. The aim of cloning for bio-recycling might involve reconstructing of some microbials which can digest the crop waste and animal manual rapidly to improve soil quality and reduce chemical fertilizing. Cloning of green algae to enhance their assimilation of N, P and CO_2, producing algae oil and protein for human beings would be meaningful.

5.3 Contribution and Impact

5.3.1 Contribution

It's very hard to evaluate the lash from gene cloning technology on the human society. People had discovered enormous of genes and their individual sequences, renewed a series of theories on molecular biology and genetics, created incredible tricks for manipulating on the tiny unseen genes and proteins, constructed varieties of **genetic modified organisms** (**GMO**), established abundant gene library and collected massive genomic information.

Based upon the cloning technology, the average human life span in the developed and rather developed countries has been elongated. New diagnostic agents had been available to detect illness which might not be easily identified by normal diagnosis. Increasingly apparently novel pharmaceuticals are so effective for curing odd diseases. Based upon the gene therapies, some of the patients who suffer from **hereditary disorders** might have a chance to be rectified. The statistics figures showed that bio-pharmaceuticals accounted for about 65% of the total marketing of biotechnology at the early 2000's, which pointed out the promising future of the application in this field.

Given that the populations in most developing countries grow rapidly, the importance of improving agricultural yield and quality will never be over-estimated. The work of cloning **super-rice** by Chinese scientists emerged a beautiful result, which has been featured on the front cover of *Nature* in May, 2003. One should make sure there will always be enough food for the people worldwide to consume and survive. The practical contribution of cloning technology to agriculture area could also be expressed by a figure, in which 28% of total bio-market was generated from transgenic plants, animals, and bio-herbicides and bio-pesticides. There are more tasks waiting for the genetic engineers to accomplish, for example, offering agricultural products rich in nutrients and bioactive components; providing cloned soil microorganisms with special functions to restore the soil fertility and clean up the pollutants.

The accelerating extinction of rare species has worried the scientists and public very much. Some of the genomes of endangered organisms have been isolated and stored in genomic banks or gene libraries in case of their extinction. The bio-rescue based on cloning seems not the best resolution for protection of the invaluable wild creature, and much more should be done on the level of remedying their living environment and maintaining the ecological harmony.

No doubt, only the fancy of Information Technology can be compared with that of gene cloning, for both brightly blaze totally the new trails and strongly accelerate the progress of science and technology and benefit our modern society.

5.3.2 The possible negative impacts

While the cloning technology is providing powerful means to improve health care, increase agriculture yield, substitute some of the chemical catalysis with bio-catalysis and provide pollution treatment, there is a growing concern among the public that about **genetically modified products (GMP)** might have diverse problems of **bio-safety, ethic wreck, trading monopoly, and warfare attempt** etc.. Undoubtedly, gene cloning technology would not maximize its great potential to benefit human beings and their society in the near further without paying attention to the possible negative impact and being aware of the public understanding.

Biosafety

In all impacts of gene cloning application, however the bio-safety issue is of paramount importance which might involve:

(1) Increasing environmental pool of antibiotic-resistant organisms. The cloning vectors used are often containing antibiotic resistant genes, when those genes carriers enter human body or are drained into water and soil, based upon the principle of Survival of the fittest in Natural Selection, the drug resistant ability of the bacteria in large would be promoted to a higher level. The antibiotics dosage spreading onto petri dish in labs is now at the level of 100g/mL, which is almost 10 times higher than that during 1970's.

(2) Bringing **potential pathogenicity**. The vectors and hosts used might carry natural or engineered pathogenic traits, and there is a potential danger of infection to human, animals and plants, causing disease, toxicity, and other problems.

(3) **Produce unsafe GM components** in food or feed. For example, Brazil GM nuts caused irritable problem for some of the consumers and the Scottish scientists once reported that the rats' immune function declined after being fed GM potato cloned with **Lectin Gene of Snowdrop**.

(4) Safety aspects associated with contamination during the industrial process might happen. The **contaminant** existing in fermentation broth would be quickly reproduced in large scale bioprocess, and the final product could be ruined by unsafe microbes.

(5) **Lost control of the mutated strains.** Since the mutated organisms would be drained into water and soil freely without strict control, those mutants would hybridize with natural species in the environment, and made the mutated traits spread at large.

Essential to the concept of biosafety are the recognition and attention of the key terms "**hazard**" and "**risk**". In the context of safety, a hazard is the potential to cause harm, while a risk defined the chance of being harmed by the hazard. Scientists and legislators are now work hard to revise the biosafety standards based upon international technical state-of-the art and relevant legislation, which always aims at preventing all kinds of hazard and risk which threaten human health, and ecological balance.

Environmental and ecological disturbance

The relationship between human beings and their living environment is becoming fragile since industrial development and agricultural exploitation had already put a heavy burden on our environment. In 1975, while the researchers were reporting the first developed cloning experiments with *E. coli* during Asiloma conference, many scientists believed that this technology was unsafe and the cloned microorganisms should be strictly controlled and prevented from being released into the environment, and they asked the reporters to make sure that there would be no engineered strain can escape from the labs into the environment with any unpredictable and even *catastrophic consequences* to upset the balance of nature. In response to those concerns, the guidelines in the form of regulations were established to prohibit unsafe manipulation and practicing. However, since 1980's, many of the regulations had been gradually relaxed on some of the identified low-risk systems.

Cloning technology is no doubt effecting the environment and eco-system, by changing many genetic traits of agricultural organisms and spreading those traits through natural hybridization in the wild field. Some reports indicated that the plants that receive foreign genes designed for prioritized growth might gain the capacity of forcible occupancy in the eco-system. it's also possible that herbicide and pesticide resistance traits incorporated into weed plants could find their way to enhance the "weediness" from the natural selection on some tough weeds. Ten years ago, there was a mighty uproar aroused from the "American Monarch Butterfly" incident, In 1999, Connell professor reported that after feeding of **monarch butterfly larva** with the **milkweed** leaves scattered with engineered maize pollen containing **Bt insecticidal crystal protein gene**, 44% of the fed larva died. Others once reported that the massive number of this insect had declined dramatically for there were no milkweed leaves for them to lay the eggs. Since the end of 1980's, there were frequent reports about the protest campaign against "molecular ranch", the local public had concerned about the negative impact from farm animal cloning. If those farm animals, carrying human protein genes and used as "bioreactor", escaped form the molecular ranches, they might freely copulate with the germ line of wild type, and make the cloned genetic traits scattered among the natural specie. How the released recombinant strains react to or interact with the native system will be a tough project to the ecologists. More strict ecological monitoring and regulation on the cloned microorganisms must be put in place, since ecological microbial communities are the balanced system.

Ethic and moral consideration

In early 1990's, some UK reports had identified the main concerns on GM food: the human genes have been cloned into some food animals without informing consumers; pig genes into sheep would offend Jews and Muslims; the genes of animal cloned into tomato would enrage vegetarians; the spent microbial cells with human genes might contaminate foods. Almost during the same period, some of the molecular ranches in California were destroyed by local people, for they worried about that normal milk product would be mixed up with the drug containing milk from GM sheep or cow, and there was no way to ensure the producer' morality.

The transgenic breeding of animal by microinjection of donor genome into the pronucleus of the fertilized eggs of the recipient was usually regarded as the most fantastic work, but some of scientists and public were strangely opposed to this technique, they considered it as an awful non-sexual reproduction which totally violates the natural evolution of animals.

Bio-warfare and bioterrorism

Since the pathogenic and toxic mechanisms had been gradually revealed, it's not difficult to clone those hazard organisms. There is a growing awareness that bioterrorism might apply bio-warfare to threaten their opponents, even the ordinary people. Potential bio-hazard agents have been assigned into three categories: 1) Those include the most serious infective viruses and most toxic producers, which might not be easily controlled; 2) Those have similar potential of large-scale dissemination but cause generally less serious illness; 3) Agents include novel infectious diseases which could emerge as the future threat.

There is also potential of targeting agricultural plants and farm animals with harmful bio-agents, which could cause devastating effects on agricultural economy and human health. People should keep eyes on any cloning which involves **carcinogenic effect** on farm animals. The effluent form industrial fermentation processes inoculated with cloned strains must be sterilized and well treated before draining into water body.

Although there is an international prohibition regulation: the Biological and Toxic Weapons Convention, the questions always arise as: How well the knowledge we have learned about the **hazard and risk** from the potential bio-weapons, how quickly these bio-weapons can be identified, and how rapidly the bio-threats can be gotten rid of. All the biotechnology people must be careful to avoid those hazard cloning.

Talking about the overall attitudes on the genetic modified organisms (GMOs), while the scientific cases are strange and persuasive, but public perception is somehow ambivalent. Some of the survey revealed that most of people have readily accepted novel medical products produced by GMOs, but they are not ready to welcome the transgenic food for daily diet. In Europe, one of the investigations on public attitudes to the application of genetic cloning has shown that the application on the item of microbial production of bio-plastics received more than 90% of the approval, Transgenic crops to grow in developing nations received only 54% of the approval, and Developing of bio-warfare was disapproved by 98.1% of the overall poll. Other investigation indicated that public also concerns about: long term impacts of transgenic food on human health; disturbance of GMO species and their mutants to the already fragile eco-system; confidential management on genetic modification; and commercial exploitation only base on the producer's profit but not the consumer's benefit.

In the future, the technology of gene cloning will still play a critical role to our society, especially in health promotion, food supply, and environmental protection. We should also keep strict control to prevent the potential negative impacts, in order to keep human society moving along a sustainable track.

Summery

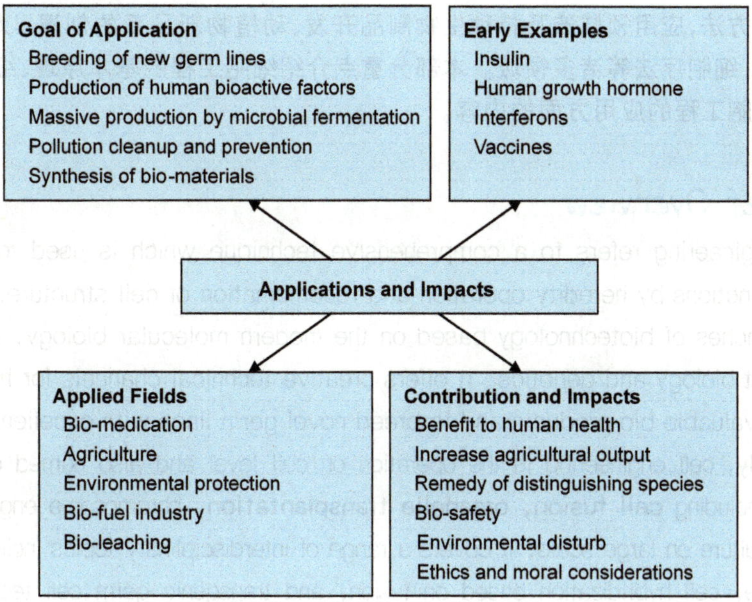

Part II Cell Engineering

Chapter 6 Introduction on Cell Engineering

细胞工程是生物技术领域的重要分支之一，建立于现代分子生物学、细胞学、发育生物学和遗传学的理论与技术基础之上，通过遗传操作和细胞结构的重组来改变生物遗传性状与功能，从而开辟出全新的技术路径，为人类提供重要的生物产品，并培育全新的生物品系。

通常细胞工程是在细胞水平上操作，也被称为细胞操作技术，包括细胞融合、器官移植、染色体基因工程和细胞组织的大规模培养。使用的材料泛指来自动物、植物和微生物的细胞、组织、器官，而一般情况下主要指动、植物细胞。细胞工程的优势在于可以将遗传物质有效的转入受体细胞形成杂合体细胞，避免了诸如分离、纯化、降解和重组等遗传操作。

历经百余年的发展，细胞工程覆盖细胞和组织培养、细胞融合杂交、转基因生殖细胞培育等基本方法，应用领域涉及特种生物制品开发、动植物新品系的创建，以及干细胞用于组织修复，细胞疗法等诸多领域。本部分重点介绍细胞工程的基本原理、细胞与组织培养技术和细胞工程的应用方面的内容。

6.1 Brief Overview

Cell engineering refers to a comprehensive technique which is used to change the biological functions by heredity operation and recombination of cell structure. It is one of the key branches of biotechnology based on the modern molecular biology, cell biology, development biology and genetics. It offers creative technical channels for human beings to produce valuable bio-products and to breed novel germ lines with excellent variety.

Generally, cell engineering is the operation on cell level and also named cell operation techniques including **cell fusion**, **organelle transplantation**, chromosome engineering, and cell tissue culture on large scale, it covers a range of interdisciplinary topics including cell and tissue culture, cell hybridization based on fusion, and transgenic germ cell techniques. The advantage of cell engineering also lies in the fact that it can raise gene transfer efficiency by transferring genetic material into the recipient cells to form **hybrid cells**, avoiding some genetic manipulation, such as separation, purification, digestion and cloning.

Its application might involve the production of important bio-medicines, exploitation of novel germ lines of animal and plant, and the uses of stem cells for tissue repairing and cellular therapy, etc.

6.2 Essentials of Cell Engineering

6.2.1 Characteristics of different cell types and their functions

The cell is the smallest unit of life. One single cell might be a whole organism in the case of bacteria, or a base unit of a larger organism such as the individual cells in human body. Although there are millions of different types of cells, the basic structure of cells has similar character. All cells have a "skin", called the plasma membrane, protecting them from the outside environment. The cell membrane is also a barrier to the substances moving into or out of the cell. At the center of the cell is the cell nucleus, which contains the cell's DNA, the genetic code of the organism. Between the nucleus and plasma membrane is the fluid called cytoplasm. According to the complexity of cell structure, the cells are generally divided into two types: the **prokaryotic cell** and the eukaryotic cell. There is a nuclear membrane surrounding the nucleus in eukaryotic cell whereas the nucleus in prokaryotic one is naked. Normally, two major types of eukaryotic cells: plant cell and animal cell, are routinely manipulated in cell engineering. Below is a labeled diagram of a model of basic animal cell to help you identify the feature of cell structures, there are also many organelles existing in the eukaryotic cell cytoplasm(Fig. 6.1).

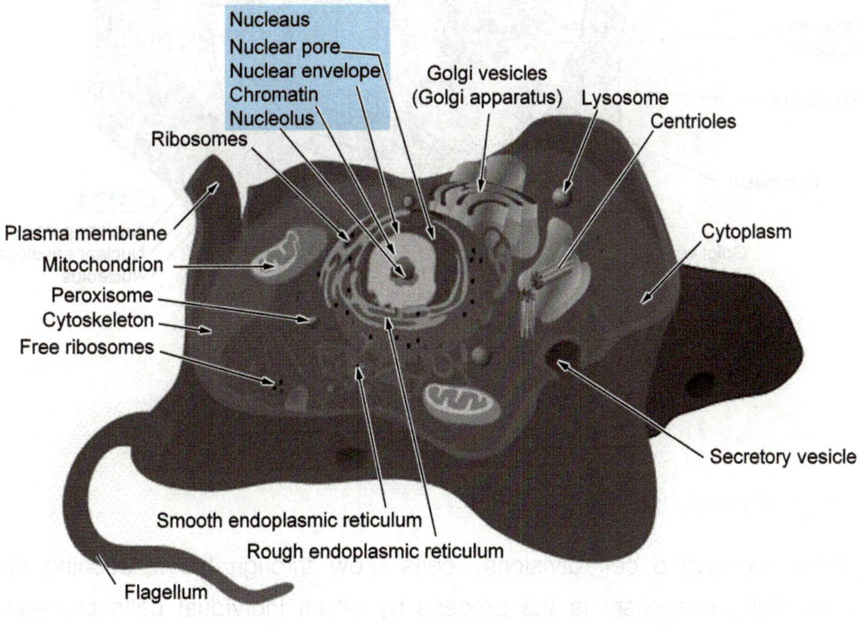

Fig. 6.1 A scheme indicating the structure of an animal cell

Despite the common features, plant cells have some additional specialized structures. Plants do not have a fixed skeleton for support body structure as many animal cells do. However plants don't just flop over in a big spongy mess. This is because of a unique cellular structure called the cell wall. The cell wall is a rigid structure outside of the cell membrane composed mainly of the polysaccharide cellulose. As pictured, the cell wall gives the plant cell a defined shape which helps support individual parts of plants (Fig. 6.2). In addition to the cell wall, plant cells contain an organelle called the chloroplast. The chloroplast allows plants to harvest energy from sunlight. Specialized pigments in the chloroplast (the major green pigment called chlorophyll) absorb sunlight and use this energy to complete the biochemical reaction:

$$6\ CO_2 + 6\ H_2O + \text{energy (from sunlight)} \longrightarrow C_6H_{12}O_6 + 6\ O_2$$

In this way, plant cells manufacture glucose and other carbohydrates that they can be stored for later use.

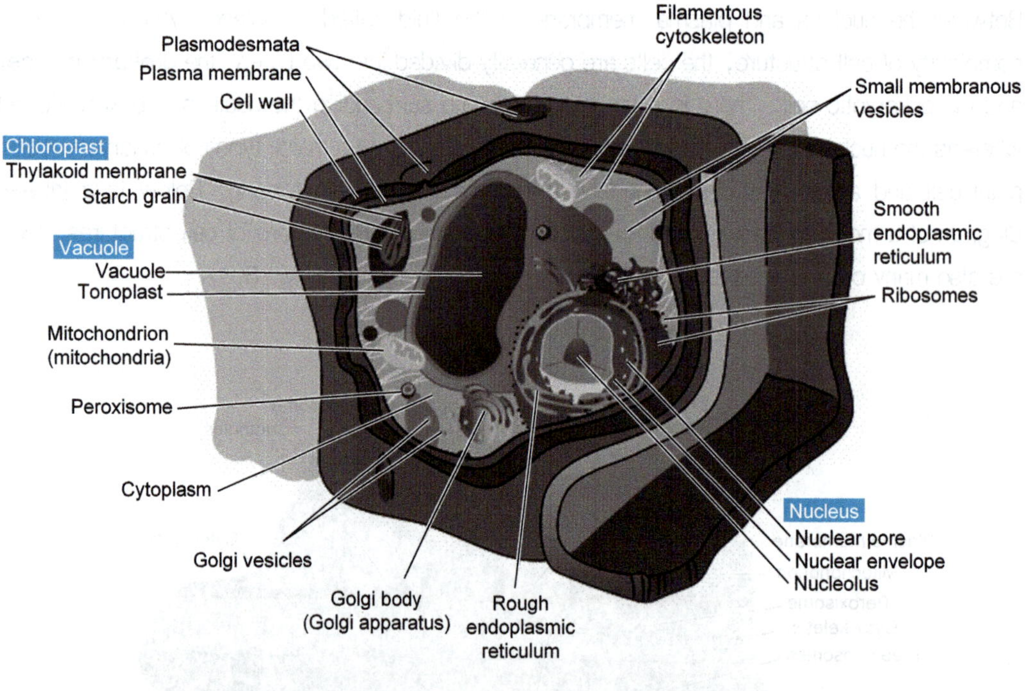

Fig. 6.2 A scheme indicating the structure of a plant cell

6.2.2 Cell metabolism and growth

Between successive cell divisions, cells grow through the functioning of cellular metabolism. Cell metabolism is the process by which individual cells process nutrient molecules into the needed biochemicals and energy. Metabolism has been divided into two distinct categories: catabolism, in which the cell breaks down complex molecules to produce energy and reducing power, and **anabolism**, in which the cell uses energy and reducing power to construct complex molecules and perform other biological functions. Polysaccharides consumed by the organism can be broken down into a less chemically complex monosaccharide called glucose. Once inside the cell, glucose is further broken down to make adenosine triphosphate (ATP), a form of energy used by many metabolic activities in cell.

Cell division involves a single cell (called a **mother cell**) dividing into two daughter cells. This leads to growth in multicellular organisms (the growth of tissue) and to procreation (vegetative reproduction) in unicellular organisms.

Prokaryotic cells divide by binary fission. Eukaryotic cells usually undergo a process of nuclear division, called mitosis, followed by division of the cell, called **cytokinesis**. A diploid cell, which contains two sets of chromosomes in the cell, may also undergo meiosis to produce **haploid** cells (with one set of chromosome), usually four. Haploid cells serve as gametes in multicellular organisms, by fusing to form new diploid cells.

6.2.3 Cell differentiation and dedifferentiation

Organisms contain many different types of cells that perform various functions. For example, some cells play the protective function; some give structural support to the body; others offer a means of transporting nutrients. Yet, these different cell types originate from one cell through cellular differentiation, by which an original cell differentiated into more specialized cells. A cell that is able to differentiate into all cell types of the adult organism is known as **totipotent**. Such cells are called embryonic stem cells in animals and **meristematic** cells in higher plants. Differentiation dramatically changes a cell's size, shape, membrane potential, metabolic activity, and responsiveness to signals.

Differentiation might be reversed in some of the organisms especially the more basal life forms such as worms and amphibians, in which a partially or terminally differentiated cell reverts to an earlier developmental stage, usually as part of a regenerative process. **Dedifferentiation** also occurs in plants, when it happens, cells in the culture would lose some of the original properties such as protein expression, or change the shape. This process is also termed dedifferentiation.

6.3 Progress Clues

The history of cell engineering can be traced to the 19th century from the establishment of cell cultures. The pioneer works in animal and plant cell cultures as well as the rapid development during the recent decades are listed in Tables 6.1, 6.2 and 6.3.

Tab. 6.1 Main events of early animal cell culture

Year	Contributors	Events
1885	W. Roux	Removed a portion of the medullary plate of an embryonic chicken and maintained it in a warm saline solution for several days, establishing the principle of tissue culture
1892	H. Driesch	Performed experiments on sea urchin embryos, by putting many embryos in a tube and shaking them to separate the cells for further cultivation
1907	R. G. Harrison	Created tissue culture technique by separating embryonic frog neural tube fragments and culturing them into fresh frog lymph successfully, which is regarded as the turning point in animal cell culture
mid-1930s	C. Lindbergh and A. Carrel	Created the "perfusion pump" which allowed living organs to exist outside of the body, which was a crucial step in organ transplants in animals and stimulated the concept of growing organs beyond the body (*in vitro* organ growing)
1938	H. Spemann	Proposed his theory of embryonic induction by organizers in his book *Embryonic Development and Induction*. Spemann was the pioneer to perform somatic cell nuclear transfer using amphibian embryos, one of the first moves towards cloning

Tab. 6.2 Early developments of plant cell culture

Year	Contributors	Events
1934～1939	R. J. Gautheret, P. A. C. Nobecourt and P. R. White	Discover the importance of auxins and B-vitamins, and laid the foundation of plant tissue culture
1954	C. Miller	Discovered kinetin which is now the well known plant hormone stimulating cell division
1957	F. K. Skoog and C. Miller	Put forth the concept of hormonal control of plant organ formation
1958	F. C. Steward and Reinert	First discovered somatic embryo production *in vitro*, and demonstrated Haberlandt's hypothesis into reality, which is also regarded as a milestone in plant tissue culture

Tab. 6.3 Consecutive rapid development

Year	Contributors	Events
1960s	G. Morel	Obtained virus-free Cymbidums by meristem culture, and vegetative propagation of orchids was established
1962	P. B. Capstick	Performed the growth of hamster kidney cells in suspension culture, which provided the basic technique of animal cell culture on a large scale
1963	E. McCulloch and J. Till	Discover the presence of self-renewing cells in mouse bone marrow
1970	J. B. Power	Demonstrated the protoplast fusion by $NaNO_3$ inducing
1972	P. S. Carlson	Used Power's method for producing the first somatic hybrid plant by fusing protoplasts of *Nicotiana glauca* and *N. langsdorffii*

Continued

Year	Contributors	Events
1975	C. Milstein and G. Köhler	Developed the hybridoma technique for the production of monoclonal antibodies
1981	M. J. Evans, M. Kaufman and G. R. Martin	Published their experimental results, in which how they isolated embryonic stem cells from mouse blastocysts and grew them in cell cultures was described
1983	R. L. Brinster and R. D. Palmiter	Created the first "transgenic mice". The methods they developed to transfer foreign genes into mammals, were used to elucidate the activity and function of genes
1988	J. Thomson	Declared the stem cell research was not limited to mice but monkeys
1996	I. Wilmut	Created Dolly, the cloned lamb using a technique called somatic nuclear cell transfer
1998	M. Shamblott and J. Thomson	Published a paper showing the discovery of human primordial germ stem cells

Above all, the recent advances in cell engineering have enabled us to engineer cells and tissues for specific purposes. For example, cells can be genetically, chemically and mechanically modified or reprogrammed to address specific biological questions. Tissues can be engineered to mimic living counterparts. It is anticipated that these approaches will lead to new cell-based therapies for age-related and degenerative conditions such as Parkinson's disease, cardiovascular disease, arthritis and bone disorders.

Summary

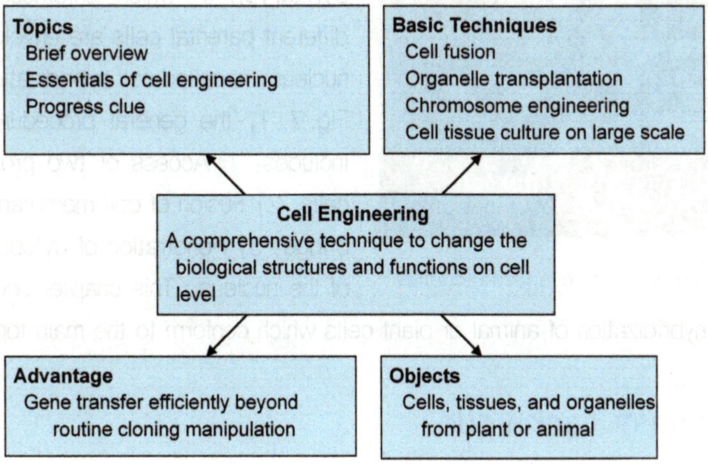

Chapter 7 Cell Fusion and Hybridization

细胞融合是指两个及以上细胞合并成为一个细胞的过程。融合的过程涉及细胞(或原生质体)的相互接近;质膜的融合形成细胞桥;细胞质的融合;细胞核的融合;进而发生的杂交。自发融合的现象存在于自然界中,人工诱导的方式更促进了种间细胞的融合,这种技术在细胞生物学、医学研究以及新品种选育等方面发挥着积极的作用。本章所讨论的主要是动物、植物细胞的融合与杂交技术,包括原生质体的制备与培养、融合方法、原生质体再生以及目的融合子的筛选。

7.1 General Aspects

Cell fusion is a specialized cellular event that two (or more) cells fuse into a single cell. Known as a phenomenon for over a hundred years, cell fusion took center stage in the early analysis of gene expression, chromosomal mapping, monoclonal antibody production, and cancer therapy. With the development of genetics, cell biology, molecular biology, biochemistry, and genomics, the molecular mechanisms of cell fusion have begun to come to light during recent years. And the technology of artificial cell fusion has been widely used in cell biology, medical research, new variety breeding, etc.

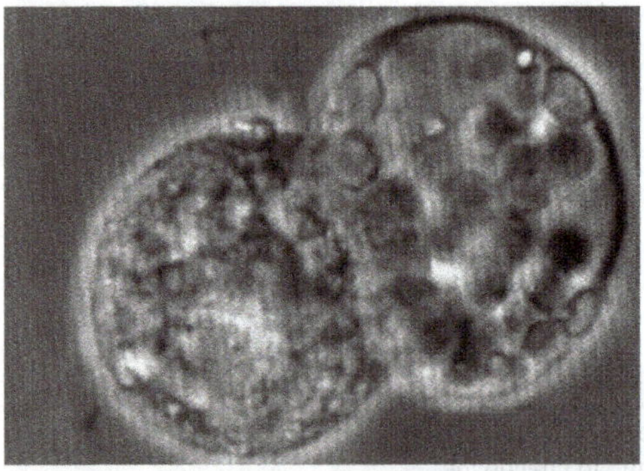
Fig. 7.1 Two different cells fused together

Cell fusion may happen in the same species or between different ones. The fusion of two or more dissimilar cells, leading to formation of a **synkaryon** in which chromosomes from two different parental cells are enveloped in a single nucleus, is called cell hybridization. As shown in Fig. 7. 1, the general procedure of cell fusion includes: 1) Access of two protoplasts or two cells; 2) Fusion of cell membrane to form a cell bridge; 3) Penetration of cytoplasm; 4) Fusion of the nucleus. This chapter only describes the cell fusion and hybridization of animal or plant cells which conform to the main topic of Part II.

7.2 Protoplast Technique

7.2.1 Isolation and purification of plant protoplast

Although the animal cell can be used in fusion directly, the protoplasts must be prepared for removing of the barrier of plant cell wall. Protoplast looks like sphere ball after the removal of the cell wall either mechanically or enzymatically. Protoplast can still carry out the metabolism, and under suitable condition the cell wall can be regenerated *in vitro*. In mechanical method, cells are kept in a suitable **plasmolyticum** (in plasmolysed cells, protoplasts shrink away from cell wall) and peeled with a fine knife, so that protoplasts are released from cells. This method is suitable for isolation of protoplasts from vacuolated cells (e. g. onion bulbs, scales, radish roots). However, this method is not suitable for isolating protoplasts from meristematic and less vacuolated cells, where it gives poor yield of protoplasts. The enzymatic method used now is almost invariable for the isolation of protoplasts (Fig. 7. 2). **Cellulase, pectinase** and **xylanase** are the appropriate polysacharide-degrading enzymes for the digestion of plant cell wall. The condition of the plant material, the formula of enzyme combination, the time, temperature, pH of the reaction must be taken into consideration in order to obtain high yield of protoplasts. Also, the process must be done in an isotonic (or hypertonic) solution to prevent rupture of the plasma membrane.

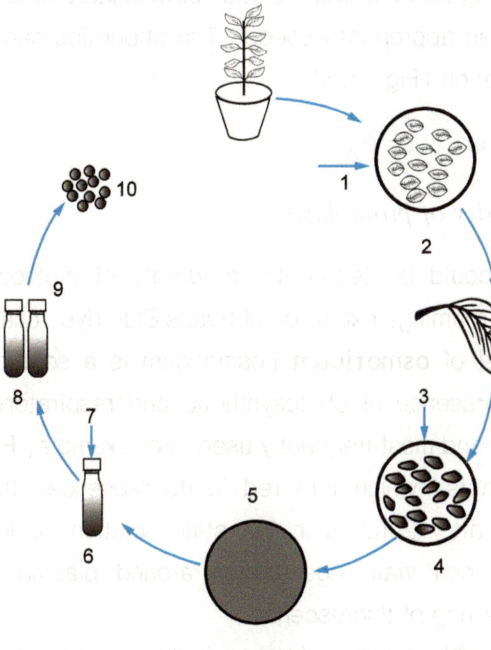

Fig. 7. 2 Diagram of the principle steps in enzymatic preparation of plant protoplasts.
1. Removal of the leaves from intact plant; 2. epidermis are peeled off with scalpel; 3~4. leaves are removed and cut into fine strips; 5. incubation in enzyme mixture; 6~7. enzyme is removed; 8~9. protoplasts are purified; 10. protoplast

The protoplasts isolated as above are present in the media together with a range of cell debris and broken cell organelles. Usually the purification of protoplasts starts from the

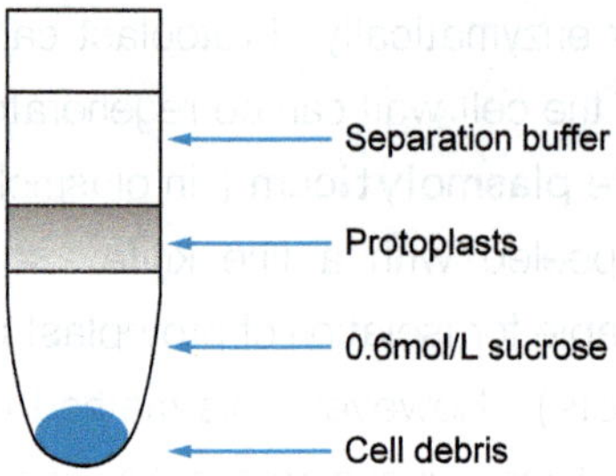

Fig. 7.3 Protoplast Separation by buoyant centrifugation

separation with a mesh to get rid of undigested cells, cell aggregates, and cell debris. The isolated protoplasts are further purified by centrifugation, flotation or both: 1) Centrifugation-the crude protoplasts suspension is centrifuged at low speed (50~100*g* for 5min). The intact protoplasts form a pellet and supernatant containing cell debris can be pipetted off. The pellet is gently resuspended in fresh isotonic (hypertonic) culture media and rewashed. This process is repeated two or three times to get relatively clean protoplast preparation. 2) Flotation-the density of protoplasts is lower than other cell debris. Therefore the protoplasts will float in solutions of 25% sucrose; the floated protoplasts can be carefully collected. More protoplasts lost after flotation. 3) Buoyant centrifugation-first the crude protoplasts suspension is centrifuged at low speed, and the supernatant is removed, then the pellets are resuspended in buffer and added in to a tube containing concentrated solution of mannitol, sorbitol or sucrose (0.3~0.6mol/L), centrifuge at an appropriate speed. The impurities can be removed off from protoplasts after centrifugation (Fig. 7.3).

7.2.2 Protoplast culture and regeneration

Viability and plating density of protoplasts

Protoplasts viability could be tested by a variety of methods which include the presence of cytoplasmic streaming; exclusion of Evans Blue dye; change in protoplast size due to change in the level of **osmoticum** (osmoticum is a solution causing change in osmotic pressure); and presence of photosynthetic and respiratory activity. The color changes of dyes are visual and most frequently used. For example, Phenosafranin (0.1%) detects the dead protoplasts, which turn red in its presence, the viable protoplasts remaining unstained even after 2 hours in the stain solution; calcofluor white (CFW, 0.1%) detects onset of cell wall regeneration around plasma membrane of viable protoplasts in the form of a ring of fluorescence.

A minimum plating density (mpd) of protoplasts is required for growth to begin, because protoplasts enrich the culture medium due to cell leakage. The suitable mpd for tobacco is ranged from 5×10^3 to 1×10^5 protoplasts/mL.

General methods for protoplast culture and regeneration

The same nutrient media that are suitable for *in vitro* cell and tissue culture are also used for protoplast culturing. However, keeping appropriate osmotic pressure at the initial stages of protoplast culture is of great importance in preserving and ensuring the stability of these cells, since they are devoid of their cell walls. Different sugars have been used to increase the **osmolarity** of culture medium for protoplasts; for example, glucose, mannitol, sorbitol, sucrose and xylose have all been successfully utilized. In some instances, the combination of two sugars may be beneficial. After the cell wall regenerated and developed into the cell colonies, the concentration of the solute used to raise the osmolarity of the culture medium can be reduced.

Methods generally used for cultivation of the isolated protoplasts are illustrated in Fig. 7.4.

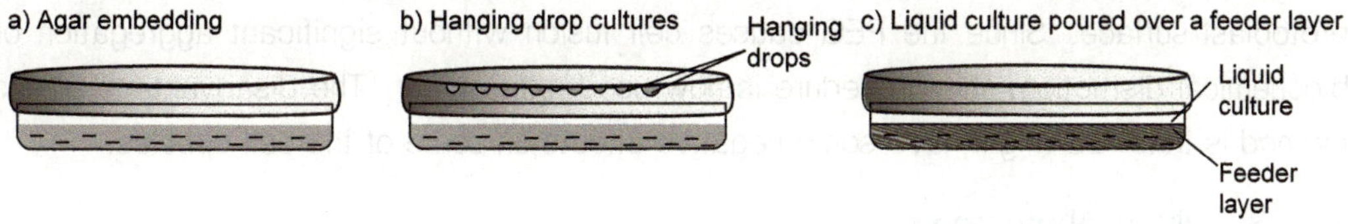

Fig. 7.4 Different methods for protoplast culture

Within 2~4 days, protoplasts start developing cell walls, which can be detected by staining with 0.1% fluorescent stain. About 3~4 weeks, the particles of **callus** form, which could be transformed to the culture medium for generation into intact plant.

7.3 Basic Methods of Cell Fusion

7.3.1 Biological methods

Virus induced cell fusion used to be the major biological method for both plant and animal cells in the earlier time of cell fusion. The invasion of virus sometimes causes the contact between host cells. If there are enough virions attached to the host cells, the aggregation of the cells will be triggered. At the site of conglutination, it's possible that the membranes of the two cells fuse to form a passage for exchanging their cytoplasm, even nucleus. If there happens the recombination of two chromosomal DNA, then the **fusant** forms. The cell fusion intermediated by virus takes the risk of viral pollution, in other cases, low concentration of mix enzymes would be used to break slightly the cell wall to facilitate cell fusion. Also in some cases, the chemical and physical methods work well in replacement of the biological methods.

7.3.2 Chemical methods

In chemical methods, chemical inducers are used to facilitate the cell fusion, which include polyethylene glycol (PEG), $NaNO_3$, high concentration of Ca^{2+} etc. 1) $NaNO_3$ induced fusion: the sodium ion in $NaNO_3$ neutralized the negative charge carried by the surface of the protoplast which leads to the aggregation of the protoplasts and thus facilitates the cell fusion. The first successful example of $NaNO_3$ facilitating cell fusion was carried out by P. S. Carlson in the fusion of tobacco leaf protoplast in 1972. Although $NaNO_3$ is harmless to the protoplast, the efficiency of this method is much lower. 2) PEG induced fusion: The mechanism of PEG fusion has not been definitively identified, but some researchers believe that the PEG, by binding a large number of H_2O molecules, effectively decreases the hydrogen bonding between the water and the lipid side group on the protoplast surface. Since the PEG causes cell fusion without significant aggregation or biochemical disruption, this procedure is now used extensively. The disadvantage of this method is that PEG might have some negative effects on some of the cell species.

7.3.3 Physical methods

The generally used physical method for fusion is electrofusion. It proposed that the electric pulse changes the charge and the redox potential of the cell membrane which leads to the adhesion of heterogeneous cells. After that the cell membranes break transiently, conjugate, and then enclose to form a fusant.

The whole electrofusion procedure can be divided into four stages: aggregation, cell membrane fusion, heterokaryon and fusant formation. This procedure can be performed in a small chamber. Under electric field, the cells will line up in a chain. The high electric pulse forces the cells contact to each other, and results in some transient pores in cell membrane, thus facilitating the fusion of the cells.

7.4 Fusant Screening

7.4.1 Screening for antibiotic resistant fusant

Antibiotic resistance can be used to screen the desired fusants. For example, the parent A confers the genetic trait of ampicillin resistance, but it is sensitive to tetracycline; In contrast, parent B is sensitive to ampicillin, but not to tetracycline. Therefore, only the hybridized fusants will grow on the medium containing both antibiotics, by this method the unfused protoplasts are all excluded.

7.4.2 Screening for nutrition complementary fusant

An organism which is unable to synthesize a particular organic compound required for its growth is defined as **auxotrophy**. Nutrition complementary of auxotrophy with different nutrition requirement can be employed in fusant screening. For example, parent A is tryptophan deficiency whereas parent B needs threonine for its growth, therefore only the hybridized fusants can grow on the minimal medium.

7.4.3 Fusant with different physical characters

According to the difference in size, color and morphology, the fusant can be distinguished under microscope from their parent cells and thus be selected. The disadvantage of this method would be that the hybrid may be damaged in the step of fusant transfer.

7.5 Application of Cell Fusion

In plant breeding experiments, **somatic cell fusion** is an important tool in the theoretical study of cytoplasmic genes and their activities. In application, somatic cell fusion overcomes sexual incompatibility barriers and increases the possibility of distantly related species crossing. It is also a routine method of recombination of plants that reproduce asexually.

Animal cell fusion can be used to study the control of cell division and gene expression, to investigate malignant transformations, to obtain viral replication, and to map the chromosome, etc. The cell fusion between immortalized cell and an antibody producing lymphocyte provides a method for monoclonal antibody production.

Cell fusion is also employed by fermentation industries, to breed the microbial strains with superior traits from both their parental lines.

Summary

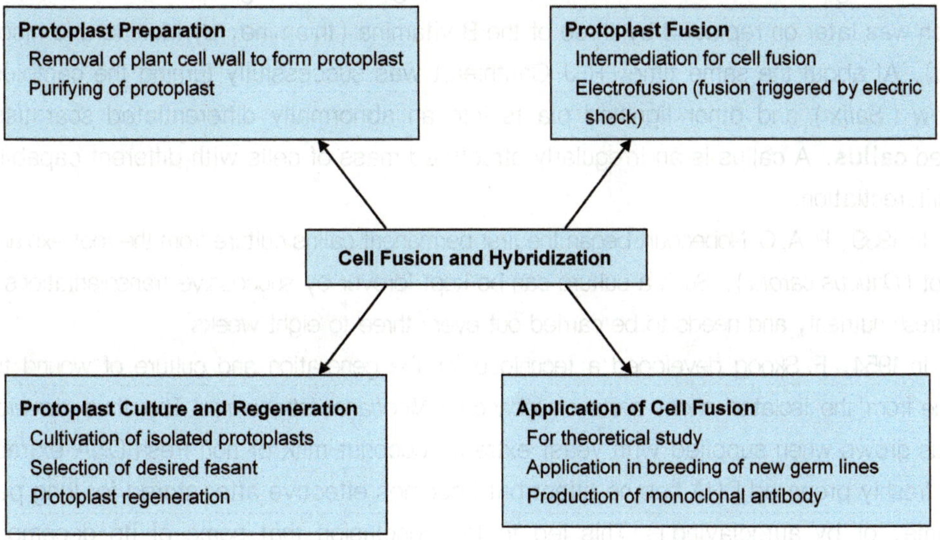

Chapter 8 Plant Tissue and Cell Culture

早期的植物组织与细胞培养以建立脱分化组织培养物为目标，目前它已成为生产细胞代谢产物，开展植物转基因、植物人工育种与快速繁殖的关键技术，涉及选择适宜的培养基和培养条件，建立愈伤组织和细胞悬浮培养体系，利用胚胎与组织培养制造人工种子和脱毒植株，采用载体转化或者基因枪方法将外源基因导入植物细胞，并再生形成完整植株等新技术体系。有关技术在植物生物制品的生产以及植物育种方面发挥着举足轻重的作用。

8.1 Plant Tissue Culture System

8.1.1 Plant cells are totipotent

In 1879, A. Hansen postulated already that plants are **totipotent**, the regeneration of a differentiated plant cell into a complete plant. He referred to the power of regeneration of Begonia which can develop new **plantules** of leaf cuttings, a feature still used by gardeners today. Successful research into plant cell culture is dependent on the development of fundamental methods for the cultivation of tissues, single cells and protoplasts. In the following, several studies that contributed essentially to the plant cell engineering are outlined.

In 1934, P. R. White cultivated tomato root tips and meristems (*Lycopersicon esculentum*) in a submerged form. The medium contained inorganic salts, glucose and yeast extract which was later on replaced by three of the B-vitamins (thiamine, pyridoxine and nicotinic acid). At about the same time, R. J. Gauthieret was successfully turning the cambium of willow (*Salix*) and other lignified plants into an abnormally differentiated scar tissue, called **callus**. A callus is an irregularly structured mass of cells with different capabilities of differentiation.

In 1939, P. A. C. Nobecourt began the first permanent callus culture from the root explants of carrot (*Daucus carota*). Such a culture can be kept forever by successive transplantations onto the fresh nutrient, and needs to be carried out every three to eight weeks.

In 1954, F. Skoog developed a technique for the generation and culture of wound tumor tissue from the isolated shoot parts of tobacco (*Nicotiana tabacum*). The thus developed callus grows when supplied with yeast extract, coconut milk or non fresh DNA extraction (for freshly prepared DNA has no effect but becomes effective after stored for long period of time, or by autoclaving). This led to the conclusion that some of its decomposed products must be required for cell growth and division. The substance was characterized as kinetin by C. O. Miller. The technique developed by F. Skoog and C. Miller proved to be ideal for the study of the regeneration capacity of callus cultures through hormonal control.

8.1.2 Composition of culture media

When initiating a plant callus culture, the first decision to be made is about the recipe of the callus culture medium. A large number of different culture media are described in literature, but only a few have found widespread usage for a range of plant species (e. g. MS, SH and GB5, see Tab. 8.1).

Tab. 8.1 Components of a few selected tissue culture media (mg/L)

Constituents	Murshige and Skoog (MS)	Schenk and Hildebrandt (SH)	Gamborg B5 (GB5)	Nitsch and Nitsch (NN)
Major inorganic nutrients				
KNO_3	1900	2500	2500	950
NH_4NO_3	1650			720
$NH_4H_2PO_4$		300		
$(NH_4)_2SO_4$			134	
$MgSO_4 \cdot 7H_2O$	370	400	250	185
$CaCl_2 \cdot 2H_2O$	440	200	150	220
$KHPO_4$	170			68
$NaH_2PO_4 \cdot H_2O$			150	
Trace elements				
$MnSO_4 \cdot H_2O$		10.0	10.0	
$MnSO_4 \cdot 4H_2O$	22.3			25
KI	0.83	1.0	0.75	
H_3BO_3	6.2	5.0	3.0	10
$ZnSO_4 \cdot 7H_2O$	8.6	1.0	2.0	10
$CuSO_4 \cdot 5H_2O$	0.025	0.2	0.025	0.025
$Na_2MoO_4 \cdot 2H_2O$	0.25	0.1	0.25	0.25
$CoCl_2 \cdot 6H_2O$	0.025	0.1	0.025	
Iron source				
$FeSO_4 \cdot 7H_2O$	27.8	15.0	27.8	27.8
Na_2EDTA	37.3	20.0	37.3	37.3
Vitamins				
Nicotinic acid	0.5	5.0	1.0	5.0
Pyridoxin-HCl	0.5	0.5	1.0	0.5
Thiamine-HCl	0.1	5.0	10.0	0.5
D-biotin				0.05
Folic acid				0.5
Myo-inositol	100	1000	100	100
Carbon source				
Sucrose	30,000	30,000	20,000	20,000
Organic nitrogen source				
Glycine	2.0			2.0

The following section outlines the components found in plant cell culture media. As with nearly all plant culture media, its components can be divided into six groups. These are: major inorganic nutrients; trace elements; iron source; vitamins; carbon source and plant growth regulators.

Major inorganic nutrients Major inorganic nutrients refer to the elements which concentrations are over 100mg/L in the media. Nitrogen is one of the important components in major inorganic nutrients. The levels of nitrogen are different between the media. MS medium has high level of inorganic nitrogen while other formulations, such as NN, have very low level. The tolerance of the inorganic nitrogen also varies between the plant species. With some species, even the ratio between NH_4^+ and NO_3^- is very critical for cell culture and plant regeneration. In general, there is a tendency to use lower levels of NH_4^+ than NO_3^- in plant tissue culture media. The amount of NH_4^+ in MS medium is almost half that of NO_3^-, while other formulations, such as SH, B5, and NN, contain even lower levels of NH_4^+. In MS and NN, inorganic nitrogen in the media is supplemented with organic nitrogen in the form of glycine since organic nitrogen has been shown to enhance somatic embryogenesis or **organogenesis** in some plant species. Phosphate element participates in the synthesis of nucleotide and protein, as well as photosynthesis, respiratory, plant maturation, etc., that's why a large amount of phosphate in form of phosphate salts should be provided in the major inorganic nutrients. Sulfur is also included in the major inorganic nutrients.

Trace elements Levels of trace elements in the media are usually lower than 100mg/L, acting as co-factors in enzymes, or promoters for cell propagation and division, for example, Mo is one of the co-factors to the enzyme with function of nitrogen-fixation.

Iron source Potassium, calcium, and magnesium are also necessary for activation of the enzymes, cell wall synthesis, chlorophyll biosynthesis and plant cell growth.

Vitamins The vitamins are necessary in plant cell metabolism as various forms of co-factors of enzymes which promote the cell growth and division. In certain cases, beneficial biological additive such as yeast extract may supply or replace the vitamins. However, use of fully-defined media is generally recommended, as the addition of biological supplements introduces possible variation due to differences in the composition of different batches.

Carbon source Carbon source is the essential nutrient to the plant cell culture, used as the backbone of metabolism and the supply of energy. It also maintains the required osmotic level in the media. The commonly used carbon source is sucrose, and its concentration is normally maintained between 1%～5%.

Plant growth regulators In most of the plant culture media, at least two classes of the growth regulators, **auxins** and **cytokinins**, are needed. Auxins and cytokinins play important roles in a number of plant activities, including: development of embryo, leaf formation, phototropism, gravitropism, apical dominance, fruit development, abscission, root initiation and development etc. Also they are the regulators to each other' synthesis according to the recent reports. In particular, auxin induces callus formation, and promotes the elongation of the cell. The most commonly used auxin include IAA, NAA, IBA and 2, 4-D; cytokinin promotes cell division and retards senescence. The commonly used cytokinin includes KT, BAP, 2iP and Zea, among them BAP and KT are the most routinely used (Tab. 8. 2). Usually, manipulation of auxin and cytokinin levels will be successful in defining a growth regulator balance necessary for the required behavior in culture. High ratio of cytokinin to auxin often induces shoots rather than roots.

Tab. 8. 2 Plant growth regulators and their abbreviations

Class	Name	Abbreviation
Auxin	*p*-Chlorophenoxyacetic acid	pCPA
	2,4-Dichlorophenoxyacetic acid	2,4-D
	Indole 3-acetic acid	IAA
	Indole 3-butyric acid	IBA
	1-Naphtalene acetic acid	NAA
	2-Napthoxyacetic acid	NOA
Cytokinin	6-Benzylaminopurine	BAP
	N-Isopentenylaminopurine	2iP
	6-Furfurylaminopurine (kinetin)	KT
	Zeatin	Zea

Plant growth hormones are normally applied in the concentration range of $10^{-7} \sim 10^{-5}$ mol/L. It is also important to remember that reference to a culture medium by its abbreviation (e. g., MS, SH) generally implies the culture medium minus growth regulators.

8. 1. 3 Preparation of plant culture media

The most efficient way of preparing plant culture media is first to make up the stock solutions of major inorganic nutrients, trace elements, iron source, carbon source, vitamins, and individual plant growth regulators. The vitamins recipe is usually stored in frozen condition (at $-20°C$) in small batches (e. g., enough for one liter of medium). Right before use, the stock solution of vitamins is thawed and mixed with other components fully. The other stocks can be kept in a refrigerator at 4℃, but should be frequently checked and discarded if precipitation occurs. The inorganic stock solutions can not be stored longer than one month. It is advisable to make up small stocks of growth regulators fresh for each batch of media, as small changes in concentration due to precipitation can seriously affect growth of the cultures. To solidify media, 0.8%～1% of agar will be added to make a gel medium. Media and glassware should be autoclaved at 121°C for 15 to 20 minutes to eliminate any germs before use.

Selection of the appropriate nutrient medium for a given species or tissue is generally achieved by empirical trials. This can be done by a systematic screen of various concentrations of the media components such as the broad spectrum screen using stock solutions of the various groups of mineral components. Thus the medium components would be grouped under four categories (major nutrients, trace elements, vitamins and phytohormones) and used in different concentrations (low, medium and high) for each category and prepare various combinations of the components. Alternatively, one can start from the standard MS medium and vary the composition in the four categories. Perhaps one of the major components having a significant effect on regeneration is the type and concentration of phytohormones in the medium. It is worth remembering that the discovery of the cytokinins led to the **centra dogma** of tissue culture, namely, endogenous ratio of cytokinin to auxin determining the nature of regeneration. Commonly, optimization of the appropriate phytohormone concentrations in the medium can also be empirically determined in the earlier set of exploratory experiments. Once an optimum medium combination is identified, it can be used as a defined medium for a specific species, and also often for closely related species of plants.

8.1.4 Different culture patterns and measurements

Explant selection

The part of the plant that is excised and put in culture is referred to as the **explant**. This may be leaf disks, cotyledons, hypocotyls, roots, shoot tips, axillary buds, or **zygotic embryos**. The selected explants have to be rendered aseptic, which often involves surface sterilization using various dilutions (10%～30% *V/V*) of Clorox® or commercial bleach that has sodium hypochlorite (5.25% *m/V*) as the active ingredient. Other chemicals such as ethanol (1～5min rinses in 70% ethanol) or mercuric chloride solution may also be used for surface sterilization. Prior to inoculating the explants onto the nutrient medium, they have to be rinsed thoroughly with autoclaved distilled water to remove traces of the chemical used and then trimmed to get rid of the cells along the edges that might have been killed by the harsh chemicals used for sterilizing. There are several simplified, modified methods of surface sterilization where the explant donor tissues, such as seed pods and floral buds, may be directly dipped in ethanol (70%). The seeds or anthers are then excised aseptically for culture. With this procedure the explants never come in contact with the chemical surface-sterilization agent, giving a better survival rate for the explants. This method is useful for intact seed pods of orchids or anthers in unopened floral buds of several species. Certain materials, especially field-grown plants, are extremely difficult to surface sterilize. In addition, selected field-grown plants with desired characteristics for clonal propagation might be located too far from the laboratory. Thus, pre-surface sterilization treatments have to be devised.

For example, in guava, scions were obtained from selected field-grown plants and grafted to seedling root-stocks. Grafted plants were kept in the laboratory for harvesting nodal explants. To remove apical dominance and encourage sprouting of the axillary buds, healthy scion branches were decapitated 5 to 8 days prior to excision of the nodal explants. The nodal segments were then surface sterilized in 80% ethanol followed by 3% to 5% Clorox solutions before successful establishment of cultures.

Once a suitable explant is selected and prepared for culture, it should be incubated as quickly as possible on an appropriate nutrient medium for growth and differentiation.

Initiation of callus

The establishment of a plant cell culture system is traditionally from the initiation of callus. After surface sterilization, the explants were cut and pressed on to the surface of a suitable solid medium. Root, shoot, and leaf material of seedlings germinated from sterilized seed can be cut and directly plated on callus induction medium. Explants are usually placed horizontally, although stem sections may produce more calli if placed vertically with one cut end in the agar. Inclusion of certain kinds of auxins and cytokinins will be necessary for callus growth, and that somewhat higher auxin concentrations may be required for callus initiation. Petri dish plates sealed with Parafilm to prevent desiccation should then be put into an incubator at approximately 25°C either in the dark or, if beneficial, under low level of illumination. If sufficient callus **subculture** has not formed within three to eight weeks, it will be necessary to re-evaluate the culture medium and conditions.

Suspension culture

Suspension cultures, in general, form readily after transfer of callus to shaken flasks of the culture medium without agar. A large inoculum may be necessary to initiate the suspension culture. If the callus culture is non-friable, very little will break off from the callus clumps, and the necessary inoculum size of freely suspended cells and small cell clusters may not be attained. In such cases, modifications should be carried out to obtain more friable callus by somewhat empirical methods. All cells are exposed uniformly to the medium in suspension compare to those in callus culture. And suspension culture is commonly used to cultivate the protoplasts for cell fusion, genetic manipulation and commercial production of secondary products.

Subculture

Subcultures are normally performed in both patterns of callus and suspension cultures. Callus subculture involves removing newly formed callus with a sterile scalpel and transferring to fresh medium. The optimum size of inoculum varies depending on the species, but the callus should not be cut into too small pieces. Most callus cultures will require regular subculture at approximately one month intervals. For newly initiated callus, it may be necessary to transfer the entire callus to the fresh medium for the first two or three subcultures. Fast growing callus cultures may require subculture at two to three week intervals.

When suspension culture has reached a suitable cell density, it can be transferred to new liquid medium for subculture. The minimum density of cells required for cell suspension cultures depends on the growth rate and the medium composition. Adding of medium in which the cell line had previously been growing (conditioned medium) can sometimes stimulate growth of cell suspensions at low inoculum density.

Measurement of growth parameters

Growth of callus and cell suspension cultures can be detected by the increase of fresh or dry cell weight, or the increase in cell number. For cell suspension cultures, increase in packed cell volume (PCV) is also a good indicator of cell growth.

Fresh and dry weight measurements

For callus culture, the weight measurement is carried out by transfer of the entire callus (scrape off the medium) to a pre-weighed weighing container and assay the fresh weight; for cell suspension culture, by transfer of the entire contents of the cell culture flask to a pre-weighed centrifuge tube, then spin down the cells and carefully pipette out of the entire medium without disturbing the pellet. Weigh the centrifuge tube with cells to measure the fresh weight, then dry the cell samples in an oven at 60°C, until no change in dry weight is observed.

Counting cell number

A haemocytometer can be used for determining the cell numbers in a fine suspension culture. Callus cultures and suspension cultures consisting of cell aggregates have to be macerated prior to the counting of cell numbers.

Measurement of packed cell volume

Measurement of packed cell volume involves: transfer the entire contents of cell culture to a centrifuge tube; determine the volume of the pellet after centrifugation; packed cell volume is expressed as a percentage of the volume of the pellet to the entire culture volume.

Counting of viable cell number

The viability of cells in culture can also be easily and reliably evaluated by specific vital cell staining which excludes the dead cells. Microscopic observation of cytoplasmic streaming in cells is also a good indicator of cell viability.

8.1.5 Properties of plant cell suspensions and organized cultures

Unlike microbial suspension cultures, plant cell suspensions generally consist of groups or aggregates of many thousands of cells, and up to 2mm or more in diameter. A fine suspension would have cell aggregates in the range of 200～500μm in diameter. As a consequence of these properties, the culture will settle out rapidly if mixing is stopped and sampling can therefore be difficult. Compared with microbial cultures, the cultures grow very slowly with doubling times of 2～6 days. Therefore the contamination of the microbes will lead to the failure of the total plant cell culture system.

Two other important characteristics are foam production and shearing sensitivity. Many plant cell suspensions in bioreactors produce foam and further develop into a matrix which can trap cells and block air exits. The formed "meringue" can be reduced by the addition of antifoam (silicon-based or polypropylene glycol), but this may affect the culture. Plant cells in suspension have been regarded as sensitive to shearing force because of their rigid cell envelop, of relatively large size, and extensive vacuole, and this property often causes the failure of plant cells to grow. There is also a need of specially designed bioreactors with mechanism of lowering shearing stress.

8.1.6 Environmental aspects of cell culture

The choice of culture vessels are varied. Erlenmeyer flasks, culture tubes, petri dishes, and specialized containers may be used to establish plant cell cultures. Cell cultures for extraction of metabolites (e. g., taxol or recombinant proteins) are generally cultivated in bioreactor vessels of desirable scale.

The influence of the headspace gas composition (or the gas space above the culture medium within the culture container) on growth and differentiation is well recorded. One of the major components in the headspace gas is ethylene, the gaseous phytohormone. It can exert an influence on regeneration even after it is released from the explant tissues. The partial pressures of carbon dioxide and oxygen in the headspace gas can also exert some influence on regeneration. The use of vented lids, specialty vented plastic film containers, or incorporation of chemicals that act as inhibitors of ethylene action into the culture medium help the explant regeneration. Once the explants are inoculated onto the nutrient medium under aseptic conditions, they are incubated at (25 ± 2)℃. Depending on the needs of the culture, incubation may be under a photoperiod for 12 to 26 hours or in darkness for a specific initial culture period. Incubation of the cultures may be done in growth chambers or on shelves fitted with illumination. Although the most common type of light source is the fluorescent bulb, more recently, light emitting diode (LED) are becoming available for this purpose. The LED light panels consume less power, generate less heat, and can be configured to deliver lights of specific wavelengths(e. g. , 460nm blue and 630nm red), so that plant cultures can be grown under the wavelengths suitable for absorption by chlorophyll.

8.1.7 Modes of regeneration

As indicated earlier, the concept of totipotency was realized in plant cells with the ability to regenerate whole new plants starting from single cells. Such a process involves dedifferentiation from partially differentiated cell/tissue type (e. g. , parenchyma) to a meristem-like state followed by redifferentiation (regeneration) into well-organized structures. The totipotency of cells in plant meristem was the basis of the concept of "stem cells" in animals that has become such an intense research area in the past decade or so. *De novo* **organogenesis** and somatic **embryogenesis** are the two major routes of regeneration of plants in tissue culture. Organogenesis can occur directly as shoot buds or roots from the cultured tissues or via an intermediate callus stage. Direct regeneration of adventitious shoots is generally preferred for clonal propagation applications, because chromosomal aberrations may be introduced during some callus induction treatments. However, the choice of the pathway of regeneration is genetically determined and, depending on the species, protocols for only one route of regeneration may be available; for example, in cereals such as rice and maize regeneration is via callus. In either case, generally a small number of cells are involved in establishing the organ or somatic embryo. In somatic embryogenesis, vegetative cells develop into plants through embryogenetic stages without the fusion of gametes. Somatic embryos are structurally similar to zygotic embryos found in seeds and possess many of their useful features, including the ability to grow into complete plants. However, somatic embryo develops from somatic cells (non-sexual) and does not involve sexual recombination which is happens in natural seed as a result of a sexual process in cross-pollinating species.

The plantlets developed from shoot buds or by germination of somatic embryos can be grown to a sufficiently large size before they are put through a hardening process and acclimation to be grown in pots.

8.2 Gene Transfer to Plants

Gene transfer of plants involves the selection of individual donor plant, purification and some time reconstruction of cloning vectors, careful introduction of donor DNA sequence usually into the cytoplasm or the nuclear genome of the recipient plant cells, and identification of the desired clones from the massive recipient cells by a well designed screening strategy. By gene transfer some special genetic traits carried originally in the parental plants would be brought to the recipient plants. Many plants including tobacco, corn, bean, etc. have been genetically engineered, and recombinant DNA technique as an important approach for breeding has been widely performed in agricultural and forestry fields. Hopefully, plant genetic engineering could also provide the salvation to those endangered rare plant species.

8.2.1 Ti plasmid

Vectors especially plasmids are most commonly used in transferring the foreign genes into the host cell. As in prokaryotic cloning, vectors should have the prokaryotic selectable (antibiotic resistance) markers for plasmid identification and maintenance in the host cell, such as in *E. coli*. The vectors must also have bacterial origins for plasmid replication, and multiple unique restriction sites for rejoining with the fragment carrying foreign genes. For the same reason, the vectors for gene transferring in plants should also have the marker genes which allow the recognition of transformed plant cells by either selection or screening. Selectable marker genes confer the plant cells with the resistance to certain antibiotics (kanamycin, methotrexate) or herbicide (glyphosphate, bialaphos). Other screenable markers (also called as reporter genes) coding for acetyl transferase, acetyl-glucuronidase, luciferase, nopaline synthase, or octopine synthase can also be used, if those genes do not exist in the recipient plant tissue.

Generally speaking, plant cells do not contain natural plasmids, however the bacterium ***Agrobacterium tumefaciens*** which primarily infects dicotyledenous plants contains **Ti plasmid**. The two letters of Ti stands for **tumor inducing**. On infection, **crown galls** (collections of plant cells growing in an undifferentiated, uncontrolled manner, as a tumor) will differentiate due to the function of the genes in the Ti plasmid. Ti plasmid is a circular plasmid with a size of 150 ~ 200kb. There are several regions commonly existing in all kinds of Ti plasmids (Fig. 8.1).

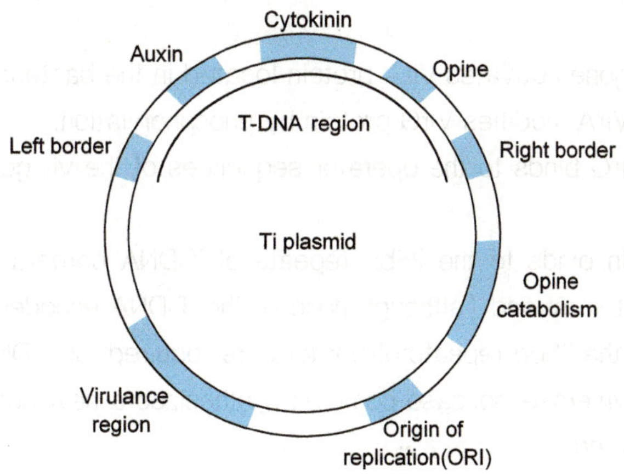

Fig. 8.1 Crucial sections in Ti plasmids

The structure and the functions of Ti plasmid are somehow complicated: the T-DNA region bracketed between the left and right borders is the part that will be integrated into the plant chromosome DNA. It also codes for one or more enzymes that direct the synthesis of unusual amino acid or sugar derivatives, collectively referred to as "**opines**". Ti plasmid is classified into different types based on the types of opine related to their genes. The different opines specified by Ti are octopine, nopaline, succinamopine and leucinopine. Several T-DNA genes encode enzymes involving cytokinins and auxins biosynthesis and the crown gall tumorigenesis is due to the "activation" of unregulated phytohormone synthesis in the transformed cells.

Another region which plays a crucial role in transfer of the T-DNA into the plant is virulence region. It is organized into six loci. The functions of each locus are listed in Tab.8.3.

Tab.8.3 Functions of each locus in virulence region of Ti plasmid

Vir locus	Functions
A	is expressed constitutively, recognizes plant metabolites and activates VirG
G	is expressed at low levels and functions as a transcriptional activator of other vir loci
D	nicks Ti plasmid at T-DNA borders, covalently attaches to "T-strand"
C	unknown function involved in host-range determination
E	ssDNA binding protein which stabilizes T-DNA during or after transfer
F	transfer apparatus

With the help of the Vir proteins, the T-DNA is transferred into plant cell genome through the wound (or crown galls), and the mechanism of transferring is usually explained as follows:

(1) Ti plasmid gene expression in vegetative *Agrobacterium* prior to the *vir* gene is induced by the plant phenolic compound acetosyringone released by plant cells at the wound site.

(2) Acetosyringone activates VirA protein located in the bacterial membrane.

(3) Activated VirA modifies VirG protein by phosphorylation.

(4) Modified VirG binds to the operator sequences of the vir genes resulting in their activation.

(5) VirD protein binds to the 25bp repeats of T-DNA borders and introduces site-specific nicks in bottom strand (although none of the T-DNA encoded genes are required for T-DNA transfer, the 25bp repeats of borders are required for T-DNA transfer).

(6) A DNA polymerase/helicase complex synthesizes a new bottom strand and then produced the top-strand.

(7) Top-strand is associated with VirE protein and translocated to the plant nuclei.

Other crucial regions including the replication origin and the sequence related to opine catabolism also exist in the Ti plasmids (Fig. 8.1).

8.2.2 Modification of botanic cloning vectors

With the specific integration function, the recombinant Ti plasmid with a foreign gene inserted in the T-DNA region helps the integration of the desired gene into the plant chromosome, where it may be expressed. However, general features of Ti plasmid such as large size and absence of unique restriction site make this simple scheme difficult in practice. Also normal shoots cannot be regenerated from oncogenic cells.

Therefore, the modification or reconstruction of plasmid vector systems is necessary in the creation of transgenic plants. It has been discovered that if the T-DNA and the remainder of the Ti plasmid are on separated molecules within the same bacterial cell, integration will still take place. Therefore the binary vector strategy has been put forward. In this strategy, one plasmid containing the recombinant T-DNA can be constructed in a standard *E. coli* plasmid with the left and right borders, selective markers for both bacteria and plant cells and multiple cloning site for carrying foreign genes. The genes for crown gall formation from the T-DNA are deleted. Another plasmid contains a modified Ti plasmid without T-DNA. Binary vector systems are now routinely used for botanic gene transfer.

8.2.3 Physical introduction of genes into host cells

The apparent recalcitrance of several agronomically important crop plants, including maize, wheat, barley, and rice, to infection by *A. tumefaciens* has driven the development of alternative methods of DNA delivery for genetic engineering. Protoplast transformation, although achievable through electroporation, **microinjection**, or polyethylene glycol fusion, is proved to be inefficient because the regeneration of plants from protoplasts is time-consuming and non-trivial. The methods of physical introduction of botanic genes into host cells were invented in 1980's. So called particle bombardment (microprojectile), in which tungsten or gold microprojectiles are coated with DNA and accelerated into the target plant tissue, has proven highly successful in a wide range of species, especially monocots like wheat or maize. The accelerated particles penetrate cell wall and membranes. The DNA separated from the metal may be integrated into the plant genome inside the nucleus.

This approach presents certain advantages over *Agrobacterium*-mediated gene delivery; many types of explants can be bombarded and yield fertile plants, and the gene to be delivered need not be cloned into a specialized transformation vector. Nonetheless, particle gun delivery of DNA is generally not the method of choice for a plant species that can be transformed by *Agrobacterium*, as the bombardment process typically results in integration of multiple copies of the DNA, as well as rearranged and/or truncated DNA

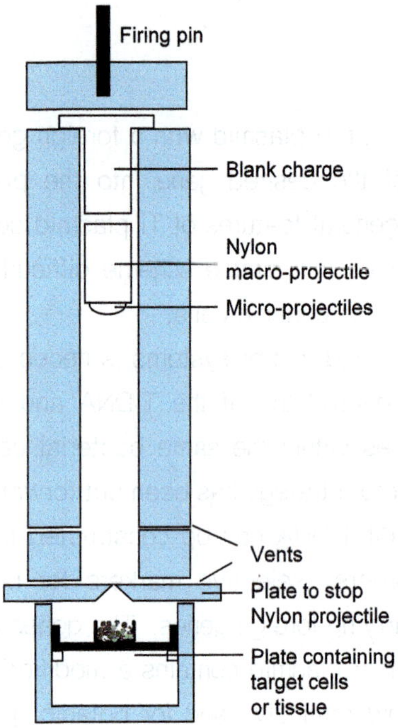

Fig. 8.2 Microprojectile acceleration device

sequences. These complex integration patterns can lead to genetic instability, due to homologous recombination among the identical copies.

Later, so called "**agrolistic**" transformation was designed to mitigate these shortcomings by combining the high efficiency of **biolistic** DNA delivery with the simpler integration pattern with characteristic of *Agrobacterium*-mediated DNA transfer (Fig. 8.2). Particle bombardment of the *virD*1 and *virD*2 genes, under the control of the CaMV35S promoter, with a target plasmid carrying the trans-gene of interest flanked by T-DNA border sequences, allows transient expression of the *vir* genes in the plant. The insertion events resulting from in *planta* VirD1/2-mediated processing and integration resemble those generated by traditional *Agrobacterium* mediated transformation.

In conclusion, *Agrobacterium*-mediated and biolistic delivery of foreign DNA are typically used to stably transform plants. Also, transient expression of genes delivered by *A. tumefaciens* on binary vectors can be used to produce recombinant proteins without the delays and technical barriers associated with stable integration.

8.2.4 Screening for desired clones

In addition to susceptibility to *Agrobacterium* infection, requirement for successful genetic modification is an efficient selection method for plant cells containing integrated trans-DNA. Identification of transformed cells was achieved by screening for nopaline production. Several key improvements have been made thereafter. Foremost among these was the use of T-DNA-derived promoters and 3′ regulatory regions (from the nopaline synthase gene) to drive in planta transcription of a bacterial antibiotic resistance gene such as chloram-phenicol acetyltransferase or neomycin phosphotransferase (nptII). Expression of these chimeric genes in the plant allowed the selection of antibiotic-resistant transformed plant cells and hence the elimination of opine synthesis genes from the transferred DNA. Phenotypically normal and fertile plants were regenerated from the resistant calli, and the resistance trait was passed to the progeny in a Mendelian fashion.

In response to the concerns about the potential for transfer of antibiotic resistance genes to gut microbes, a number of antibiotic-free marker systems have been developed. Herbicide resistance genes are also in wide-spread use as selectable markers, although they too pose perceived dangers to health and the environment. Rather than killing non-transformed cells (negative selection), one can also use a positive selectable marker that confers on transformed cells a growth or metabolic advantage. For example, introduction of the phosphomannose isomerase gene rescues plants from the growth inhibition associated with mannose. The desire to transform recalcitrant plant species has driven the development of other positive selection markers, including native plant genes conferring resistance to bacterial pathogens. Other positive selection markers, such as the agrobacterial or plant cytokinin synthesis isopentyl transferase (ipt) genes, promote regeneration of shoots from transformed calli or explants in the absence of critical growth regulators. Finally, marker-free transformation was achieved in potato using a virulent *A. tumefaciens* and PCR screening for successfully altered shoots.

Non-antibiotic/herbicide resistance markers address concerns about potential health and ecological risks, but they still suffer from other shortcomings: the selection scheme can have negative consequences for plant cell proliferation and differentiation, and multiple transgenes cannot be stacked through sequential retransformations using the same marker gene. These constraints have spurred the development of various methods to remove the marker gene after transformation. For example, the selectable marker is inserted into a transposable element, allowing transposition-mediated loss of the marker after selection of the transformed plants.

8.2.5 Regenerate whole plants from transformed cells

After the transformation of the foreign genes into the cell, the regeneration of cell into whole plant is a necessary step in transgenic plant construction. Transformation without regeneration, and regeneration without transformation, are both of limited value. Hormone-induced regeneration of transgenic plants from transformed explants can occur via organogenesis or somatic embryogenesis. Most economically important plants, especially monocots, are regenerated using the latter approach, since callus is easily initiated from the scutellum of immature embryos. Delivery of the foreign DNA directly into meristematic tissue or immature embryos has also been found to limit somaclonal variation because it shortens the time in tissue culture. A vacuum infiltration method, in which agrobacteria are applied to entire flowering Arabidopsis, was developed to avoid altogether the requirement for plant tissue culture or regeneration. More recently, this approach has been further simplified; in the "floral dip" process only the developing floral tissue is submerged into the suspension of agrobacterial cells, and the labor-intensive vacuum infiltration step is avoided.

8.3 Artificial Seeds and Virus-free Plants

8.3.1 Artificial seeds

The idea that basic plant materials can be combined with a nutrient filled environment,

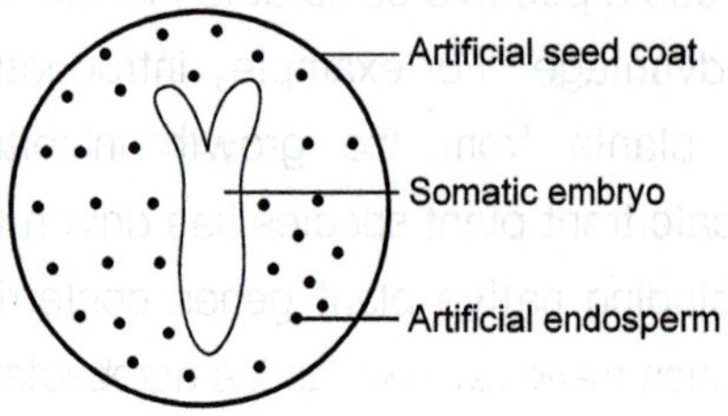

Fig. 8.3 Structure of artificial seed

preserved and subsequently capable of revitalizing into a viable living plant connotes endless possibilities for a world where arable agricultural land is diminishing at an alarming rate annually. The case for artificial seeds is an attractive one. An artificial seed is often defined as a novel analogue to botanic seed consisting of a somatic embryo surrounded by an artificial coat (Fig. 8.3). This definition is primary based on the similarity of somatic embryo with zygotic embryo in morphology, physiology and biochemistry.

In some cases, shoot and artificial endosperm are also enclosed in the coat, therefore an alternative definition to an artificial seed is provided as "artificial seed comprises a capsule prepared by coating a cultured matter, a tissue piece or an organ which can grow into a plant body and nutrients with an artificial film". The somatic embryo or plant material suitable for propagation into a plant is at the core of the artificial seed concept. And the choice of the accompanying matrix materials used for combining with the biological material is equally vital.

8.3.2 Plant materials used for encapsulation

Somatic embryos can be produced clonally in large numbers and make them attractive materials for encapsulation. Artificial seeds, consisting of somatic embryos enclosed in a protective coating, have been proposed as a low cost, high volume propagation system. Apical and axial shoot buds were used for encapsulation of several plant species ranging from mustard, medicinal plant to tree species.

8.3.3 Matrix material selection for artificial seeds

The matrix is responsible for the immediate surrounding of the plant material. Factors will include mediating the environment such as temperature and humidity to protect the biological material as well as provide a nutrient reservoir. Therefore, the role of matrix materials has an important bearing on the ultimate viability of the artificial seed. The concept of matrix materials has over the years developed into a rather sophisticated interplay that focuses on the germination viability of artificial seeds. Among the matrix materials, the most common one used to generate the artificial endosperm is based on sodium alginate that in the presence of divalent calcium ions forms a **hydrogel**. In the hydrogel approach, somatic embryos are mixed with a polymeric solution that when introduced drop-wise into a separate solution containing divalent metal ions, initiated a crosslinking reaction to form the hydrogel that in the process encapsulates the embryo. With this advancement, the true seed character of an embryo in a simulated endosperm was realized. The nature of the hydrogel being soft ensures that little pressure acts on the embryo therefore minimizing harm to the embryo. Nutrients and other materials such as bactericide and growth hormones for example could be included. Distinct beadlike artificial seeds of good quality could be obtained with an 80% or higher rate of germination.

8.3.4 Virus-free plants in breeding

Organogenesis or embryogenesis allows the production of large numbers of uniform individuals of horticultural species from limited starting material or virus-free plants in breeding. In conventional gardening, viruses and other pests would be transmitted when using rooted cuttings. In these mother plants, however, usually the uppermost shoot apex (<0.5mm), generally without conductive elements, is free of infections, and its use for *in vitro* propagation can yield virus-free offspring. This is enforced by a heat shock treatment before explantation (up to 40℃). Still, often a stepwise treatment of several subcultures is necessary. For this work, a very sensitive control method is required to screen the plant material, often available only in tiny amounts. To control virus infection, immunological methods are commonly applied (e.g., an ELISA test). In Fig. 8.4, a schematic summary of the production of healthy offspring from an infected "mother plant" is given.

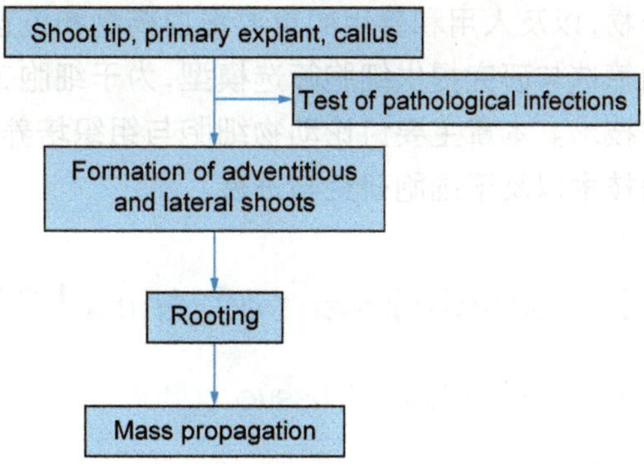

Fig. 8.4 Scheme for the production of *Pelargonium* mother plants and cuttings, by means of tissue culture methods, and used in gardening

Summary

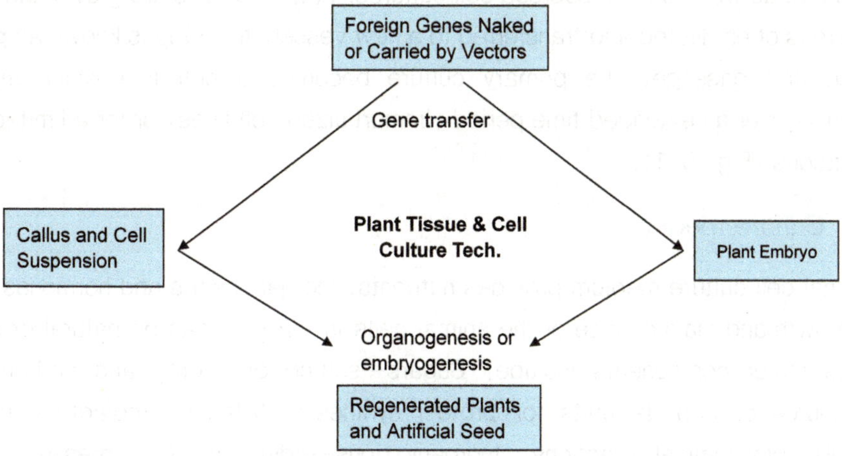

Chapter 9 Animal Tissue and Cell Culture

1907 年 R. Harrison 的青蛙组织培养开启了动物细胞与组织培养的先河，该技术模拟体内生理环境，分离动物细胞，使之在体外生存与繁殖。历经数十年的发展，动物细胞与组织培养已成为现代生物技术的重要分支之一，用于生产原核细胞宿主不能合成的代谢产物，以及人用和兽用的重要蛋白药物和疫苗。同时动物细胞与组织培养也为新型药物的筛选与研究提供细胞筛选模型，为干细胞工程、胚胎工程、克隆动物的培育等提供了支撑技术。本章主要讨论动物细胞与组织培养体系的建立、转基因技术、单克隆抗体和杂交瘤技术以及干细胞研究与进展。

9.1 Establishment of Animal Tissue Culture System

9.1.1 Initiation of tissue culture

The initiation of animal tissue culture is involved in the dissociation of the organized tissue. According to the degree of dissection, the primary culture (cells obtained from animal tissue) can be divided into three main types: 1) Organ culture—the tissue explant retains, at least in part, its architectural features; 2) Primary explant culture— a fragment of tissue is placed at a liquid-solid interface, where, following attachment, outgrowth and migration of cells occurs in the plane of the solid substrate; 3) Cell culture—the tissue, or outgrowth, from the primary explant is mechanically or enzymatically dispersed into a cell suspension. The cell suspension can be cultured as an adherent monolayer or as a non-adherent suspension.

In general, the phrase "tissue culture" is broadly used to describe three-dimensional organ culture as well as the dispersed cell cultures. After the cells are grown and divided, the culture is often diluted and transferred to a new vessel, this step is known as **passing**. After the first passage, the primary culture becomes a cell line which may grow continuously over an extended time period (immortalized cell lines) or for a limited number of generations (Fig. 9.1).

9.1.2 Culture media

Animal cell culture medium provides nutrients, inorganic salts and hormones required for the growth and maintenance of the animal cells *in vitro*. It can be natural or synthetic one. The major components include: Sugars—source of energy and carbon; Amino acids—source of peptide units for protein synthesis; Vitamins—cofactors in various intracellular biochemical reactions; Inorganic ions—adjust osmotic pressure; provide

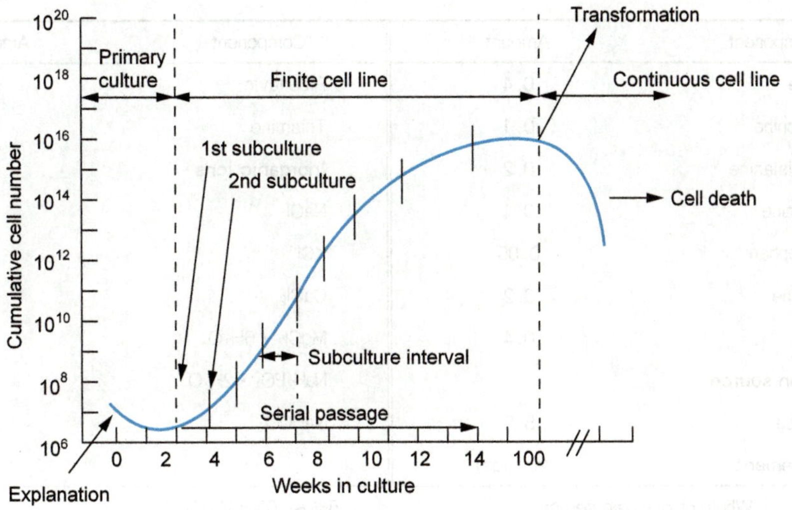

Fig. 9. 1 Development curve of cell lines

electrolyte balance or buffering capacity; maintain the activity of protein; Supplement—supra-physiologic hormones for stimulating cell growth; or non-physiologic indicator dyes, and some time antibiotics for prevention of contamination. Serum is the most effective and commonly used supplement in animal tissue culture; it contains important growth factors, adhesion factors, minerals, lipids, and hormones that would be critical to a successful tissue culture. However sera are made in different lots or batches, and therefore have some disadvantages, such as: 1) variability in composition; 2) high cost and periodic scarcity; 3) source of contamination (mycoplasma, bacteriophages, viruses) and toxins; 4) prevents definition of nutritional environment; 5) interference with detection and purification of cell products. The quality of the lots of sera is usually controlled based on a well-run experimental study in the cell-culture lab. Also, serum-free medium has been developed to avoid the above problems.

An example of medium composition for animal cell culture is listed in Tab. 9. 1.

Tab. 9. 1 Eagle's minimum essential medium (MEM)

Component	Amount	Component	Amount
L-amino acids		**Vitamins**	
Arginine	0.6	Choline	8.3
Cystine	0.1	Folic acid	2.3
Glutamine	2.0	Inositol	11.0
Histidine	0.2	Nicotinamide	8.2
Isoleucine	0.4	Pantothenate	4.6
Leucine	0.4	Pyridoxal	6.0

Continued

Component	Amount	Component	Amount
Lysine	0.4	Riboflavin	0.27
Methionine	0.1	Thiamine	3.0
Phenylalanine	0.2	**Inorganic ions**	
Threonine	0.4	NaCl	116
Tryptophan	0.05	KCl	5.4
Tyrosine	0.2	$CaCl_2$	1.8
Valine	0.4	$MgCl_2 \cdot 6H_2O$	1.0
Carbon source		$NaH_2PO_4 \cdot 2H_2O$	1.1
Glucose	5.5	$NaHCO_3$	23.8
Supplement	(depends)		
Whole or dialyzed serum		5%～10% (V/V)	

Note: All values are millimolar, except for vitamin amounts which are micromolar.

9.1.3 Physical condition for animal tissue and cell culture

Temperature

Most mammalian cell and its derived cell lines are routinely cultured at 37℃; this temperature was chosen because it is the critical body temperature of *Homo sapiens*. However, there is no reason to expect 37°C to be optimal for all types of cells. Insect, worm, and fish cells require significantly lower temperature. Growing cells at too high a temperature is more detrimental than too low a temperature. Even fairly short exposures to temperature 3°C higher than the standard may lead to the death of the entire animal cell culture. In contrast, cells can tolerate low temperature and survive for year in **cryostorage**.

pH

Regulation of extracellular and intracellular pH is also essential for survival of tissue and cells. The pH is not only important for maintaining the appropriate ion balance, but also for maintaining the optimal function of enzymes. The mammalian cell types may have an optimal pH between 7.1 and 7.3. The regulation of pH is done through a variety of buffering systems such as phosphate buffer system.

Osmolality

Without cell wall, animal cells are very sensitive to the osmolality of the medium. Altering the osmolality significantly will almost always affect cell growth and function in some manner. Most cells grow well in the range of 290 to 310 mOsm. The components in the medium such as sugar, ions and amino acids contribute to the final osmolality of the medium. Usually the alteration of osmolality of medium can be obtained by adding various concentrations of salts.

Gases

Oxygen and CO_2 are the main gaseous components in the culture environment that affect the cell culture. All mammalian cells need oxygen for their metabolism. The oxygen deprivation causes poor cell growth. In contrast, oxidative damage is one of the constant threats to cell health. Usually, oxygen levels in the incubator are maintained by keeping 90%～95% air in the incubator. CO_2 is also a metabolite from the cell. The excessive release of CO_2 leads to change of pH of the medium. CO_2 level should be controlled at the range from 0 to 10% CO_2 in air mixtures.

9.1.4 Techniques for culture of animal cells

Typically, cells are grown in the form as monolayers attached to adhesive substrates. These cells require attachment for growth and are often referred to as "anchorage dependent". Other cells and cell lines (hematopoietic, rodent ascites tumor, etc.) can be grown in suspension and are therefore referred to as "anchorage independent". Therefore the culture methods can be classified into **adherent culture**, **immobilized culture** and **suspension culture**. Fig. 9.2 shows the relationship among the culture type, matrix type, deep culture technique and the types of bioreactor.

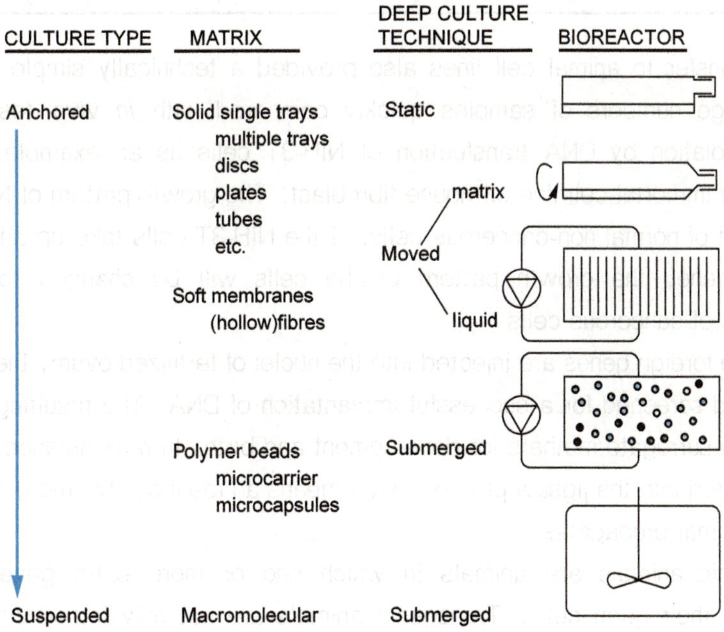

Fig. 9.2 Techniques for cultivation of animal cells

All the bioreactors used would be strictly aseptic, and except the hormones and some of the organic supplement, other components of medium are usually sterilized before use. In the suspension culture, the types and timing of the stirring would be vital factors to the suspended cells, and the shearing force, if over employed, would break the cells and cause failure of the cultivation. Many anchorage-dependent cells exhibit a characteristic growth curve, progressed through a lag phase, a log phase, and then a plateau phase. Anchorage-independent cell lines go through a similar growth cycle. The limiting factors on cell density are media components and secreted factors.

9.2 Gene Transfer to Animal Cell

The genetic engineered animal cell lines are used increasingly for large-scale culture of proteins of animal origin, since these cells might produce many of post-translational modifications of animal proteins which a bacterial expression system usually cannot provide (Tab. 9. 2).

Tab. 9. 2 Post-translational modification of proteins by different host species

Native and/or desired property	Ability of organism to mimic protein natural properties				
	E. coli	Yeast	*Aspergillus*	Insect	Mammalian cell
Glycosylation	No	Yes	Yes	Yes	Yes
Accurate secretion	No	Yes	Yes	Yes	Yes
Folding	Yes/No	Yes/No	Yes/No	Yes	Yes
Phosphorylation	No	Yes	Yes/No	Yes	Yes

Gene transfer to animal cell lines also provided a technically simple procedure for screening large numbers of samples quickly compared with *in vivo* tests. Take the **oncogene** isolation by DNA transfection of NIH-3T cells as an example. NIH-3T cell belongs to an immortal cell line of mouse fibroblast. The growth pattern of NIH-3T cells in culture is that of normal non-cancerous cells. If the NIH-3T cells take up and express the foreign oncogene, the growth pattern of the cells will be changed to one that is characteristic of cancerous cells.

After the foreign genes are injected into the nuclei of fertilized ovum, then they can be cultivated and screened for a successful implantation of DNA. The resulting embryos are transferred to surrogate mothers for development and birth. In rare instance (when all the pieces are fitted into the jigsaw puzzle (might means an ideal combination) an offspring of transgenic animal propagates.

Transgenic animals are animals in which one or more extra genes have been introduced to their germ cells. Transgenic animals are not only important tools for the biological research, but also one kind of living bioreactors for producing valuable products (see also Chapter 10). A cow that synthesizes human ion-binding proteins and goats that manufacture medications to relieve cystic fibrosis are examples of living bioreactors, and sometimes the ranches which grow those transgenic animals are referred as "molecular ranches".

9.2.1 Vector based trans-infection

Vectors for carrying genes into animal cells have been constructed based on some related animal viruses (for example, *Baculovirus* only infects insect cells). One of the major proteins encoded by the virus genome is polyhedron, which accumulates in a very large amount in the nuclei of infected cells. Since the gene has an extremely active promoter, a foreign gene can be inserted under this viral promoter and transferred to the infected insect cells in culture to let over expressed. Simian virus 40 (SV40) was one of the first vectors for the transfer of genes into mammalian cells. The small size of genome and the packing constraints limited the large fragments of foreign DNA to be transferred into SV40. The reconstruction of SV40 with the deletion of unnecessary segment of SV40 genome and incorporation with bacterial plasmid genes results in a shuttle vector for easy selection. This technique was proved very successful and later become a common strategy for animal gene transfer (Fig. 9.3). Retroviruses have also been used in transferring genes into mammalian cells. Retroviruses have a single-stranded RNA genome, which is copied into dsDNA by its own revetro-transcriptase, then stably integrated into the host genome during the late period of infection.

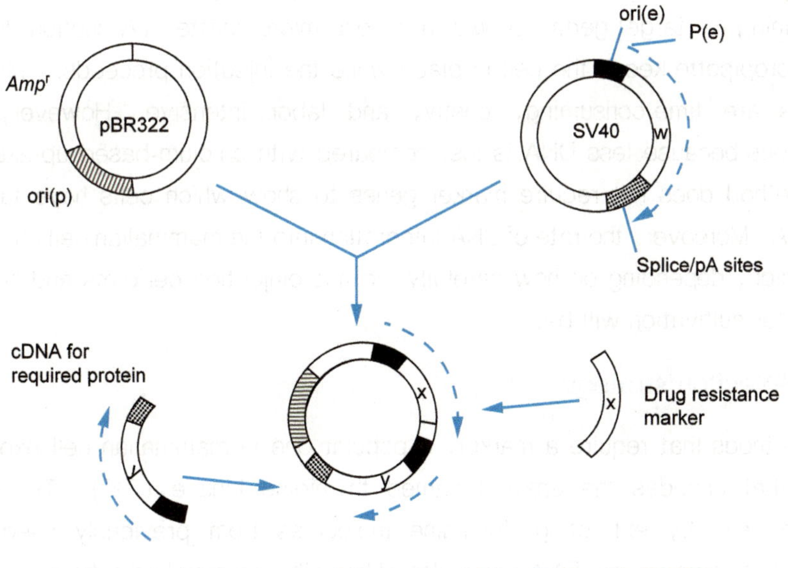

Fig. 9.3 Construction of shuttle vector from SV40 and pBR322

9.2.2 Direct gene transfer

Animal genes can also be introduced into culture eukaryotic cells transiently or permanently without the intermediation by a vector. These methods include **calcium phocphate precipitation**, **microinjection**, and **electroporation** etc. In the method of calcium phosphate precipitation, the foreign DNA fragments are mixed with calcium phosphate to form the complex of DNA-calcium precipitate. When the complex is mixed with cultivated mammalian cells, some of the cells take up the precipitate efficiently, possibly because of **endocytosis** functions. Endocytosis is the process in which a cell forms a membrane around the food particles in its environment and then pinches off the new vesicle into its cytoplasm. For example, in the screening for the presence of an oncogene, the foreign DNA, as a form of DNA-calcium phosphate precipitate, is poured onto NIH-3T cells, it will be pinched off by endocytosis and fortunately expressed in the new host cells. By comparison with the growth patterns of the transfected cells, the oncogene could be isolated.

The microinjection now widely used is a much simpler and more direct process, as Fig. 9.4 displays. A thin glass pipette is placed in a flame and its tip becomes liquefied. Quickly, the technologist draws out the tip to form an ultrathin tip about one-half to a few micrometers in diameter (roughly a hundredth the diameter of a human hair). The micropipette is actually a syringe with an ultrathin tip. It is placed in a microinjection apparatus and filled with DNA fragments or with vectors containing genes. The experimenter controls the micropipette while watching through the oculars of a microscope and restraining the target germ cell with a second micropipette. (A suction device in the second micropipette keeps the cell in place while the injection proceeds.) Microinjection procedures are time-consuming, costly, and labor intensive. However, they are advantageous because less DNA is lost compared with calcium-based uptake methods, and the method does not require marker genes to show which cells have taken up the foreign DNA. Moreover, the rate of DNA integration into the mammalian cell nucleus can be relatively high, depending on how carefully the microinjection performs and how well the conditions for cultivation will be.

9.2.3 Selection strategy

For methods that require a marker, a popular one in mammalian cell experiments is the gene that encodes the enzyme named thymidine kinase (TK). TK functions in mammalian cells by extracting thymidine molecules from previously used DNA and reincorporating them to new DNA molecules. Normally, mammalian cells do not need TK, because they have an alternate method to accomplish the same goal. By use of a biochemical inhibitor, DNA technologists can kill the cells that rely on TK, and let the cells

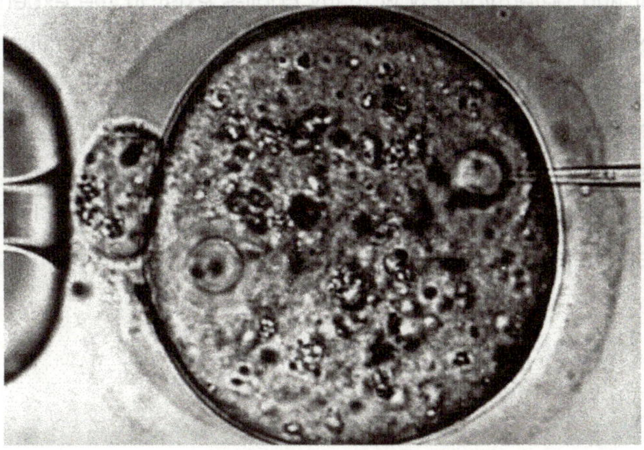

Fig. 9.4 Extra-DNA is microinjected into the nucleus of a fertilized mouse egg. The egg is held in place by a suction pipette shown at the left, while the ultrathin tip is penetrating the egg at the right

with thymidine metabolism in "alternative pattern" remain. The cells that survive the selection process (the "alternative" cells) obtain their thymidine molecules by synthesizing them, not by salvaging them.

To synthesize thymidine, the cells must obtain raw products from the local environment. However, if hypoxanthine and aminopterin are present in the environment, the two chemicals will inhibit thymidine synthesis causing cells stop growing. Moreover, if thymidine is present in the environment, one would expect the cells to take it up and use it. This does not happen, however, because the cells lack the TK necessary to salvage thymidine and incorporate it. A laboratory medium named **HAT medium**, which contain hypoxanthine (H), aminopterin (A), and thymidine (T) has been developed and put into routinely uses.

In practice, the DNA technologist inserts a gene coding TK enzyme into the DNA vector first, and transfers the recombinant DNA into the mammalian cell. If the donor DNA can be successfully inserted into the recipient genome, then the enzyme TK will be expressed, and thymine will be synthesized and incorporated into new DNA molecules to allow the cell growth. Thus, the growing cells represent the ones with the TK gene. By contrast, the cells without TK are finally eliminated.

Other gene markers were also developed, such as *Alu* marker and varieties of fluorescent markers. *Alu* is a repetitive sequence of base pairs, which was named after the related restriction enzyme. Nearly half a million copies of the *Alu* sequences exist in the human genome; the number of repetitive copies in the cells of other mammals varies. By using radioactive gene probes to pinpoint the number of repetitive sequences present, DNA technologists can determine which type of DNA is present. For example, the presence of human DNA in a mammalian cell genome can be determined by locating the human *Alu* marker and determining how many copies exist in the experimental cells. The fluorescent markers are also excellent indicators for their distinct luminary effects. A clone with the desired gene arranged next to the fluorescent marker will be quickly spotted out from the massive individuals. The fluorescent markers from firefly genome have been successfully transformed into tobacco plants; during the night, the transgenic tobacco plants can be easily screened out by examining the luminary effect. In 2009, there was a report on the Korean scientist research work with the transgenic fluorescent puppies.

9.3 Animal Hybrids and Hybridoma

9.3.1 Antibody

The existence of antibodies in the vertebrates forms one of the most important human immune response systems. The host can be induced to have a highly efficient defense response, in the presence of a specific foreign molecule which might be a protein or carbohydrate as a single molecule or attached to cells. Antibodies are **immunoglobulins** (Igs) produced by B lymphocytes (humoral immunity), or by T lymphocytes (cell-mediated immunity). Antibodies and the receptors on the surface of T cells recognize and

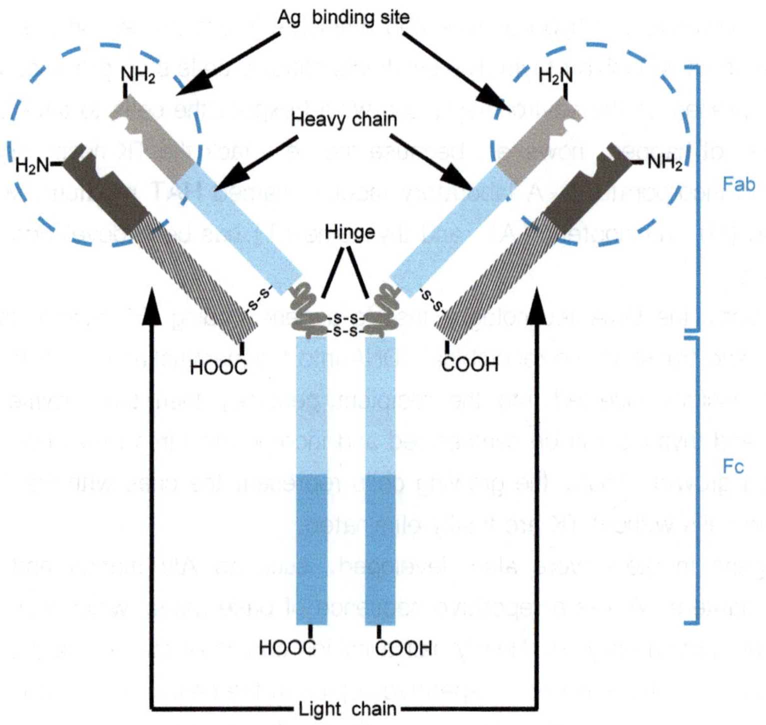

Fig. 9.5 Molecular structure of immunoglobulin (antibody)

bind to specific foreign molecules called antigens, resulting in a variety of responses depending on the nature of target. Immunoglobulins are large Y-shaped proteins made up of two heavy and two light peptide chains, which create two functionally important areas, as shown in Fig. 9.5. Each tip of the "Y" of an antibody (Fab) contains a **paratope** that is specific for one particular **epitope** of antigen and the Fc portion at the other end of the molecule which mediates the function of the antibody. Polyclonal antibodies are secreted in the presence of antigens or an antigen with different epitopes. Polyclonal antibodies are a combination of immunoglobulin molecules which are difficult to be separated from each other.

9.3.2 Monoclonal antibody production

In 1975, the J. F. Köhler and C. Milstein reported the production of mouse monoclonal antibdies (mAbs) by hybridoma technology, which have had a profound impact on medicine by providing an almost limitless source of therapeutic and diagnostic reagents. The procedure of hybridoma is shown in Fig. 9. 6. Monoclonal antibodies are typically made by fusing myeloma cells with the spleen cells from a mouse that has been immunized with the desired antigen. Polyethylene glycol is used to fuse adjacent plasma membranes, and the selective medium is used to let only the fused cells be capable of growing on it. This is because myeloma cells have lost the ability to synthesize hypoxanthine-guanine-phosphoribosyl transferase (HGPRT), an enzyme necessary for the salvage synthesis of nucleic acids. The absence of HGPRT is not a problem for these cells unless the **de novo** purine synthesis pathway is also disrupted. By exposing cells to aminopterin (a folic acid analogue, which inhibits dihydrofolate reductase, DHFR), they are unable to use the de novo pathway and become fully auxotrophic for nucleic acids requiring supplementation to survive.

The HAT medium (see also 9. 3. 3) is selective for the hybrid (hybridoma) cells. Unfused myeloma cells cannot grow because they lack the related metabolic ability, and thus cannot replicate their DNA. Unfused spleen cells cannot grow indefinitely because of their limited life span. Only fused hybrid cells, referred to as hybridomas, are able to grow indefinitely in the media because the spleen cell partner supplies HGPRT and the myeloma partner has traits that make it immortal (as it is a cancer cell).

This mixture of cells is then diluted and clones are grown from single parent cells on microtitre wells. The antibodies secreted from the different clones are then assayed for their ability to bind to the antigen (with a test such as in ELISA or Antigen Microarray Assay) or immuno-dot blot. The most productive and stable clone is then selected for future use. Although the hybridomas can be grown indefinitely in a suitable cell culture medium, the medium must be enriched during *in vitro* selection to favor the hybridoma growth. This can be achieved by the use of a layer of feeder fibrocyte cells or supplement

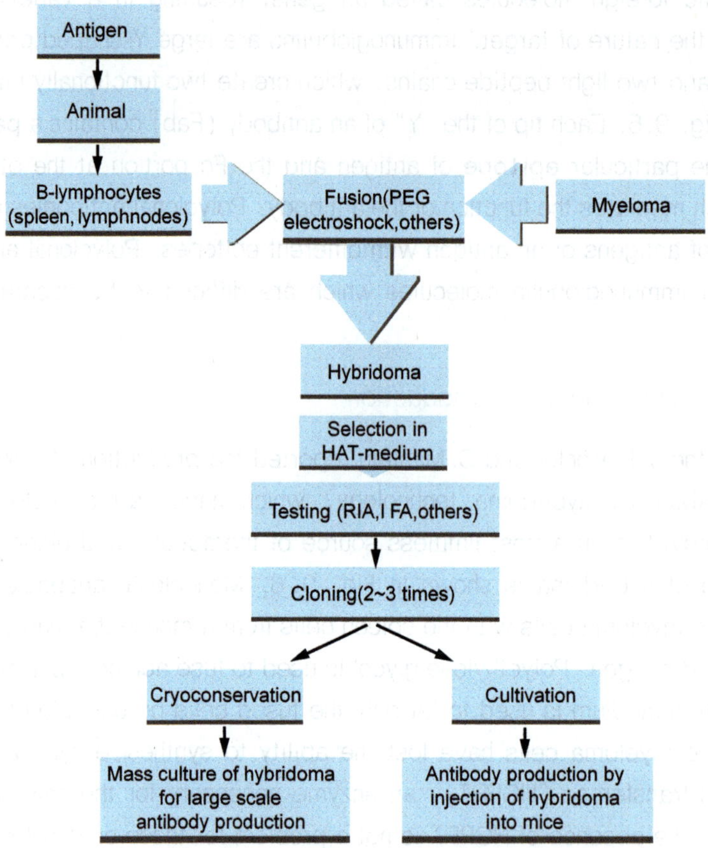

Fig. 9.6 The hybridoma procedure

medium such as briclone. Culture-medium conditioned by macrophages can also be used. The hybridomas can also be injected into mice (in the peritoneal cavity, surrounding the gut). There, the animals produce tumors which secrete an antibody-rich fluid called ascites fluid for use of making monoclonal antibodies.

9.3.3 Monoclonal antibody purification

After obtaining either a media sample of cultured hybridomas or a sample of ascites fluid, the desired antibodies must be extracted. The contaminants in the cell culture sample would consist primarily of media components such as growth factors, hormones, and transferrins. In contrast, the *in vivo* sample is likely to have host antibodies, proteases, nucleases, nucleic acids, and viruses. In both cases, cytokines would also be secreted from the hybridomas. In addition, there may be bacterial contamination, such as **endotoxins** existing in the medium samples. There are no standardized procedure for monoclonal antibody purification for different cases, one method (*in vivo* or *in vitro*) may be preferable to the other, in practice.

Cells, cell debris, lipids, and clotted material are first removed, typically by centrifugation followed by filtration with a 0.45μm filter. These large particles can cause a phenomenon called *membrane fouling* in later purification steps. In addition, the concentration of product in the sample may not be sufficient, especially in cases where the desired antibody is produced by a low-secreting cell line. The sample is therefore condensed by ultrafiltration or precipitated out using ammonium sulfate. Antibodies precipitate usually at low salt concentrations, while most other protein molecules precipitate at rather higher salt concentrations. The appropriate level of salt is added in order to achieve the best separation. Excess salt must then be removed by a desalting step such as dialysis.

Most of the charged impurities are usually anions such as nucleic acids and endotoxins. These are often separated by ion exchange chromatography. Either cation exchange chromatography is used at a low enough pH that helps the desired antibody bind to the column while anions flow through, or anion exchange chromatography is performed at a high enough pH, which allow the desired antibody flows through the column while the anion impurities bind to it. Various proteins can also be separated out along with the anions based on their isoelectric point (*pI*). For example, albumin has a *pI* of 4.8, which is significantly lower than that of most monoclonal antibodies (*pI* of 6.1). In other words, at a given pH, the average charge of albumin molecules is likely to be more negative. Transferrin, on the other hand, has a *pI* of 5.9, so it cannot easily be separated out by this method. A differential *pI* value of 1 (at least), is necessary for achieving a good separation.

Transferrin can be removed by size exclusion chromatography. The advantage of this purification method is that it is one of the more reliable chromatography techniques. Since we are dealing with proteins, properties such as charge and affinity are not consistent and vary with pH as molecules are protonated and deprotonated, while size stays relatively constant. Nonetheless, the method might have some disadvantages as low resolution, low capacity and longer elution times.

A single-step and much quicker method of separation is Protein A/G affinity chromatography. The antibody selectively binds to Protein A/G, so a high level of purity (generally >80%) is obtained. However, this method may be problematic for antibodies that are easily damaged, as harsh conditions are generally used. A low pH can break the bonds to remove the antibody from the column. In addition to possibly affecting the product, low pH can cause Protein A/G itself to leak off the column and appear in the eluted sample. Gentle elution buffer systems that employ high salt concentrations are also available to avoid exposing sensitive antibodies to low pH. Cost is also an important consideration with this method because immobilized Protein A/G is a more expensive resin.

The final purity can be analyzed using a chromatogram. Any impurities will produce peaks, and the volume under the peak indicates the amount of the impurity. Alternatively, gel electrophoresis and capillary electrophoresis can be carried out. Impurities will produce bands beyond the major Ig band of varying intensity, thus be measured.

9.3.4 Chimeric antibodies

Early on, a major problem for the therapeutic use of monoclonal antibodies in medicine was that initial methods used to produce them yielded mouse, not human antibodies. While structurally similar, differences between the two are sufficient to invoke an immune response occurring when murine monoclonal antibodies were injected into humans and resulted in their rapid removal from the blood, systemic inflammatory effects, and the production of human anti-mouse antibodies (HAMA). In an effort to overcome this obstacle, approaches using recombinant DNA have been explored since the late 1980s. In one approach, mouse DNA encoding the binding portion of a monoclonal antibody was merged with human antibody-producing DNA in living cells, and the expression of this chimeric DNA through cell culture yielded partially mouse, partially human monoclonal antibody. The terms "chimeric" and "humanized" monoclonal antibody have been used to descript this product, for reflecting the combination of mouse and human DNA sources in the process.

Several studies have demonstrated that substitution of mouse with human Fc dramatically reduce the anti-antibody response (AAR). While with mAb 84% marked, 7% tolerable and 9% negligible, as to HAMA responses to chimeric Abs there were only 40% marked, 27% tolerable, and 33% negligible human **antichimeric antibodies** (HACA) responses. To counteract both HAMA and HACA responses associated with murine and chimeric antibodies, humanization was done by grafting mouse complementarity-determining regions (CDRs), the hypervariable regions that bind to the specific antigen, into human antibody backbone. In terms of immunogenicity of humanized mAbs, there is a dramatic decrease in marked anti-humanized antibody (HAHA) responses to only 9%. Nevertheless, the tolerable and negligible responses are similar to the chimeric mAbs with 36% and 55% incidence, respectively. Another problem arising from using the humanization approach was that the antibody variant in which only CDRs had been replaced had reduced affinity. This was overcome by the conversion of human amino acids in framework region to their corresponding mouse sequences. This was followed by a "resurfacing" approach in which only the surface exposed mouse residues are replaced by human amino acids.

Hybrid cultures as well as other permanent lines of cells are usually stored and saved under deep freezing (−196°C). One of the problems of hybrid cultures maintenance and cultivation, especially interspecific, is the segregation of chromosomes. Therefore, the control of desirable correlation saving between hybridization partners' chromosomes is obligatory.

9.4 Approaches of Stem Cell Research

9.4.1 Background of stem cells

Stem cells (SC) are the types of self-renewing and with potential to differentiate along one or two lineages. Historically, it has been thought that stem cells have capacity of division without limitation in a life period. There are different kinds of stem cells classified by the names: Unipotent (produce one cell types); pluripotent (produce many cell types); or totipotent (produce all types of cell). Stem cells can also reconstitute a damaged tissue and maintain the normal tissues (Fig. 9.7).

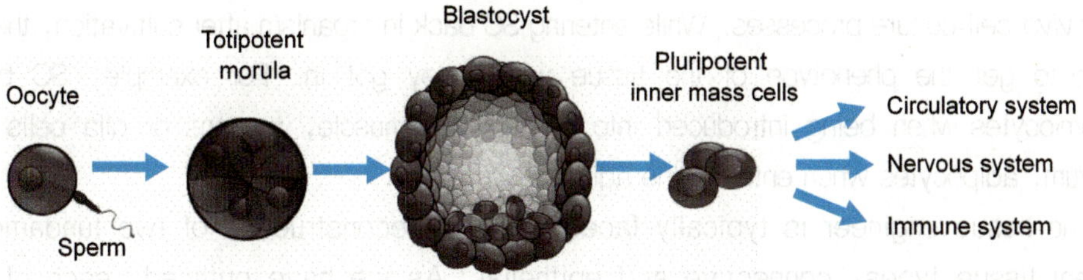

Fig. 9.7 Stem cells with potential to differentiate

In mammals, there are two broad types of stem cells: embryonic stem cells (ESC), which are isolated from the inner cell mass of blastocysts, and adult stem cells or regional stem cells (RSC), which are not differentiated cells discovered in specialized tissues. They have the possibility of self-reproducing into all the cell types of the tissue they are originating from. Three types of ESC have been described by scientists of different countries: 1)ESC obtained from the embryoblast of pre-implantation embryos of mammals; 2)Cells of embryonic carcinoma (ECC); 3)Primordial gametes of embryos (PGCE). The ESC of hamsters, rabbits, sheep, pigs, cattle, and primates are obtained from three embryonic layers(ectoderm, endoderm and mesoderm). The lines of murine permanent ESC had been intensively studied, from each SC in the organism two daughter cells appear after mitosis, one of which is the complete copy of maternal and is able to self-renewal, and the second is initially determined and possesses the certain potency to differentiate.

RSC are not numerous and are dispersed in the tissues. The origin of the RSC from different tissues of adult organism is still not clearly known. The adult stem cells differentiating firstly form the cells of intermediate type are called the precursors. These are partly differentiated cells, which divide and give the origin to differentiated ones. The adult SC are characterized as pluripotent cells which become the main feature of all the SC, that means the most part of all RSC (except the unipotent SC of cornea) can give the origin to the cells of different types. For example, blood SC can differentiate into different cell elements of blood. The RSC investigation was lasting many years, and the endotheliocytes, hepatocytes, and neurocytes, the cells of skeleton and muscles, the cells of skin, pancreas, and others were isolated and identified, and their differentiation potential was studied.

9.4.2 Stem cells and tissue building

Stem-cell commitment along differentiated lineages underlies the organ function, repair, and formation, the processes of cell proliferation, cell death, cell motion, and cell differentiation together determine the fate of the committed cell. Stem-cell depletion may lead to full or partial loss of organ function. Stem cells are thus the starting material for many key processes in tissue engineering and must be considered as an important source of regenerative potential *in vivo* and for *ex vivo* cell-culture processes. While entering SC back in organism after cultivation, there is a chance to get the phenotype of the tissue which they got in. For example, SC become cardiomiocytes when being introduced into the cardiac muscle, neurons or glia cells in the cerebrum, adipocytes when entering the adipose tissue.

The tissue engineer is typically faced with the reconstruction of two fundamentally different tissue types: connective and epithelial. As we have outlined, each of these tissues has been found to have stem cells. There are direct descriptions of hematopoietic and mesenchymal stem cells, and interestingly they both reside in the bone marrow. Epithelia in the intestinal lining, the skin, and the cornea have also been described as having stem cells. In the case of the liver, recent evidence suggests that a stem cell, that can repopulate the liver after injury, also resides in the bone marrow. Thus, the bone marrow may serve as an attractive target to tissue engineers for generating the new tissues. It is important to note that while a single mesenchymal stem cell (that can form cartilage, bone, fat, etc.) has been isolated, a typical "epithelia" stem cell has not been isolated for there are many different types of epithelium.

9.4.3 Isolation of stem cells for scientific and clinical purposes

Several methods have been developed to isolate stem cells or enrich the stem-cell content of a cell population. These include affinity columns, magnetic columns, and counter-flow elutriation. The difficulty with positively identifying the presence of stem cells is the requisite functional assay, namely, the reconstitution of tissue *in vivo*.

Ex vivo growth and manipulation of stem cells

Being able to grow stem cells (SC) *ex vivo*, relies on how to let the cells divide without differentiation, and theoretically provide an inexhaustible source of stem cells. However, as previously stated, adult stem cells appear to age, and this capability may never be realized. Embryonic stem cells can be propagated *in vivo*; however, the differentiation cues to efficiently produce populations of specialized cells have not been elucidated in most cases.

Engraftment potential of cultured stem cells

The procedure of isolating stem cells and culturing them for transplantation purposes has a number of attractive attributes from a clinical standpoint. Unfortunately for unknown reasons, often the stem cells differentiate and lose their engraftment potential quickly in culture. One strategy for preventing such problem is to mimic the *in vivo* microenvironment of the stem cell. Soluble growth factors, together with adequate nutrients and signals from supportive cells of mesenchymal origin, seem to prolong the culture period without differentiation and loss of pluripotency. Indeed, the strategy of using cell-cell communication has been employed with both hematopoietic stem cells and embryonic stem cells in the culture system. Recently, the discovery and extensive research on stem cells brings good news to those patients who need special therapeutic treatment. According to some very new reports, Chinese scientists are exploiting some innovative techniques to allow the neural stem cells to grow on chitosan backbone, by which to help the injured neural tissue regenerate. The news casts a light on repairing of neuro-injury, and also indicates there is a bright future of applying stem cells to routinely therapeutic treatment.

Summary

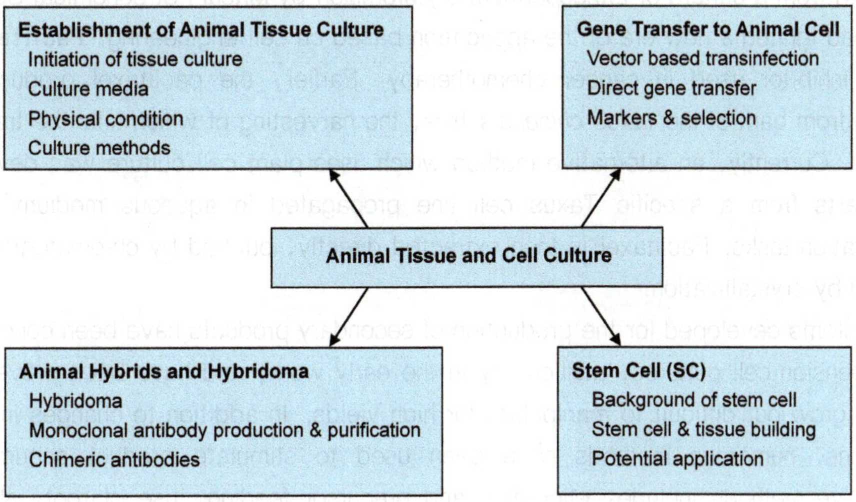

Chapter 10 Application Examples from Cell Engineering

作为现代生物技术体系的重要组成部分之一，细胞工程为细胞生物学、分子生物学、生物化学等生命科学的理论研究提供了全新的实验材料与研究路径。此外细胞工程的应用领域也非常广泛，在讨论了细胞融合、杂交、动植物细胞培养、转基因技术等基础理论和方法之后，本章主要列举细胞工程应用的一些例子，特别介绍生物产品、农业育种、新药以及研究性物种建构等方面的应用。

10.1 Bioproducts from Cell Engineering

10.1.1 Secondary metabolites form plant cells

Higher plants are still major sources of pharmaceuticals, colors, dyes, and flavors. The plants grown in large scale plantations might also be of value, particularly to the pharmaceutical industry, (in the way of producing useful compounds in a factory setting). Cultured plant cells were not only developed as a tool for studying plant metabolism, physiology, and tissue development but have also been used as a possible source of natural products especially those pharmaceuticals with a high value. Many natural products have been detected in plant cell cultures, either callus or suspension, but only a few have been at levels which would make an industrial process possible. This probably reflects the lack of knowledge of the pathways and controls on synthesis of many natural products. One notable success was announced in 1983, the commercial production of **shikonin** from a culture of *Lithospermum erythrorhizon* by Mitsui Petrochemical Company, which had ignited a new era of the application based on cell engineering. **Paclitaxel** is a mitotic inhibitor used in cancer chemotherapy. Earlier, the paclitaxel produced was derived from bark of the *taxus chinensis* tree, the harvesting of which kills the tree in the process. Currently, an alternative method which uses plant cell culture was developed. This starts from a specific Taxus cell line propagated in aqueous medium in large fermentation tanks. Paclitaxel is then extracted directly, purified by chromatography and isolated by crystallization.

Systems developed for the production of secondary products have been concentrated on suspension cell cultures, particularly in the early work, and have been proved to be easy to grow but difficult to manipulate for high yields. In addition to changes in cultural conditions, numerous methods have been used to stimulate product accumulation. Alternative methods include: elicitation and precursor feeding; use of root, shoot, or embryo cultures; use of transformed cultures; and immobilized cultures.

It has been observed that secondary product accumulation can often by initiated if the culture is allowed to differentiate into roots, shoots, or embryos. Immobilization of plant cells increases the production of secondary products in some cases and has a number of other advantages over the use of suspension cultures, for example, the biocatalyst would be easily recovered and reused, easily separated from the catalytic medium. Immobilization of plant cells is also beneficial to the continuous operation. In principle, the nearly unlimited enzymatic potential of plant cell cultures can be employed for bioconversions. The best known example is the conversion of β-methyldigoxin by *Digitalis lanata* cells.

10.1.2 Products from cell hybridization

Among the bioproducts from cell hybridization, monoclonal antibodies (mAb) would be the most successful examples and have particular usage. They are also very useful in immuno-histochemistry, which detect antigen in fixed tissue sections; in immuno-fluorescence test, which detect the substance in a frozen tissue section or in live cells, and in Western blot test and immuno dot blot tests, which detect the protein on a blot membrane.

One possible treatment for cancer involves monoclonal antibodies (mAb) that bind only to cancer cell-specific antigens and induce an immunological response against the target cancer cell. Since mAb could also be modified for delivery of a toxin, radioisotope, cytokine or other active conjugate; it is also possible to design biospecific antibodies that can bind with their both Fab regions to the target antigen and to a conjugate or effector cells.

10.1.3 Transgenic animal for medicine production

As living bioreactors, transgenic animals can be genetically altered to produce valuable products. The methods involve microinjection of DNA into fertilized egg cells of embryos and transferal to surrogate females, where the animal develops. In this way a transgenic pig has been produced that can synthesize human hemoglobin for use as a blood substitute. A transgenic cow has also been bred with the ability to produce **lactoferrin**, an iron-building milk protein and a potential antibacterial agent. A transgenic sheep is now available that can synthesize a protein to keep cell membranes more elastic and help treat emphysema patients. And finally, a transgenic goat has been developed with the capability of synthesizing the protein which lacks in the patients suffering from cystic fibrosis.

Transgenic methods can also be used to develop animals with better traits. For example, the nutritional value of milk can be upgraded by enhancing the ability of a cow's mammary gland to secrete high quality casein. Another possibility is to develop transgenic animals that can resist viral infection by modifying animal genes to produce viral glycoproteins. Animals can also be transgenically treated to increase their production of the antiviral substance, such as **interferon**. It may also be possible to introduce transgenic snails and mosquitoes to crowd out the natural vectors that transmit epidemic disease.

10.2 Breeding of New Agronomic Varieties

During the last decades, a tremendous progress has been made in the development of transgenic plants using the various techniques of genetic engineering. The plants, in which a functional foreign gene has been incorporated by any biotechnological methods that generally are not present in the plant, are called transgenic plants. According the uncompleted statistic figures, transgenic crops are cultivated world-wide on about 148 million acres (587 million hectares) of land by about 5.5 million farmers, in 2002, but now, in USA the output of transgenic soy beans is around 93% and corn around 90%. Transgenic plants have many beneficial traits like insect resistance, herbicide tolerance, delayed fruit ripening, and improved oil quality etc.

Some of the commercially grown transgenic plants in developed countries have been put into large scale plantation, for example, 'Roundup Ready' soybean, 'Freedom II squash', 'High-lauric' rapeseed (canola), 'Flavr Savr' and 'Endless Summer' tomatoes. In 1995, the granted registrations on genetically engineered species involved Bt gene containing insect resistant 'New Leaf' (potato), 'Maximizer' (corn), 'BollGard' (cotton), and other brands in USA. More and more traits (only a few of them are listed as follows), have been transferred into different agronomic plants to make them new varieties.

10.2.1 Abiotic stress tolerance

Special strategies had to be developed to overcome the problems caused due to biotic stresses (viral, bacterial infections, pests and weeds) and abiotic stresses (physical actors such as temperature, humidity, salinity etc). The plants show their abiotic stress response reactions by the production of stress related osmolytes like sugars (e.g. trehalose and fructans), sugar alcohols (e.g. mannitol), amino acids (e.g. proline, glycine, betaine) and certain proteins (e.g. antifreeze proteins). Transgenic plants have been produced which overexpress the genes for one or more of the compounds mentioned above. Such plants show increased tolerance to environmental stresses. Resistance to abiotic stresses includes stress induced by herbicides, temperature (heat, chilling, freezing), drought, salinity, ozone and intense light. These environmental stresses result in the destruction, deterioration of crop plants which leads to low crop productivity.

Other abiotic stress is also known as water deficit stresses. The plants produce osmolytes or osmoprotectants to overcome the osmotic stress. The attempts involve using some genetic engineering strategies to build up the ability of osmoprotectants in the plants. The biosynthetic pathways for the construction of many osmoprotectants have been established and genes coding the key enzymes have been isolated. For example, glycine betaine is a cellular osmolyte which is produced by the participation of a number of key enzymes like choline dehydrogenase and choline monooxygenase etc. In other cases, the choline oxidase gene from *Arthrobacter* sp. was transferred to produce transgenic rice with high levels of glycine betaine, which offers rice better tolerance against water deficit stress. Scientists also developed cold-tolerant genes in *Arabidopsis* when this plant was gradually exposed to slowly declining temperature. By introducing the coordinating gene (it encodes a protein which acts as transcription factor for regulating the expression of cold tolerant genes), expression of cold tolerant genes was triggered giving protection to the plants against the dropping temperatures.

10.2.2 Biotic stress tolerance

Several strategies have also been developed to build resistance in the plants against biotic stresses such as:

Herbicide resistance

Weeds are unwanted plants which lower down the crop yields by competing with crop plants for light, water and nutrients. Several biotechnological strategies for weed control are used, e.g. the over-production of herbicide target enzyme (usually in the chloroplast) in the plant which makes the plant insensitive to the herbicide. This is done by the introduction of a modified gene that encodes a resistant form of the enzyme targeted by the herbicide in weeds and crop plants. **Roundup Ready Crop** plants tolerant to herbicide-Roundup have already been used. The biological manipulations using genetic engineering to develop herbicide resistant plants are involving: 1) over-expression of the target protein by integrating multiple copies of the gene or by using a strong promoter; 2) enhancing the plant detoxification system which helps in reducing the effect of herbicide; 3) detoxifying the herbicide by using a foreign gene; and 4) modification of the target protein by mutation.

More examples are: Glyphosate resistance—Glyphosate is a glycine derivative herbicide which is found to be effective against worst weeds by being the competitive inhibitor of the enzyme 5-enoyl-pyruvylshikimate 3-phosphate synthase (EPSPS) in the shikimic acid pathway. Due to its structural similarity to the substrate phosphoenol pyruvate, glyphosate binds more tightly with EPSPS and thus blocks the shikimic acid pathway. Certain strategies were used to provide glyphosate resistance to plants: 1) It was found that EPSPS gene was over expressed in Petunia due to gene amplification. EPSPS gene was isolated from Petunia and introduced in to the other plants. These plants could tolerate glyphosate at a dose of 2~4 times higher than that required to kill wild type plants. 2) By using mutant EPSPS genes—A single base substitution from C to T resulted in the change of an amino acid from proline to serine in EPSPS. The modified enzyme cannot bind to glyphosate and thus provides resistance. 3) The detoxification of glyphosate by introducing the gene (isolated from soil organism—*Ochrobactrum anthropi*) encoding glyphosate oxidase into crop plants. The enzyme glyphosate oxidase converts glyphosate to glyoxylate and aminomethylphosponic acid. The transgenic plants exhibited very good glyphosate resistance in the field. And phosphinothricin resistance—Phosphinothricin is a broad spectrum herbicide and is rather effective against broad-leafed weeds. It acts as a competitive inhibitor of the enzyme glutamine synthase which results in the inhibition of the enzyme glutamine synthase and accumulation of ammonia and finally causes the death of the herb. The disturbance in the glutamine synthesis also inhibits the photosynthetic activity. The enzyme phosphinothricin acetyl transferase (which was first observed in *Streptomyces* sp. in natural detoxifying mechanism against phosphinothricin) acetylates phosphinothricin, and thus inactivates the herbicide. The gene encoding phosphinothricin acetyl transferase (bar gene) was introduced in transgenic maize and oil seed rape to provide agronomic plants with the resistance against phosphinothricin.

Insect resistance

A variety of insects, mites and nematodes significantly reduce the yield and quality of the crop plants. The conventional method is to use synthetic pesticides, which also have severe effects on human health and environment. The transgenic technology uses innovative and eco-friendly methods to improve pest control. About 40 genes obtained from microorganisms, higher plants and animals have been used to offer insect resistance in crop plants.

The first genes available for genetic engineering of crop plants for pest resistance were Cry genes (popularly known as Bt genes) from a bacterium *Bacillus thuringiensis*. These are specific to particular group of insect pests, and are not harmful to other useful insects like butter flies and silk worms. Many transgenic crops (e.g. cotton, rice, maize, potato, tomato, brinjal, cauliflower, cabbage, etc.) carrying Bt genes have been developed; this has proved to be an effective way of controlling the insect pests and has reduced the pesticide use. The most notable example is Bt cotton (which contains CryIAc gene) that is resistant to a notorious insect pest Bollworm (*Helicoverpa armigera*). There are certain other insect resistant genes from other microorganisms which have been used for similar purpose. Isopentenyl transferase gene from *Agrobacterium tumefaciens* has been introduced into tobacco and tomato; the transgenic plants were found to reduce the leaf consumption by tobacco hornworm and decrease the survival of peach potato aphid.

Certain genes from higher plants were also found to result in the synthesis of products possessing insecticidal activity. One of the examples is the Cowpea trypsin inhibitor gene (CpTi) which was introduced into tobacco, potato, and oilseed rape for developing transgenic plants. It was observed earlier that the wild species of cowpea plants growing in Africa were resistant to the attack form a wide range of insects. The insecticidal protein as a trypsin inhibitor was capable of destroying insects belonging to the orders of Lepidoptera and Orthoptera etc. Cowpea trypsin inhibitor (CpTi) has no effect on mammalian trypsin, thus it is non-toxic to mammals.

Pathogens resistance

Anti-viral species There are several strategies for engineering plants for viral resistance, and these utilise the genes from virus itself (e.g. the viral coat protein gene). The virus-derived resistance has given promising results in a number of crop plants such as tobacco, tomato, potato, alfalfa, and papaya. The induction of virus resistance is done by employing virus-encoded genes-virus coat proteins, movement proteins, transmission proteins, antisense RNAs, and ribozymes. The virus coat protein-mediated approach is most successful to provide virus resistance to plants. It was in 1986, transgenic tobacco plants expressing *Tobacco mosaic virus* (TMV) coat protein gene were first developed. These tobacco plants exhibited high levels of resistance to TMV. The transgenic plants providing coat protein-mediated resistance to virus are also including rice, potato, peanut, sugar beet, alfalfa etc. The viruses being manipulated include *Alfalfa mosaic virus* (AlMV), *Cucumber mosaic virus* (CMV), *Potato virus X* (PVX), *Potato virus Y* (PVY) etc.

Resistance against fungal and bacterial invasions As a defense strategy against the invading pathogens (fungi and bacteria), the plants accumulate low molecular weight proteins which are collectively known as pathogenesis-related (PR) proteins. Several transgenic crop plants with increased resistance to fungal pathogens are being raised with genes coding for the different compounds. One of the examples is the glucanase enzyme that degrades the cell wall of many fungi. The most widely used glucanase is beta-1, 4-glucanase. The gene encoding beta-1, 4 glucanase has been isolated from barley, introduced, and expressed in transgenic tobacco plants. This gene provides good protection against soil-borne fungal pathogen *Rhizoctonia solani*. Also, Lysozyme degrades chitin and peptidoglycan of cell wall, and in this way fungal infection can be reduced. Transgenic potato plants with lysozyme gene have been developed for providing resistance to *Erwinia carotovora*.

10.2.3 Delayed fruit ripeness

The gas hormone, ethylene, regulates the ripening of fruits, therefore, ripening can be slowed down by blocking or reducing ethylene production. This can be achieved by introducing ethylene forming gene(s) in a way that will suppress its own expression in the crop plant. Such fruits ripen very slowly (however, they can be ripened by additional ethylene application) and this helps export the fruits to longer distances without spoilage due to longer-shelf life. The most common example is the "Flavr Savr" (a genetically modified tomato), which were commercialized in U. S. A in 1994. The main technique used was the antisense RNA approach. In the normal tomato plant, the PG gene (coding for the enzyme polygalacturonase) encodes a normal mRNA that produces the enzyme polygalacturonase which is involved in the fruit ripening. The cloned PG complimentary DNA encodes antisense mRNA, which is complimentary to the normal (sense) mRNA. The hybridization between the sense and antisense mRNAs renders the sense mRNA ineffective. Consequently, polygalacturonase can not be translated because of lacking template molecule, hence causing the delay in fruit ripening. Transgenic plants with antisense gene of ACC oxidase (another enzyme involved in the biosynthetic process of ethylene) have been developed. In these plants, production of ethylene was reduced by about 97% with a significant delay in the fruit ripening. The bacterial gene encoding ACC deaminase (an enzyme that acts on ACC and removes amino group) has been transferred and expressed in tomato plants which showed 90% inhibition in the ethylene biosynthesis.

10.3 Transgenic Animals as Models for Medical Research

10.3.1 Custom-made animals

A wide variety of experiments are currently in progress to assess the various uses of transgenic animals. Among the earlier transgenic animals were the so-called **custom-made animals**. In 1983, R. L. Brinster and R. D. Palmiter inserted into mice the growth hormone gene and achieved an animal of bigger-than-normal size. The initial success, however, was tempered when it became clear that the bigger is not necessarily the better. The transgenic mice encountered a number of physiological problems such as reproductive failure and organ diseases. Also, the mice did not convert food to muscle any more efficiently than normal mice, even though they were of larger size. Thus, the initial expectation of leaner cattle and meatier pigs remained elusive.

10.3.2 Mouse with human immune system

So called **Human mouse** is a mouse with partial human immune system. This animal was not manipulated from the microinjection with human genes, so in strictly sense, it is not a product of DNA cloning technology. Instead, it was exposed as an embryo to human cells that grew together with the mouse tissues and became part of the mouse's system. Having a mouse with a human immune system is a valuable asset to many forms of research, particularly to AIDS researches which are seriously hampered by the lack of ideal animal models for laboratory tests. For example, no animal except the chimpanzee contractining AIDS. The use of an animal species rather distanced to human being would be safer; the mice with human immune system will help researchers to determine whether an experimental vaccine is eliciting a response. Moreover, pharmaceutical companies will be able to observe the effects of new drugs on the ability of human immunodeficiency virus (HIV) to infect immune system cells.

10.3.3 The Alzheimer's mouse

Alzheimer's symptom is a brain disease marked by severe memory loss, decline in judgment and thinking, and mood swings. Postmortem analyses of brains of Alzheimer's patients consistently reveal some "plaques" of dead nerve cells enmeshed in fibers of a protein called amyloid. Several years ago, researchers purified amyloid and determined its amino acid sequence. Soon thereafter, a precursor protein was identified, and its gene sequence was subsequently determined. Through all these experiments, the question of amyloid's involvement in Alzheimer's disease was still left unresolved. Was it a cause of the disease or was it debris left by the disease? Experiments reported in 1991 shed a light on the puzzle, when a group of DNA technologists led by Jon Gordon of *New York's Mount Sinai Hospital* engineered a transgenic mouse having the signs and symptoms of Alzheimer's disease. Gordon's team introduced the amyloid precursor gene to the fertilized egg cells of the mice and observed that adult mice were synthesizing human amyloid protein. As amyloid concentrations built up in the animal brain, there appeared the typical brain configurations of amyloid plaques seen in Alzheimer's patients. The model replicated the disease process accurately and offered a practical understanding of the disease. Now those transgenic mice are used as the model to show the relation of amyloid protein to human Alzheimer's disease.

10.3.4 Knockout mouse

Mice are currently the most closely related laboratory animal species to humans for which the knockout technique can easily be applied. They are widely used in knockout experiments, especially those investigating genetic questions that relate to human physiology.

One of the hottest commodities in contemporary molecular genetics is the so-called knockout mouse in which one or more genes has been eliminated. The loss of gene activity often causes changes in the mouse's phenotype, which includes appearance, behavior and other observable physical and biochemical characteristics. Knockout mice are important animal models for studying the role of genes which have been sequenced but whose functions have not been determined. By causing a specific gene to be inactive in the mouse, and observing any differences of normal behavior or physiology, researchers can infer its probable function.

Along with the methodological development of animal cell engineering, either the theory or the practice is frequently renewed. It's not surprising that there are so many research works come out with almost incredible results, and we have confidence to wait for more intensive applications. Meanwhile, there are always debates on the ethic effects of animal cell engineering, for example, many people consider that it's not moral to hybridized human and animal cells together which might create a potential "monster", and to propagate animal progeny with only maternal genomic transform which totally disturb the dual-sexual propagation system. The studies of "**Stem Cell Engineering**" are also opposed by many scientists and philosophers. It's necessary to ask all the scientists involved in this field to uphold justice and be fully responsible to our society and the ecosystem.

Summary

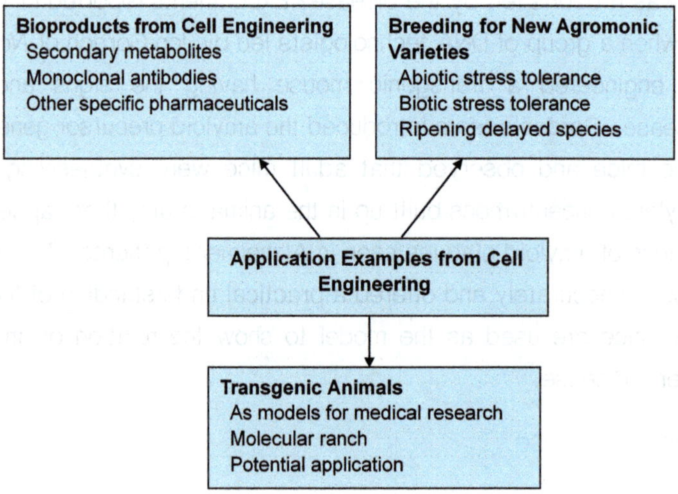

Part III Enzyme Technology

Chapter 11 Fundamental Nature of Enzyme

酶是催化化学反应的各种生物催化剂的总称，一般情况下，酶是蛋白质，但新近发现一些剪切核酸分子的生物催化剂是具有三维结构的 RNA 分子，称之为核酶(ribozyme)。几乎所有的生物都含有酶或利用酶，酶对于生命活动中的物质代谢和能量代谢起到不可替代的作用。本章介绍酶的基本性质、用途、大致分类、反应动力学以及酶的来源等基本知识。

11.1 Definition, Basic Nature and Classification of Enzyme

11.1.1 What is enzyme?

Enzymes are amazing and important substances in nature. First, it's important to realize that enzymes are expressed by living cells for **catalyzing** the biochemical reactions necessary for metabolism. If one were able to look inside a cell, he would see different activities going on. There would be some molecules joining together, others breaking apart or changing into different molecules. These activities keep the cell alive and enzymes make these activities possible. That's why every cell of every living creature on the Earth must synthesize enzymes. As a matter of fact, in some of the sophisticated lives, the species and the amount of enzymes vary all the time.

In a broad sense, there are three categories of enzymes. One undertakes the synthetic works to help join specific molecules together to make a different compound; one plays degradation works to break specific molecules apart into separate molecules; and other may alters the residues of some molecules to form new substances. Enzymes may play important roles outside the living cells as well, if they were obtained gently from the cells without losing the catalytic activities. One of the best examples of this is the enzymatic digestive system. For instance, enzymes in human digestive system break food down into small molecules that can be absorbed by our body. The enzyme extraction forming our digestive system could also work in intro to break down starch, proteins and fats into small units.

Additionally, a great number of enzymes have been identified and characterized; and due to their significantly different bio-functions, they have been further arranged as Oxidoreductases, Transferases, Hydrolases, Lyases, Isomerases, and Ligases. It is interesting that some enzymes are composed of proteins, carbohydrates, lipids and even nuclear acids, with very complex structures and particular activities.

For many years, the chemical manipulation of DNA lagged behind that of proteins. The chemical complexity and variety of proteins, with up to 20 different naturally occurring amino acids as the basic units, served to make them more amenable to increasingly sophisticated methods of analysis. On the other hand, DNA, composed of only four different nucleotides, appeared dauntingly large, with few structural features to make it yield to available methodology.

Paradoxically, this very lack of variety in the nature of the constituent nucleotides of DNA has permitted the revolution in genetic engineering, in which the enzymology of DNA processing has played a prominent part. For example, the discovery and purification of DNA restriction endo-nucleases enabled DNA to be cleaved at defined sites to form a fixed length of DNA fragments with specific ends; DNA ligase permits the DNA fragments to be rejoined selectively; and DNA polymerases allow DNA strand to be synthesized in a complementary form and sequenced at astonishing speed.

The most incredible discovery involved in enzymology was the revealing of the phenomenon of RNA "self splicing". The intron sections in eukaryotic primary mRNA molecules were found to be spliced without the action of protein-like catalysts, which finally brought about the theory of "Ribozyme", meaning that certain RNA molecules have special catalytic roles on eukaryotic RNA processing.

11.1.2 Basic properties of enzyme

Specific: An enzyme is able to do specific catalytic job; for example, a protease breaks protein down but not touch fat or starch. Generally speaking, one enzyme always performs only one catalytic function, which means a special enzymatic reaction would cause almost no side effects. It also explains why there are so many different types of enzymes existed in one single life. To date, thousands of different enzymes have been well identified, some of them are proteins of tertiary structure, without or with co-enzyme factors, and some might have quaternary structure composed form many subunits and have complicate functions. Furthermore, more than one different catalytic site was detected in some enzymes, with significantly different activities [e.g. glutamine synthetase (GS) and DNA polymerase I].

Reusable: It is important to know that enzyme can perform the same job over and over again, even millions of times under the optimal conditions, without being consumed in the process. When performing the *in vitro* enzymatic reactions, it's better to follow the kinetic models and ensure the reaction going on under the optimal temperature and pH condition. It's worth to mention that the immobilization of enzyme molecules would elongate the duration of these bio-catalysts.

Efficient: Not only do enzymes work hard, they also work at a dashing speed. For instance, there is an enzyme in human liver that helps hydrogen peroxide break down into water and oxygen to protect human health. What an amazing fact is that one molecule of the enzyme can break 5 million molecules of hydrogen peroxide in a single minute. The generation period of *E. coli* may be as short as 20 minutes, under suitable growth condition; one can see how efficient the enzymes work for the cells.

Mild: Enzymes are bio-macromolecules working only under mild physiological conditions. In rare cases, few thermal-enzymes might work at rather high temperature (e.g. 80～90℃), but most of normal enzymes are not tolerated to violent conditions. The mildness of enzymatic reaction offers great advantages for its application, for example, chemical hydrolysis of starch needs high temperature and low pH, which usually causes energy consumption and reactor corruption, but hydrolysis of the same starch by amylase only happens at 37℃ and at rather neutral pH.

Industrial application of enzymes would be benefited by their properties of specialty, reusing ability, efficacy and mildness etc. In other cases, some special reactions could not be carried out by chemical catalysts, but only by enzymes. Even in the cases of either the chemical or enzymatic catalysts can be used, more and more chemical catalytic processes are expecting to be replaced by enzymatic treatment. On the other hand, enzymes are biochemical protein or RNA molecules, they are easy to lose the activity under unfavorable conditions such as high temperature and extreme pH, offering the optimal catalytic conditions and maintaining the stability of enzymes will be the critical tasks to the biotechnologists.

11.1.3 Classification of enzymes

There are approximately 3000 enzymes which have been well characterized. The International Union of Biochemistry and Molecular Biology have developed a nomenclature for enzymes, the EC numbers; each enzyme is described by a sequence of four numbers preceded by "EC". The first number broadly classifies the enzyme based on its mechanism.

EC 1 Oxidoreductases

These enzymes catalyze oxidation and reduction reactions involving the transfer of hydrogen atoms or electrons, play vital parts in the life and existence of organisms. The following are of particular importance in the design of enzyme electrodes. This group can be further divided into 4 main classes.

Dehydrogenases catalyze hydrogen transfer from the substrate to a nicotinamide adenine dinucleotide cofactor (NAD^+). An example of this is lactate dehydrogenase which catalyzes the following reaction:

$$\text{Lactate} + NAD^+ \text{===} \text{Pyruvate} + NADH + H^+$$

Oxidases catalyze hydrogen transfer from the substrate to molecular oxygen producing hydrogen peroxide as a by-product. An example of this is FAD dependent glucose oxidase which catalyses the following reaction:

$$\text{D-glucose} + O_2 \text{===} \text{Gluconolactone} + H_2O_2$$

Peroxidases catalyze oxidation of a substrate by hydrogen peroxide. An example of this type of enzyme is horseradish peroxidase which catalyzes the oxidation of a number of different reducing substances (dyes, amines, hydroquinones, etc.) and the concomitant reduction of hydrogen peroxide. The reaction below illustrates the oxidation of neutral ferrocene to ferricinium in the presence of hydrogen peroxide:

$$2[Fe(Cp)_2] + H_2O_2 + 2H^+ \text{===} 2[Fe(Cp)_2]^+ + 2H_2O$$

Oxygenases catalyze substrate oxidation by molecular oxygen. The reduced product of the reaction in this case is water and not hydrogen peroxide. An example of this is the oxidation of lactate to acetate catalyzed by lactate-2-monooxygenase.

$$\text{Lactate} + O_2 \text{===} \text{Acetate} + CO_2 + H_2O$$

EC 2 Transferases

These enzymes transfer C, N, P or S containing groups (alkyl, acyl, aldehyde, amino, phosphate or glucosyl) from one substrate to another. Transaminases, transketolases, transaldolases and transmethylases belong to this group. This group could be further divided into several sub-classes.

Methyltransferase is a type of transferase enzyme that transfers a methyl group from a donor to an acceptor. Methylation often occurs on nucleic bases in DNA or amino acids in protein structures. Methytransferases use a reactive methyl group bound to sulfur in S-adenosyl methionine (SAM) as the methyl donor (e.g., DNA methyltransferase, Histone methyltransferase, 5-Methyltetrahydrofolate-homocysteine methyltransferase, *O*-methyltransferase).

Acyltransferase is a type of transferase enzyme that acts upon acyl groups (e.g., Glycerone phosphate *O*-acyltransferase, Lecithin-cholesterol acyltransferase).

Glycosyltransferases are enzymes (EC 2.4) that act as a catalyst for the transfer of a monosaccharide unit from an activated nucleotide sugar (also known as the "glycosyl donor") to a Glycosyl acceptor molecule, usually an alcohol. The result of glycosyl transfer can be a carbohydrate, glycoside, oligosaccharide, or a polysaccharide. Some glycosyltransferases catalyse transfer to inorganic phosphate or water. Glycosyl transfer can also occur to protein residues, usually to tyrosine, serine, or threonine to give O-linked glycoproteins, or to asparagine to give N-linked glycoproteins. **Phosphotransferases** are a category of that catalyze phosphorylation reactions. The general form of the reactions they catalyze is:

$$A — —P + B \longrightarrow B — —P + A$$

Where P is a phosphate group, A and B are the donating and accepting molecules, respectively.

Sulfurtransferase is a transferase enzyme that acts upon atoms of sulfur. An example is thiosulfate sulfurtransferase that catalyzes the chemical reaction:

Thioulfate + Cyanide ⇌ Sulfite + Thiocyanate

EC 3 Hydrolases

These enzymes catalyze cleavage reactions or the reverse fragment condensations. According to the type of bond cleaved, a distinction is made among peptidases, esterases, lipases, glycosidases, phosphatases and so on. Examples of this class of enzyme also include: cholesterol esterase, alkaline phosphatase and glucoamylase.

EC 4 Lyases

These enzymes cleave various bonds by means other than hydrolysis and oxidation; they non-hydrolytically remove groups from their substrates with the concomitant formation of double bonds or alternatively add new groups across double bonds.

EC 5 Isomerases

These enzymes catalyze isomerization changes within a single molecule (intramolecular rearrangements) and are subdivided into:

Racemases catalyze the stereochemical inversion around the asymmetric carbon atom in a substrate having only one center of asymmetry.

Epimerases catalyze the stereochemical inversion of the configuration about an asymmetric carbon atom in a substrate having more than one center of asymmetry, thus inter-converting epimers.

***Cis/trans* isomerases** play vital parts in organic chemical reactions as geometric isomerism or configuration isomerism or E/Z isomerism. It is a form of stereoisomerism describing the orientation of functional groups within a molecule. In general, such isomers contain double bonds, which cannot freely rotate, but they can also arise from ring structures, wherein the rotation of bonds is greatly restricted. Cis and trans isomers occur both in organic molecules and in inorganic coordination complexes.

Mutases are enzymes that catalyze the shifting of a functional group from one position to another within the same molecule. Examples of this are bisphosphoglycerate mutase which appears in red blood cells, and phosphoglycerate mutase which acts in glycolysis. In glycolysis, it changes 3-phosphoglycerate to 2-phosphoglycerate. In particular, it moves phosphate groups within a single molecule, for instance, phosphoglyerate mutase.

EC 6 Ligases

In biochemistry, ligases (from the Latin verb ligate— "to bind" or "to glue together") are a group of enzymes that can catalyze the joining of two large molecules by forming a new chemical bond, usually with accompanying hydrolysis of a small chemical group dependent on one of the larger molecules. The reaction is usually accompanied by the consumption of a high energy compound such as ATP and other nucleoside triphosphates. An example of this type of enzyme is pyruvate carboxylase which catalyses the following reaction:

$$\text{Pyruvate} + HCO_3^- + \text{ATP} \rightleftharpoons \text{Oxaloacetate} + \text{ADP} + P_i$$

Besides, DNA ligases are very useful groups which catalyze the forming of bonds between two neighbor bases in the DNA chain.

11.2 Determination of Enzyme Activities

The theory of enzyme-catalyzed reactions proposed by Michaelis and Menten in 1913 is based on the assumption that the enzyme (the catalyst, E) and the substrate (the reactant, S) form a complex (ES) by a reversible reaction. The complex is then converted into the product (P) with the reaction rate k_2, when practically no product is present.

$$S + E \underset{k_{-1}}{\overset{k_1}{\rightleftharpoons}} ES \xrightarrow{k_2} P + E \qquad (11.1)$$

Under commonly used conditions of enzyme activity measurement, constant of ES can be considered to be sufficiently constant during the observed reaction period (steady state assumption, Briggs and Haldane):

$$\frac{d[ES]}{dt} = 0 = k_1[E][S] - k_{-1}[ES] - k_2[ES] \qquad (11.2)$$

By using a term for the total concentration of enzyme

$$[Et] = [E] + [ES] \qquad (11.3)$$

one obtains

$$0 = k_1([Et] - [ES])[S] - k_{-1}[ES] - k_2[ES] \qquad (11.4)$$

or

$$[ES] = \frac{k_1[Et][S]}{k_{-1} + k_2 + k_1[S]} \qquad (11.5)$$

By introducing the Michaelis constant K_m

$$K_m = \frac{k_{-1} + k_2}{k_1} \qquad (11.6)$$

The reaction rate is a function of [ES].

$$v = k_2[ES] \tag{11.7}$$

The maximum reaction rate v_{max}, is reached, when all of the enzyme is saturated with substrate.

$$[ES] = [Et]$$

$$v = k_2[Et] \tag{11.8}$$

So-called Michaelis & Menten equation is obtained,

$$v = \frac{v_{max}[S]}{K_m + [S]} \tag{11.9}$$

Where,

v_t = the velocity at any time;

$[S]$ = the substrate concentration at this time;

v_{max} = the highest under this set of experimental conditions (pH, temperature, etc.);

K_m = the Michaelis constant for the particular enzyme being investigated.

above equation shows the dependency of the reaction rate on the substrate concentration (first-order reaction). The plot of this relationship is shown in Fig. 11.1:

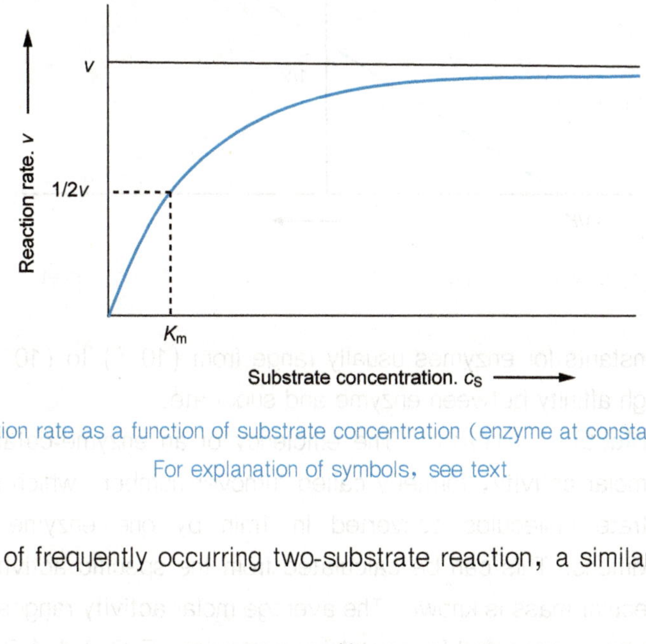

Fig. 11.1 Reaction rate as a function of substrate concentration (enzyme at constant concentration). For explanation of symbols, see text

In the case of frequently occurring two-substrate reaction, a similar derivation leads to the formula:

$$v = \frac{v_{max}}{1 + \frac{(K_m)_{S_1}}{[S_1]} + \frac{(K_m)_{S_2}}{[S_2]} + \frac{(K_m)_{S_1S_2}}{[S_1][S_2]}} \tag{11.10}$$

This is called a second-order reaction. If, however, the concentration of the second substrate is kept at a level many times of the respective Michaelis constant, then the third and fourth term of the denominator are practically zero and the equation is identical with Equation (11.9), allowing the same evaluation as with a one-substrate reaction.

Michaelis constant As can be derived from Equation (11.9), the Michaelis constant equals the substrate concentration at half the maximal reaction rate. The value K_m can be obtained by plotting the experimentally measured reaction rate against the various substrate concentrations (Fig. 11.2). A more convenient way is the plot according to Lineweaver and Burk, using a reciprocal of the Michaelis & Menten equation.

$$\frac{1}{v} = \frac{K_m}{v_{max}[S]} + \frac{1}{v_{max}} \qquad (11.11)$$

A schematic plot is shown in Fig. 11.2. The intersections with abscissa and ordinate allow the determination of the values for K_m and v_{max}. The Michaelis constant approaches the dissociation constant K_s of the enzyme ± substrate complex and is therefore valuable for estimating individual reaction kinetics.

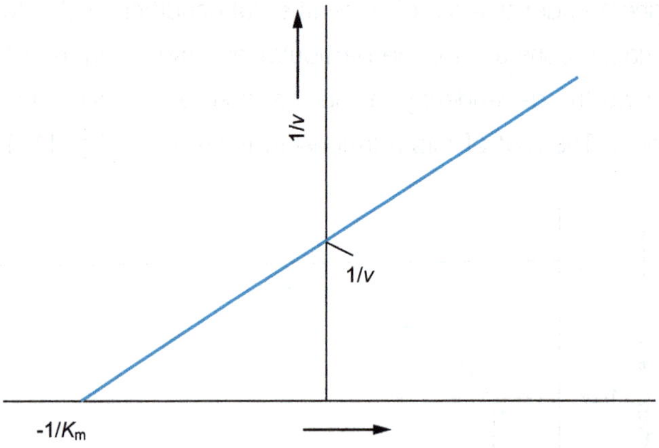

Fig. 11.2 Lineweaver ± Burk graph for determination of the kinetic parameters of an enzyme

Michaelis constants for enzymes usually range from (10^{-2}) to (10^{-5}) mol/L; a low K_m indicates a high affinity between enzyme and substrate.

Molar or molecular activity The efficiency of an enzyme-catalyzed reaction is indicated by the molar activity, formerly called turnover number, which is defined as the number of substrate molecules converted in 1min by one enzyme molecule under standardized conditions. This can be calculated from the specific activity of a particular enzyme if its molecular mass is known. The average molar activity ranges from 10^3 to 10^4; peak values have been measured for acetylcholinesterase (E.C.3.1.1.7) [9000-81-1] at 1.106 and for catalase (E.C.1.11.1.6) [9001-05-2] at 5.106.

Enzyme assays

Reaction rate as a measure of catalytic activity. As biological catalysts, enzymes increase the rate of a reaction or permit it to proceed. Therefore, the conversion rate of substrates, v, is measured to determine the catalytic activity. The respective formulae for enzyme kinetics are given in Equations (11.9) and (11.10).

If in Equation (11.9) the substrate concentration is kept at a level considerably above the **Michaelis constant** ($[S] > K_m$), then the enzyme is saturated with substrate and the reaction proceeds with the constant and the maximum rate $v = v_{max}$ (zero order reaction). Consequently, catalytic activity is linearly dependent on the amount of enzyme used. This situation is attempted to be reached in enzyme activity measurements.

Definition of units for expression of enzymatic activity

Originally, units were defined by the investigator who first discovered and described the enzyme.

Therefore, in the older literature, enzyme activity was expressed in arbitrary units, e.g., changes in absorbance, increase of reducing groups, amount of converted substrate expressed in milligrams or micromoles. These parameters were related to various time units, such as 1min, 30min, or 1h.

To obtain standardized values for each enzyme, in 1961 the Enzyme Commission of the International Union of Biochemistry defined the International Unit U as the activity of an enzyme which, under optimized standard conditions, catalyzes the conversion of 1mol of substrate per minute. With respect to basic SI units, the Expert Panel on Quantities and Units (EPQU) of the International Federation of Clinical Chemists (IFCC) and the Commission on Quantities and Units in Clinical Chemistry (CQUCC) of IUPAC defined the base unit katal as the catalytic amount of any enzyme that catalyzes a reaction (conversion) rate of 1mol of substrate per second in an assay system. This, however, is not in common use.

The temperature must be stated for each assay. As a general rule, the rate of enzymatically catalyzed reactions is about doubled by a temperature increase of 10℃ in the range of 0～40℃. The temperature must be controlled precisely and kept constant to achieve reproducible results.

However, for practical reasons, many enzyme reactions cannot be monitored by measuring the stoichiometric amount of substrate consumed or product formed. Therefore, the catalytic activity of a particular enzyme may be impossible to express in International Units.

The definition of units for most enzymes used in molecular biology is rather arbitrary. For example, the unit of a DNA restriction endo-nuclease (E.C. 3.1.21.3), is defined as the catalytic activity of the enzyme that yields a typical cleavage pattern, detectable after electrophoresis, with a precise amount (usually 1μg) of a particular DNA under defined incubation conditions. Other parameters that define the activity of such enzymes are the degradation of a nucleic acid or the incorporation of nucleotides into a nucleic acid, expressed in micrograms, nanomoles, absorbance units, or number of base pairs. For practical reasons, different time periods such as 1, 10, 30, or 60min, may be chosen as references.

11.3 Factors Affecting Enzyme Activity

Knowledge of enzyme kinetic theory is important in enzyme analysis, in order both to understand the basic enzymatic mechanism and to select a suitable method for enzyme analysis. The conditions selected to measure the activity of an enzyme would not be the same as those selected to measure the concentration of its substrate. Several factors affect the rate at which enzymatic reactions proceeded-temperature, pH, enzyme concentration, substrate concentration, and the presence of any inhibitors or activators.

11.3.1 Enzyme concentration

In order to study the effect of increasing the enzyme concentration upon the reaction rate, the substrate must be present in an excess amount; i. e., the reaction must be independent of the substrate concentration. Any change in the amount of product formed over a specified period of time will be dependent upon the level of enzyme present. Graphically this can be represented in Fig. 11.3.

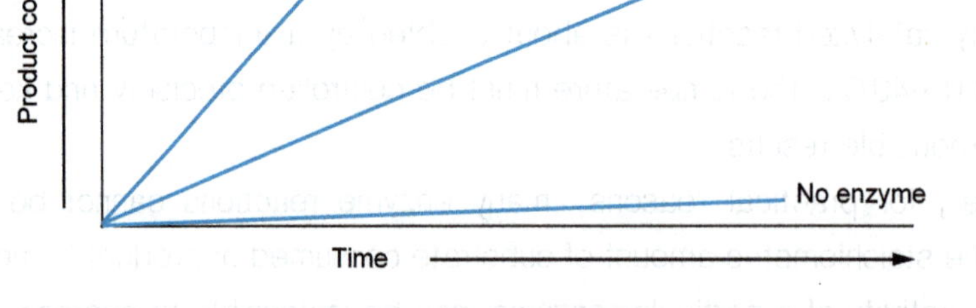

Fig. 11.3 "Zero order" reaction rate is independent of substrate concentration

These reactions are said to be "zero order" because the rates are independent of substrate concentration, and are equal to some constant k. The formation of product proceeds at a rate which is linear with time. The addition of more substrate does not serve to increase the rate. In zero order kinetics, allowing the assay to run for double time results in double the amount of product (Tab. 11.1).

The amount of enzyme present in a reaction is measured by the activity it catalyzes.

The relationship between activity and concentration is affected by many factors such as temperature, pH, etc. An enzyme assay must be designed so that the observed activity is proportional to the amount of enzyme present in order that the enzyme concentration is the only limiting factor. It is satisfied only when the reaction is under "zero order".

Tab. 11.1 Reaction orders with respect to substrate concentration

Order	Rate equation	Comments
Zero	Rate = k	Rate is independent of substrate concentration
First	Rate = $k\,[S]$	Rate is proportional to the first power of substrate concentration
Second	Rate = $k\,[S]\,[S] = k\,[S]^2$	Rate is proportional to the square of substrate concentration
Second	Rate = $k\,[S_1]\,[S_2]$	Rate is proportional to the first power of each of two reactants

In Fig. 11.4, activity is directly proportional to concentration ranged within the area AB, but not within area BC. Enzyme activity is generally greatest when substrate concentration is unlimited.

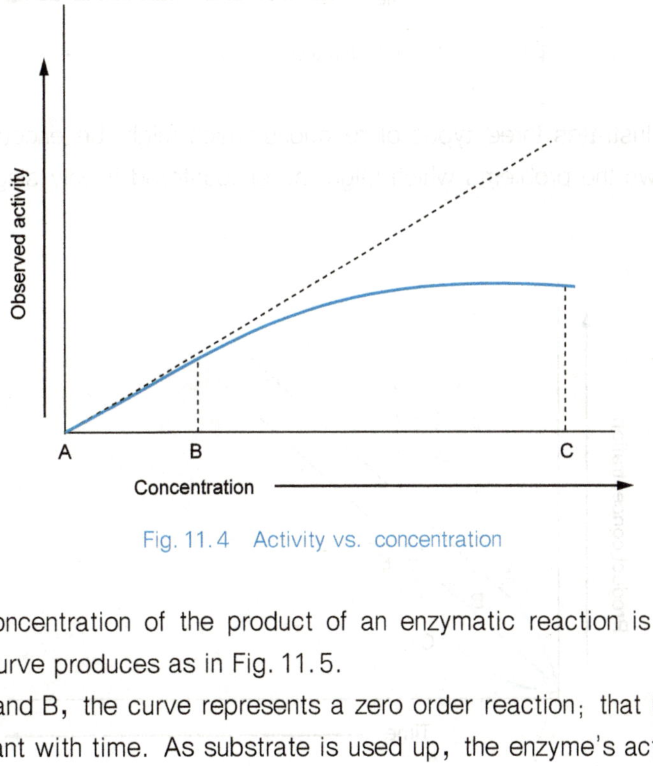

Fig. 11.4 Activity vs. concentration

When the concentration of the product of an enzymatic reaction is plotted against time, a similar curve produces as in Fig. 11.5.

Between A and B, the curve represents a zero order reaction; that is, one in which the rate is constant with time. As substrate is used up, the enzyme's active sites are no longer saturated, substrate concentration becomes rate limiting, and the reaction becomes first order between B and C.

To measure enzyme activity ideally, the measurements must be made in that portion of the curve where the reaction is under zero order. A reaction is most likely to be in this situation initially since substrate concentration is the highest. To be certain that a reaction has reached this part, multiple measurements of product (or substrate) concentration must be made.

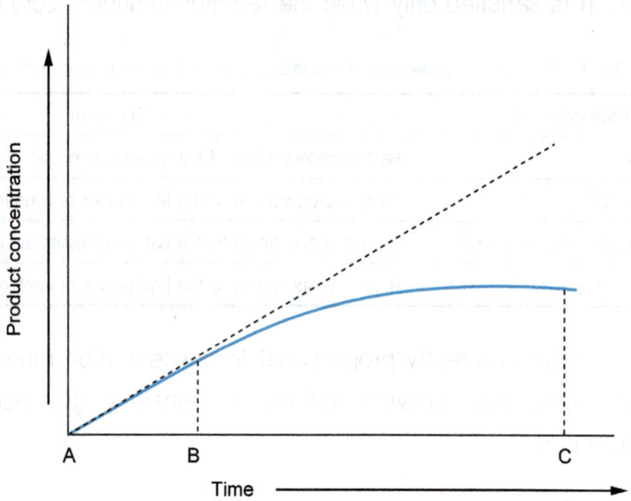

Fig. 11.5 Reaction rate limited by substrate concentration

Fig. 11.6 illustrates three types of reactions which might be encountered in enzyme assays and shows the problems which might be encountered if only single measurements are made.

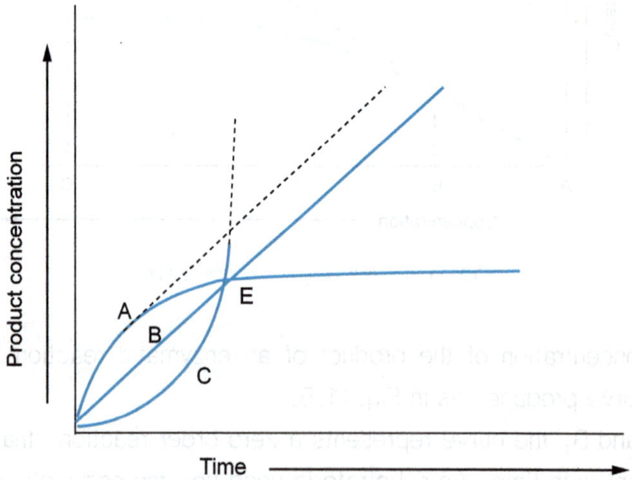

Fig. 11.6 Leading, lagging and linear reactions

B is a straight line representing a zero order reaction which permits accurate determination of enzyme activity for part or all of the reaction time. A represents the type of reaction that was shown in Fig. 11.5. This reaction is zero order initially and then slows, presumably due to substrate exhaustion or product inhibition. This type of reaction is sometimes referred to as a "leading" reaction. True "potential" activity is represented by the dotted line. Curve C represents a reaction with an initial "lag" phase. Again the dotted line represents the potentially measurable activity. Multiple determinations of product concentration enable each curve to be plotted and true activity determined. A single end point determination at E would lead to the false conclusion that all three samples had identical enzyme concentration.

11.3.2 Substrate concentration

It has been shown experimentally that if the amount of the enzyme is kept constant and the substrate concentration is then gradually increased, the reaction velocity will increase until it reaches a maximum. After this point, increases in substrate concentration will not increase the velocity (delta A/delta T). This is represented graphically in Fig. 11.7.

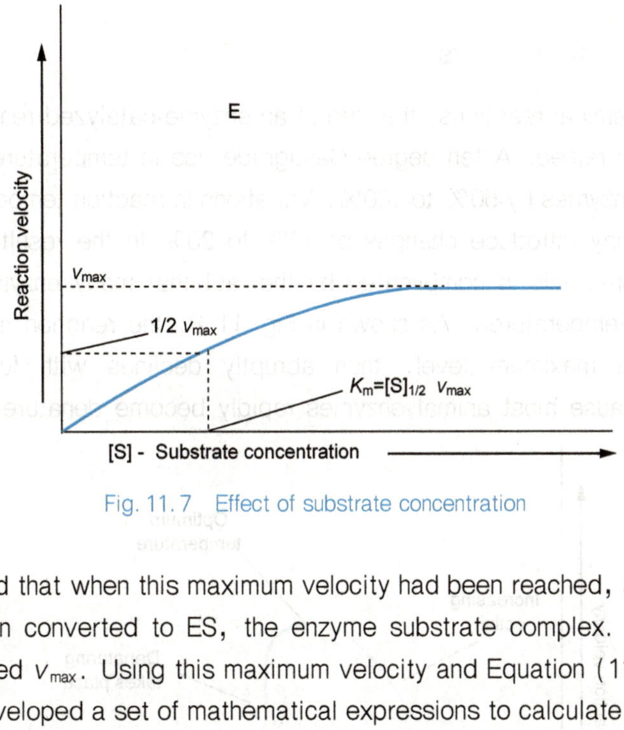

Fig. 11.7 Effect of substrate concentration

It is theorized that when this maximum velocity had been reached, all of the available enzymes has been converted to ES, the enzyme substrate complex. This point on the graph is designated v_{max}. Using this maximum velocity and Equation (11.6).

Michaelis developed a set of mathematical expressions to calculate enzyme activity in terms of reaction speed from measurable laboratory data.

The Michaelis constant K_m is defined as the substrate concentration at 1/2 the maximum velocity. This is shown in Fig. 11.7. Using this constant and the fact that K_m can also be defined as:

$$K_m = \frac{K_{-1} + K_2}{K_1}$$

K_1, K_{-1} and K_2 are the rate constants from Equation (11.6). Michaelis developed the following expression for the reaction velocity in terms of this constant and the substrate concentration.

$$V_t = \frac{V_{max}[S]}{K_m + [S]}$$

Michaelis constants have been determined for many of the commonly used enzymes. The value of K_m tells us several things about a particular enzyme.

(1) A small K_m indicates that the enzyme requires only a small amount of substrate to become saturated. Hence, the maximum velocity is reached at relatively low substrate concentrations.

(2) A large K_m indicates the need for high substrate concentrations to achieve maximum reaction velocity.

(3) The substrate with the lowest K_m upon which the enzyme acts as a catalyst is frequently assumed to be enzyme's natural substrate, though this is not true for all enzymes.

11.3.3 Temperature effects

Like most chemical reactions, the rate of an enzyme-catalyzed reaction increases as the temperature is raised. A ten degree Centigrade rise in temperature will increase the activity of most enzymes by 50% to 100%. Variations in reaction temperature as small as 1 or 2 degrees may introduce changes of 10% to 20% in the results. In the case of enzymatic reactions, this is complicated by the fact that many enzymes are adversely affected by high temperatures. As shown in Fig. 11.8, the reaction rate increases with temperature to a maximum level, then abruptly declines with further increase of temperature. Because most animal enzymes rapidly become denatured at temperatures

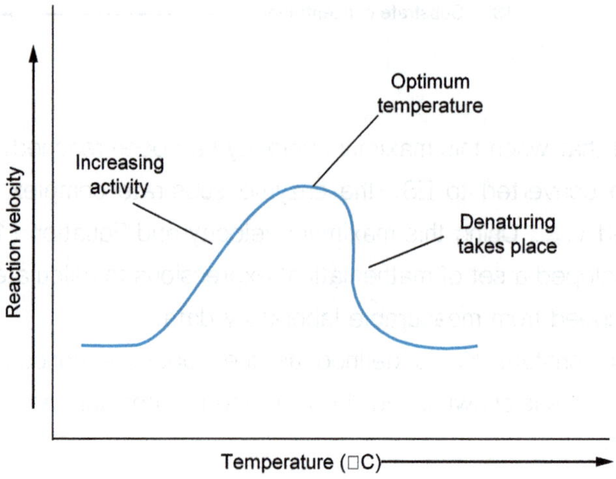

Fig. 11.8 Effect of temperature on reaction rate

above 40℃, most enzyme determinations are carried out somewhat below that temperature.

Over a period of time, enzymes will be deactivated at even moderate temperatures. Storage of enzymes at 5℃ or below is generally the most suitable. Some enzymes lose their activity when frozen.

11.3.4 Effects of pH

Enzymes are affected by changes in pH. The most favorable pH value-the point where the enzyme is most active-is known as the optimum pH. This is graphically illustrated in Fig. 11.9.

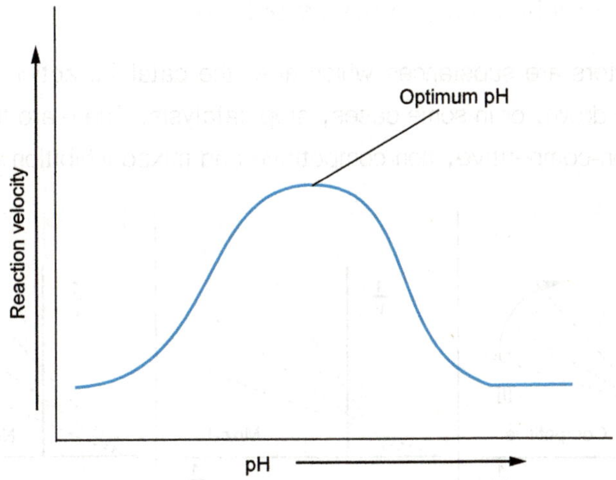

Fig. 11.9 Effect of pH on reaction rate

Extremely high or low pH values generally result in complete loss of activity for most enzymes. pH is also a factor in the stability of enzymes. As with activity, for each enzyme there is also a region of pH optimal stability.

The optimum pH value will vary greatly from one enzyme to another, as Tab. 11.2 shows:

Tab. 11.2 pH for optimal activity

Enzyme	pH optimum
Lipase (pancreas)	8.0
Lipase (stomach)	4.0~5.0
Lipase (castor oil)	4.7
Pepsin	1.5~1.6

Continued

Enzyme	pH optimum
Trypsin	7.8~8.7
Urease	7.0
Invertase	4.5
Maltase	6.1~6.8
Amylase (pancreas)	6.0~7.0
Amylase (malt)	4.6~5.2
Catalase	7.0

11.3.5 Effects of inhibitors on enzyme activity

Enzyme inhibitors are substances which alter the catalytic action of the enzyme and consequently slow down, or in some cases, stop catalysis. There are three common types of enzyme inhibition-competitive, non-competitive and mixed inhibition (Fig. 11.10).

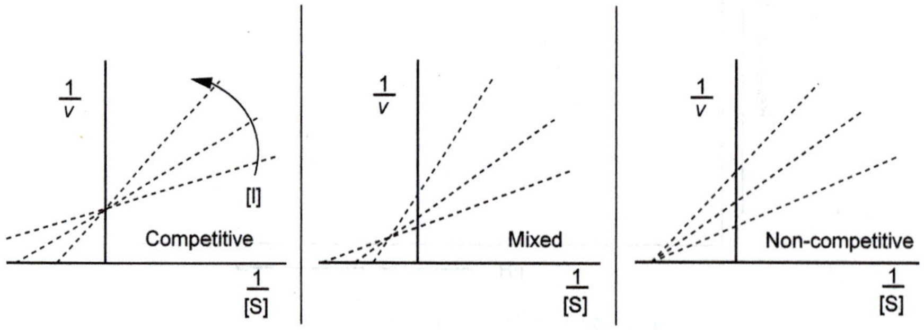

Fig. 11.10 Lineweaver-Burk plots of different types of reversible enzyme inhibitors. The arrow shows the effect of increasing concentrations of inhibitor

Most theories concerning inhibition mechanisms are based on the existence of the enzyme substrate complex ES. As mentioned earlier, the existence of temporary ES structures has been verified in the laboratory.

Competitive inhibition occurs when the substrate and a substance resembling the substrate are both added to the enzyme. A theory called the "lock-key theory" of enzyme catalysts can be used to explain why inhibition occurs.

The lock and key theory utilizes the concept of an "active site." The concept holds that one particular portion of the enzyme surface has a strong affinity for the substrate. The substrate is held in such a way that its conversion to the reaction products is more favorable. If we consider the enzyme as the lock and the substrate the key (Fig. 11.11) — the key is inserted in the lock, is turned, and the door is opened and the reaction proceeds. However, when an inhibitor which resembles the substrate is present, it will compete with the substrate for the position in the enzyme lock. When the inhibitor wins, it enters the lock position but is unable to open the lock. Hence, the observed reaction is slowed down because some of the available enzyme sites are occupied by the inhibitor. If a dissimilar substance which does not fit the site is present, the enzyme rejects it, accepts the substrate, and the reaction proceeds normally.

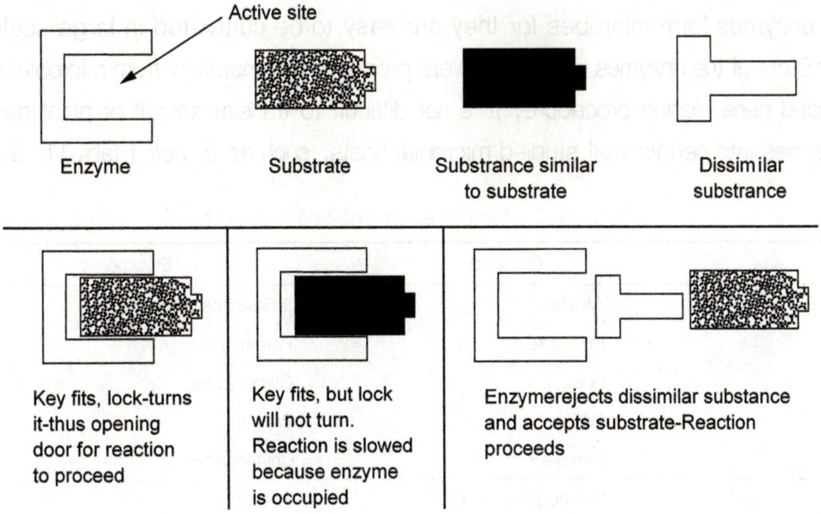

Fig. 11. 11 Lock-key theory of competitive inhibition

Non-competitive inhibition is a form of mixed inhibition where the binding of the inhibitor to the enzyme reduces its activity but does not affect the binding of substrate. As a result, the extent of inhibition depends only on the concentration of the inhibitor. V_{max} will decrease due to the inability for the reaction to proceed as efficiently, but K_m will remain the same as the actual binding of the substrate, by definition, will still function properly.

In **mixed inhibition**, the inhibitor can bind to the enzyme at the same time as the enzyme's substrate. However, the binding of the inhibitor affects the binding of the substrate, and vice versa. This type of inhibition can be reduced, but not overcome by increasing concentrations of substrate. Although it is possible for mixed-type inhibitors to bind in the active site, this type of inhibition generally results from an allosteric effect where the inhibitor binds to a different site on an enzyme. Inhibitor binding to this allosteric site changes the conformation (i. e. , tertiary structure or three-dimensional shape) of the enzyme so that the affinity of the substrate for the active site is reduced.

11.4 Resources of Enzyme

Build-up of new tissue, replacement of old tissue, conversion of food to energy, disposal of waste materials, reproduction to produce offspring-all these activities characterized as "life" are catalyzed by enzymes. Animals, plants and microbes are independently living organisms, no doubt they have a complete set of enzymes for metabolism. Viruses might also produce their own enzymes, even they live in parasitized form with ability of modifying and making use of the host enzymes, for example, the DNA ligase expressed from *E. coli* phage T4 used to be a tool for DNA ligation. Originally, people extract enzymes from the cells of different bio-species, but the most efficient way to obtain enzymes form microbes for they are easy to be cultivated in large scale. So far, more than 99% of the enzymes in quantity were produced commercially from microbial cells. With the advanced gene cloning procedure, it's not difficult to transfer animal or plant genes coding useful enzymes into certain well studied microbial hosts, such as *E. coli* (Tab. 11. 3).

Tab. 11.3 Examples of enzyme resources

	Organism	Enzyme
Plant	Alfalfa	β-Galactosidase
	Horseradish	Peroxidase
	Beet	Glycosylase
	Papaya	Papain
	Papaya	Chymopapain
Animal	Bovine pancrease	Trypsin
	Bovine pancrease	Ribonuclease
	Bovine pancrease	Rnase
	Bovine stomach	Rennet
	Snake venom	Thrombin-like enzyme
	Leech	Thrombin
Bacterial	Bacillus subtilis	α-Amylase
	Aspergillus	β-Amylase
	Aspergillus	Acidic proteinase
	Myrothecium	Cellulase
	Candida	Lipase
Phage	*E. coli* phage T4	DNA ligase
	Phage	Lyase
Recombinant	*E. coli*	DNA polymerase I
	E. coli	Peroxidase
	E. coli	Kinase
	E. coli	*Taq* DNA polymerase
	E. coli	Carboxypeptidase

11.5 Uses of Enzyme

Many people believe enzyme technology is fairly new. However, that's not the case. Enzymes have been used by man from the dawn of civilization. As long as people have been eating bread and cheese, and drinking wine and beer-they have been using enzymes. That's because enzymes help to make these products.

Cheese Take cheese for example. Most likely, cheese was discovered by ancient hunters who stored milk in the stomachs of slaughtered calves. When exposed to heat, the milk inside the container would turn into a solid cheese. The reason for this is simple. Calves have an enzyme called chymosin in their stomachs. At ambient temperature chymosin will break down milk protein, causing milk to be separated into curd and whey. If you strain away the watery whey you are left with a soft tangy curd or cheese. Chymosin derived from calf stomachs has been used in cheese making ever since.

Detergent In more recent times, other commercial uses of enzymes have been found. Laundry detergents are good examples. As mentioned, there are certain enzymes which dissolve proteins and others which dissolve fat. Proteins and fat make up two of the major causes of stains on clotting. Grass, blood and egg are all protein stains. Lipstick, frying oil, butter, sauces, and tough stains on cuffs and collars are fatty stains. Without enzymes, removing of these stains is difficult and requires a lot of washing detergents working at high temperatures.

Enzymes can remove these stains better, faster and in a way that is a lot kinder to the clothing and the environment. The enzymes in detergent dissolve stains in the same natural way they help digest food in your stomach-specifically and efficiently. Just a tiny amount of enzymes can dissolve stains that would require much larger amounts of detergent alone. Because of this they can reduce the amount of detergents actually found in laundry detergent. And since enzymes work best in mild conditions, washing machines using enzymatic detergents can be set at lower temperatures, reducing electricity consumption by as much as a third. In the future, enzymes could replace more and more of the harsh chemicals found in detergents.

Enzymes are used in the chemical industry and other industrial applications when extremely specific catalysts are required. However, enzymes in general are limited in the number of reactions they have evolved to catalyze and also by their lack of stability in organic solvents and at high temperatures. Consequently, protein engineering is an active area of research and involves attempts to create new enzymes with novel properties, either through rational design or in vitro evolution. These efforts have begun to be successful, and a few enzymes have now been designed "from scratch" to catalyze reactions that do not occur in nature (Tab. 11.4).

Tab. 11.4 Applications of enzymes in industry

Application	Enzymes used	Uses
Food processing	Amylases from fungi and plants	Production of sugars from starch, such as in making high-fructose corn syrup. In baking, catalyze breakdown of starch in the flour to sugar. Yeast fermontation of sugar produces the carbon dioxide that raises the dough
	Proteases	Biscuit manufacturers use them to lower the protein level of flour
Baby foods	Trypsin	To predigest baby foods
Brewing industry	Enzymes from barley are released during the mashing stage of beer production	They degrade starch and proteins to produce simple sugar, amino acids and peptides that are used by yeast for fermentation
	Industrially produced barley enzymes	Widely used in the brewing process to substitute for the natural enzymes found in barley
	Amylase, glucanases, proteases	Split polysaccharides and proteins in the malt
	Betaglucanases and arabinoxylanases	Improve the wort and beer filtration characteristic
	Amyloglucosidase and pullulanases	Low-calorie beer and adjustment of fermentability
	Proteases	Remove cloudiness produced during storage of beers
	Acetolactatedecarboxylase (ALDC)	Increases fermentation efficiency by reducing diacetyl formation
Fruit juices	Cellulases, pectinases	Clarify fruit juices
Dairy industry	Rennin, derived from the stomachs of young ruminant animals(like calves and lambs)	Manufacture of cheese, used to hydrolyze protein
	Microbially produced enzyme	Now finding increasing use in the dairy industry
	Lipases	Is implemented during the production of Roquefort cheese to enhance the ripening of the blue-mold cheese
	Lactases	Break down lactose to glucose and galactose
Meat tenderizers	Papain	To soften meat for cooking
Starch industry	Amylases, amyloglucosideases and glucoamylases	Converts starch into glucose and various syrups
	Glucose isomerase	Converts glucose into fructose in production of high-fructose syrups from starchy materials. These syrups have enhanced sweetening properties and lower calorific values than sucrose for the same level of sweetness

Continued

Application	Enzymes used	Uses
Paper industry	Amylases, xylanases, cellulases and ligninases	Degrade starch to lower viscosity, aiding sizing and coating paper. Xylanases reduce bleach required for decolorizing; cellulases smooth fibers, enhance water drainage, and promote ink removal; lipases reduce pitch and lignin-degrading enzymes remove lignin to soften paper
Biofuel industry	Cellulases	Used to break down cellulose into sugars that can be fermented (see cellulosic ethanol)
	Ligninases	Use of lignin waste
Biological detergent	Primarily proteases, produced in an extracellular form bacteria	Used for presoak conditions and direct liquid applications helping with removal of protein stains from clothes
	Amylases	Detergents for machine dish washing to remove resistant starch residues
	Lipases	Used to assist in the removal of fatty and oily stains
	Cellulases	Used in biological fabric conditioners
Contact lens cleaners	Proteases	To remove proteins on contact lens to prevent infections
Rubber industry	Catalase	To generate oxygen from peroxide to convert latex into foam rubber
Photographic industry	Protease (ficin)	Dissolve gelatin off scrap film, allowing recovery of its silver content
Molecular biology	Restriction enzymes, DNA ligase and polymerases	Used to manipulate DNA in genetic engineering, important in pharmacology, agriculture and medicine. Essential for restriction digestion and the polymerase chain reaction. Molecular biology is also important in forensic science

Summary

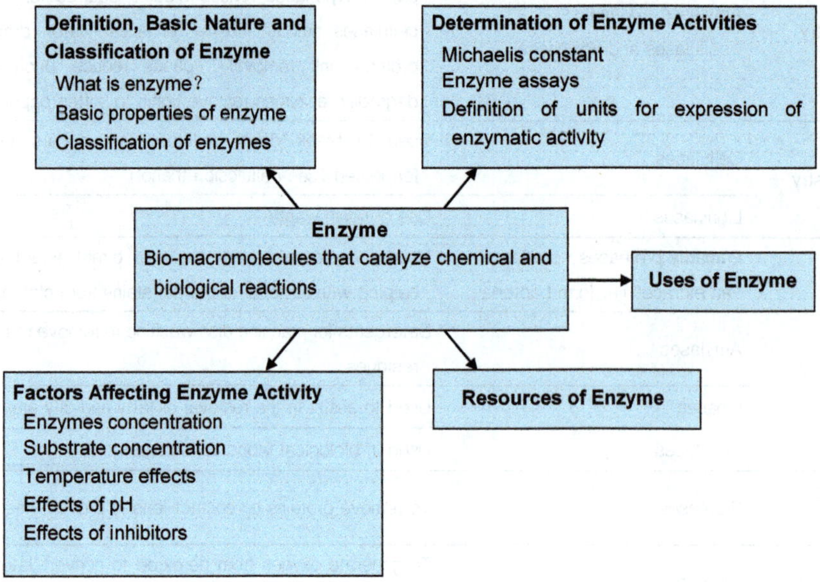
Definition, Basic Nature and Classification of Enzyme
What is enzyme?
Basic properties of enzyme
Classification of enzymes
Determination of Enzyme Activities
Michaelis constant
Enzyme assays
Definition of units for expression of enzymatic activity
Enzyme
Bio-macromolecules that catalyze chemical and biological reactions
Uses of Enzyme
Factors Affecting Enzyme Activity
Enzymes concentration
Substrate concentration
Temperature effects
Effects of pH
Effects of inhibitors
Resources of Enzyme

Chapter 12 Enzyme Biosynthesis and Products

酶是基因表达的产物，在细胞内合成。酶的来源主要分两类，即天然的生物酶和经过基因工程改造的酶分子。了解酶的生物合成过程及机理，选择有利于酶合成的原料，确立先进的生物发酵技术，大批量培养合成酶的生物细胞或经基因改造后的高产株，有助于提高酶的产率，规模化开展酶的生物合成。随着酶的规模化生物合成技术的发展，酶的应用范围不断扩展。

12.1 Bio-resources of Enzyme

12. 1. 1 Early exploitation of enzyme

The production of enzymes is central to the modern biotechnology industry. The traditional industrial enzymes continue to have expanding markets. Man has utilized enzymes throughout the ages either in the form of vegetables rich in enzymes, or as microorganisms used for a various purposes, for instance in brewing, baking or in cheese production. However, it was not until the 19th century that the various biological conversions were ascribed to the action of enzymes. An enzyme preparation was extracted from malt by A. Payen and F. Persoz in 1833, and termed "diastase", which converts gelatinized starch into surges, primarily maltose.

The history of modern enzyme production really began in 1874 when the Danish chemist C. Hansen first produced rennet by extracting it from dried calves' stomachs with saline solution. Apparently, this was the first enzyme preparation of relatively high purity used for industrial purposes.

In 1894, E. Fischer extracted invertin from yeast by a normal procedure, and showed that it hydrolyzed the α-, but not the β-methyl-D-glucoside. In contrast, the emulsion, a commercial preparation from Merck Ltd., hydrolyzed the α-, but not the β-methyl-D-glucoside. Fischer therefore deduced the famous picture of enzymatic mechanism as the "lock and key", which he considered a precondition for the potential of an enzyme to have a chemical effect on the substrate. In this way he assumed that the "geometrical form of the (enzyme) molecule concerning its asymmetry, corresponds to that of the natural hexoses" (sugars).

In 1898, C. Hill performed the first enzymatic synthesis of isomaltose with yeast extract (α-glycosidase) acting on a 40% glucose solution. In 1900, J. H. Kastle and A. S. Loevenhart showed that the hydrolysis of fat and other esters by lipases was a reversible reaction, and that enzymatic synthesis could occur in a dilute mixture of alcohol and acid.

This principle was subsequently utilized in 1902, in the synthesis of numerous glycosides by E. Fischer and coworkers, and in 1913 by E. Bourquelot and coworkers. G. Bertrand In 1897 observed that only when certain dialyzable substances exist with the enzymes, then the catalytic reaction happened, and thus he termed coenzymes.

In 1890s, J. Takamine began isolating bacterial amylases in famous Miles Laboratories. In 1895, A. Boidin discovered a new process for the manufacture of alcohol, termed the "Amylo process", which comprised cooking of the cereals, inoculation with a mold which synthesizing enzymes for saccharification, and subsequent fermentation with yeast. The early applications and patents on enzyme applications in the food industry (which numbered about 10 until 1911) have been reviewed by S. L. Neidleman (1991).

At the beginning of the 20th century, plant lipases were produced and utilized for the production of fatty acids from oils and fats, typically in the scale of 10 tons per week. Likewise, in the chill-proofing of beer, proteolytic enzymes have been used for classifying the product successfully since 1911 in the USA. Lintner, as early as 1890, noted that wheat diastase interacts in dough making, and studied the effect extensively. As a result, the addition of malt extract came into practice, and in 1922, American bakers used 30million pounds (13.5×10^6 kg) of malt extract valued at US$ 2.5million.

During the 1960s, enzyme production gained speed only in modest proportions, as reflected by the growing sales of bacterial amylases and proteases. Indeed, the annual turnover of the enzyme division of Novo Industri (now Novozymes), the leading enzyme manufacturer at the time, did not exceed $ 1million until 1965. However, with the appearance of the detergent proteases, the use of enzymes increased dramatically, and during the late 1960s, the protease-containing detergent became quite popular. At the same time, an acid/enzyme process to produce dextrose using glucoamylase was practiced increasingly in starch processing.

The main industrial processes with free or immobilized enzymes as biocatalysts are listed in Tab. 12. 1.

Tab. 12. 1 Industrial products from enzymatic process

Product	Enzyme	Free or immobilized enzyme	Companies
>10,000,000 t/a			
Hydrofluorocarbons(HFCS)	Amylase	Free	Several
	Glucoamylase	Free	
	Glucose isomerase	Immobilized	
Ethanol(gasoline additive)	Amylase	Free	Several
	Glucoamylase	Rree	
>10,000 t/a			
Acrylamide	Nitrilase	Immobilized cells	Nitto, DSM

Continued

Product	Enzyme	Free or immobilized enzyme	Companies
>10,000,000 t/a			
6-aminopenicillanic acid	Penicillin amidase	Immobilized	Several
Cacao butter	Lipase	Immobilized	Fuji Oil, Unilever
Isomaltulose	Sucrose mutase	Immobilized cells	Sudzucker
>1000 t/a			
7-aminocephalosporanic acid (7-ACA)	(*R*)-amino oxidase	Immobilized	Several
	Glutaryl amidase	Immobilized	
7-aminodesacetoxycephalosporanic acid (7-ADCA)	Glutaryl amidase	Immobilized	DSM
(*S*)-aspartic acid	Aspartase	Immobilized	Tanabe
Aspartame	Thermolysin	Immobilized	Toso, DSM
(*S*)-methoxy isopropyl amine	Lipase	Immobilized	BASF
(*R*)-pantothenic acid	Aldolactonase	Immobilized	Fuji chem. Ind
(*R*)-phenylglycine	Hydantoinase	Immobilized	Several
	Carbamoylase	Immobilized	
(*S*)-amino acid	Aminoacylase	Free	Degussa, Tanabe
>10 t/a			
Amoxicillin	Penicillin amidase	Immobilized	DSM
Cephalexin	Penicillin amidase	Immobilized	DSM
(*S*)-DOPA	β-tyrosinase	Immobilized	Ajinomoto
Human insulin	Carboxypeptidase A	Free	Aventis
	Lysyl endopeptidase	Free	Several
	Trypsin	Free	BASF
Sterically pure alcohols and amines	Lipase	Immobilized	BASF
(*R*)-mandelic acid	Nitrilase	Immobilized	BASF

The primary strategy in production of any enzyme relates to the choice of source. In most cases, the desired activity can be obtained from several sources. In some cases, the choice of source has been more restricted, for example, the enzyme rennet was until recently obtained from the stomach of suckling calves; the corresponding microbial enzyme led to an off-flavor problem in the cheese produced. Today, recombinant DNA technology allows the microbial cells to produce the calf enzyme.

Production of a new microbial enzyme usually starts with screening of microorganisms for desirable activity using appropriate selection procedures. The harsh environment to which several enzymes are subjected during process applications has given impetus to screening of extremophiles for enzymes having desirable features of activity and stability. The level of enzyme activity produced by an organism from a natural environment is often low and needs to be elevated for industrial purposes. Increase in enzyme levels is often achieved by mutation of the primitive producers. An alternative strategy that has gained favor, is to produce the enzyme from a recombinant organism of choice whose growth conditions are well optimized and whose GRAS status is confirmed. Random or site-directed mutagenesis, with the purpose of engineering on the activity and stability properties of an enzyme prior to its production, is becoming a common practice. The microbial strains used for enzyme production are grown in fermenter using an optimized cultivation medium. Both solid state-and submerged fermentation are employed commercially; however the latter is preferred in many countries because of a better handle on aseptic conditions and process control. The enzymes produced by the microorganism may be intracellular or secreted into the extracellular medium.

During the early part of the last century, in the East Asia, an age-old tradition was involving the use of mould fungi called koji in the production of certain foodstuffs and flavor additives based on soy-protein, which formed the basis of the development of fermentation process for the industrial production of fungal amylase. The process included the culture of *Aspergillus oryzae* on moist rice or wheat bran, which is still used as a digestive aid.

The value of the industrial enzymes market was estimated to $ 2 billion, and has increased at an average annual rate of 3～5 percent during the past decade. A number of companies are competing in the industrial enzymes business, Novozymes dominating with 45 percent of sales, followed by Danisco that holds a 20 percent share of the market.

12.1.2 Enzyme from plants and animals

Despite the advantages of microorganisms as enzyme sources, some enzymes are still economically produced from plant and animal sources. This is possible because of sufficiently high amounts of these enzymes in such sources and also as a means to convert inexpensive, renewable material like agricultural and slaughter waste into value added products.

Examples of the enzymes isolated from plants include several proteinases (papain, ficin and bromelain) and peroxidases. Glycolate oxidase is a peroxisomal oxidase that has been isolated from many green plants, including spinach leaves, peas, pumpkins and cucumbers. An investigation of glycolate oxidase production and substrate specificity to glycolic acid indicated that the enzyme from spinach leaves produced the highest yields and specific activities in terms of glyoxylic acid production. In many plants, the release of Hydrogen acid (HCN) from damaged plant tissues is believed to serve as a defensive strategy against herbivoral attack or as a nitrogen source for the biosynthesis of L-asparagine. Hydroxynitrile lyases and mandelonitrile lyases (HNLs, MDLs) are responsible for these reactions and their preferred natural substrates, cyanohydrins, are separated from the catalysts by compartmentalization on tissue. In synthetic chemistry, HNLs are mainly employed for the reverse reactions.

As mentioned previously, rennet has been among the most industrially significant enzymes obtained from animal tissue. The other enzymes obtained from animal sources e. g. proteases, like trypsin, chymotrypsin and urokinase, and also lactate dehydrogenase, peroxidase and lysozyme, etc., have diverse applications in industry, analysis, and therapy. The pancreas also makes amylase (alpha amylase) which hydrolyzes dietary starch into disaccharides and trisaccharides, which then are subsequently converted by other enzymes to glucose to supply the body with energy. Dehydrogenases became very important catalysts for organic syntheses due to their strong capability to convert carbonyl groups to optically pure alcohols. Bovine liver glutamate dehydrogenase (GDH) and horse liver alcohol dehydrogenase (HLADH) are examples from higher animal resources, also a variety of tissue-specific lipases and esterases have been isolated from animals. Porcine pancreatic lipase is the cheapest and one of the most widely used lipases for kinetic resolution of non-natural substrates. Pig liver esterase (PLE) plays a crucial role in the enantio-selective hydrolysis of prochiral or racemic esters.

12.1.3 Microbial resources

Microorganisms represent an attractive source of enzymes, simply for the easy cultivation in large quantities in a relatively short period by established methods of fermentation. The organism preferred usually has predominant property of high yields of enzyme in shortest possible fermentation time. The producers used in industry can be rationally modified by routine mutation or genetic manipulation to up grade the production yield. Both prokaryotic and eukaryotic producers are employed to produce useful enzymes.

Prokaryotic enzymes

Prokaryotic enzyme producers are mostly welcomed by industries, for those microbes have rather simpler life cycle, shorter growing period, and easier to be controlled. Some prokaryotes produce many different of enzymes, for example, the Gram positive bacteria, Bacillus genus, is the most active group for industrial production of α-amylases, proteases, lipases and cellulases.

Amylase is an enzyme that catalyses the breakdown of starch into sugars. Food those contain much starch but little sugar, such as rice and potato, taste slightly sweet as they are chewed because amylase turns some of their starch into sugar in the mouth.

Many species of actinomycetes are also the producers of those enzymes but with different peptide sequences and 3-D structures compared with the bacterial species. The enzymes of some prokaryotic strains sampled from salsa or specific submarine volcanic area are found to have thermophilic properties and could be developed into very useful industrial producers.

In addition, a great deal of enzymes have been discovered and extracted from either gram positive or negative bacteria, and some of them are genetic engineered for special uses. Most of the restriction enzymes were reported to be primarily discovered in prokaryotes, e. g. , *Eco*RI, *Eco*RII, *Eco*RV* and *Eco*P15I from *Escherichia coli*, *Bam*HI from *Bacillus amyloliquefaciens*, *Hin*dIII from *Haemophilus influenzae*, *Taq* I from *Thermus aquaticus*, *Not* I from *Nocardia otitidis*, *Hin*fI from *Haemophilus influenzae*, *Sau*3A from *Staphylococcus aureus*, *Pov* II* from *Proteus vulgaris*, *Sma* I* from *Serratia marcescens*, *Hae* III* from *Haemophilus aegyptius*, *Hga* I from *Haemophilus gallinarum*, *Alu* I* from *Arthrobacter luteus*, *Kpn* I from *Klebsiella pneumoniae*, *Pst* I from *Providencia stuartii*, *Sac* I from *Streptomyces achromogenes*, *Sal I* from *streptomyces caespitosus*, *Spe* I from *Sphaerotilus natans*, *Sph* I from *Streptomyces phaeochromogenes*, *Stu* I from *Streptomyces tubercidicus*, and *Xba* I from *Xanthomonas badrii*.

Eukaryotic enzyme

A great number of enzymes from eukaryotic microorganisms have been extracted, characterized and applied for industrial uses:

Glycoside hydrolases: Fungi also produce glycoside hydrolases which have been widely used in food and chemistry industry. So called amylases are glycoside hydrolases and act on α-1, 4-glycosidic bonds. Specific amylase proteins are designated by different Greek letters. The α-amylases are calcium metalloenzymes, completely have no function in the absence of calcium. By acting at random locations along the starch molecules, α-amylase breaks down long-chain carbohydrates, ultimately yielding maltotriose and maltose from amylose, or maltose, glucose and "limit dextrin" from amylopectin. Because it can act anywhere on the substrate, α-amylase tends to be faster-acting than β-amylase. Working from the non-reducing end, β-amylase catalyzes the hydrolysis of the second α-1, 4 glycosidic bond, cleaving off two glucose units (maltose) at a time. During the ripening of fruit, β-amylase breaks starch into maltose, resulting in sweet taste of the riped fruit. In addition, α-amylase could also be extracted from some fungi. The species of *Aspergillus* and *Rhizopus* are also suitable producers for providing glucoamylases, which hydrolyze starch molecules from the non-reducing end at α-1.4 or α-1.6 glycosidic bonds and release glucose.

Proteinases which refer to a series of enzymes which hydrolyze the protein molecules into shorter peptide or finally into amino acids. There are two different forms of digestion, the endo-proteases cut protein at internal of the polypeptide, and exo-proteases release the amino acid one by one from the ends of peptide chain. Some species of *Mucor* and *Aspergillus* are excellent proteinase producers. Many traditional fermented foods, for example, soy sauce and fermented Doufu are the products from the fungal proteases having action on the soy bean material. Proteinases are also used in leather industry for removing off hairs form the animal skin.

Cellulases, which refer to a class of enzymes produced chiefly by fungi, (also bacteria, and protozoans) that catalyze cellulolysis. *Trichoderma*, one of the famous fungi genus, is known as an ideal producer of cellulases. The origin of this specie was isolated from a badly rotten soldier's belt during the Vietnam War. American microbiologists revealed that it was the cellulase secreted by *Trichoderma reesei* causing the cotton made belt become rotten. More cellulases were subsequently founded, and later put into production. Cellulase has been used for commercial coffee processing; it performs hydrolysis of cellulose during drying of beans. Furthermore, cellulases are widely used in textile industry and in laundry detergents. They have also been used in the pulp and paper industry for various purposes, and they are even used for pharmaceutical applications. In the approaches of fermentation of celluloid biomass into biofuels, cellulases would play important role for primary degradation of cellulose into glucose, although this process is relatively experimental at present. It is worth noting that, the bulk of cellulase used in industries is from fungal species.

Pectinases is a general term for pectin hydrolases, such as pectolyase, pectozyme and polygalacturonase, commonly referred to as pectic enzymes in brewing. These enzymes break down pectin, a polysaccharide substrate that is found in plant cell walls. One of the most studied and widely used commercial pectinases is polygalacturonase. The enzymes are useful because pectin is the jelly-like matrix which helps cementing plant cells together, in which other cell wall components, such as cellulose fibrils, are embedded. Therefore pectinases are commonly used in processes involving the degradation of plant materials, such as speeding up the extraction of the juice from fruits, including apples and sapota. Pectinases have also been used in wine production since the 1960s. In brewing, pectinase exerts dual functions, first it helps breakdown the plant (typically fruit) material and so helps the extraction of flavors from the mash; secondly the presence of pectin in finished wine causes a haze or slight cloudiness, helps the removing of unstable substance form the wine. Pectinases can be extracted from *Aspergillus niger*. The fungus produces these enzymes to break down the middle lamella in plants so that it can extract nutrients from the plant tissues and insert fungal hyphae.

Viral enzymes

Viruses are parasites, which can neither grow nor replicate except within a host cell, but usually use parts of the machinery of the cell to make their own enzymes and run into reproduction period. Since viruses have very short life cycle, the viral enzymes must be very active, and that's why which draws great attention from the bio-engineers. Many viral enzymes are now becoming commercial products, for example: T4 DNA ligase the commonly used viral enzyme is purified from *E. coli* host cells infected with bacteriophage

T4. The reaction occurs between the 3′ hydroxyl group of one DNA strand and the phosphate group of the partner strand. ATP provides the energy for the reaction. DNA ligation can occur efficiently only between compatible ends of DNA fragments, and this specificity gives the investigator some control over how different DNA fragments ligate together in a reaction mix. Since all the cloning works would be involved with DNA cutting and ligation, T4 DNA ligase is now used broadly in all the molecular biology labs and the related industrial sectors. Reverse transcriptase is a significant enzyme firstly discovered in virus by Howard Temin at the University of Wisconsin-Madison, and independently by David Baltimore in 1970 at MIT, also known as RNA-dependent DNA polymerase, which means a DNA polymerase enzyme that transcribes single-stranded RNA into single-stranded DNA. As the matter of fact, then it is also a DNA-dependent DNA polymerase which synthesizes the second strand of DNA, while degrading the primitive RNA strand via its RNase H activity. The final product is a double stranded cDNA with one end having hairpin structure. Normal transcription involves the synthesis of RNA from DNA; hence, reverse transcription is the reverse of this.

The enzyme is encoded and used by reverse-transcribing viruses during their process of genomic replication. Reverse-transcribing RNA viruses, such as retroviruses, use the enzyme to reverse-transcribe their RNA genomes into DNA, which is then integrated into the host genome and replicated along with it. Reverse-transcribing DNA viruses, such as the hepadnaviruses, can allow RNA to serve as a template in assembling, and making DNA strands. HIV infects humans with the use of this enzyme. Without reverse transcriptase, the viral genome would not be able to incorporate into the host cell, resulting in the failure of the ability to replicate. Now, the reverse transcriptase from avian species has been commercially exploited into a useful enzymatic tool.

12.2 About Enzyme Synthesis

Except the ribozyme, most of the enzymes are proteins which have been translated in cell cytoplasm. Even a simple prokaryotic cell possesses hundreds of different enzymes to maintain its metabolism and reproduction. Some of the knowledge about the enzyme biosynthesis are introduced briefly as follows: **Constitutive and inducible enzymes. Constitutive enzymes** are produced constitutively by the cell under all physiological conditions. Therefore, they are not controlled by substrate induction or repression. Constitutive enzymes are produced in constant amounts without regard to the physiological demand or the concentration of the substrate. They are continuously synthesized because their critical role in maintaining cellular processes or structure is indispensable. E. g., enzymes in tricarboxylic acid cycle including citrate synthase, aconitase, isocitrate dehydrogenase, α-oxoglutarate dehydrogenase, succinatedehydrogenase, malate dehydrogenase etc., could be regarded as constitutive enzymes. **Inducible enzymes** are those expressed only under certain conditions in which it is clear of adaptive value, as opposed to constitutive enzymes which being produced all the time. The Inducible enzymes are employed to deal with the changing environment, for example, different sorts of substrate, changed concentration of nutrients, and toxic material existed. The enzymes have hydrolysis function are normally inducible, only when the substrate existed around their living environment, the cell starts to synthesize the related hydrolysases; it seems very reasonable to have this mechanism for highly efficient utilizing of substrate. In the prokaryotic Operon Model, inducer or repressor plays switch role to control the genes coding for inducible enzymes in the operon, to turn "on" or "off". The inducer causes the gene to turn on (controlled by the amount of reactant which turns the gene on). Then there's the repressor protein that turns genes off. The inducer can even remove this repressor, turning genes back on. An example of inducible enzyme is COX (cyclooxygenase)-2 which is synthesized in macrophages to produce prostaglandin E2 while the constitutive enzyme COX-1 (another isozyme in COX family) is always produced in a variety of organisms in body (as in stomach). *E. coli* β-galactosidase is a typical inducible enzyme, which can be induced by native inducer lactose or artificial inducer IPTG (isopropyl β-D-1-thiogalactopyranoside). The gene from *E. coli* lac operon had been cloned into many vectors for construction of genetic marker and inducible control.

12.2.1 Endo-, exo-, and membrane enzymes

Endoenzymes, or intracellular enzymes, are the enzymes functioning within the cell in which they are produced. Because the majority of enzymes fall within this category, the term is used primarily to define the enzymes which can not be secreted outside of the cell. Since all the enzymes are biosynthesized within the cell, although some of them might later be secreted out of the cell, It is possible for a single enzyme to have both endoenzymatic and exoenzymatic functions. Endoenzymes are usually binding with organelles in the cells, e.g., the enzymes in tricarboxylic acid cycle and the oxidative phosphorylases are almost always associated with mitochondria; additionally, the enzymes working for protein synthesis are usually associated with ribosome in endoplasmic reticulum. Exoenzymes, or extracellular enzymes, are enzymes that are secreted by a cell and that work outside of the cell. They are usually used for breaking up large molecules that would not be able to enter the cell otherwise, e.g., the enzymes having digestion functions on macromolecules, including pepsin, amylase, lipase, chymotrypsin and trypsin etc. In the enzyme production sectors, this term is also often referred to the hydrolytic digestive enzymes secreted by microbes. The secretion mechanism of prokaryotic exoenzymes is now quite clear, it has been found that the translational peptide of the pro-enzyme molecules is longer than the secreted one, the N-terminal section of the peptide, called signal peptide, which usually consists of few alkaline amino acids at the very end and around 20 hydrophobic amino acids in the form of α-coil. The signal peptide is associated with membrane by forming a special passageway, and the rest part of peptide can be secreted out. There is another enzyme, called signal peptidase, inlayed on the membrane, and cutting accurately between the signal peptide and the secretive segment. Membrane enzymes, means directly the membrane located enzymes, or enzymes associated with the membrane of a cell or an organelle, like signal peptidase. More than half of all the proteins interact with membranes. Membrane enzymes produce a variety of substances essential for cell function, E. g., Apyrase (adenosine diphosphatase) is a calcium-activated plasma membrane-bound enzyme (magnesium can also activate it) (EC3.6.1.5) that catalyses the hydrolysis of ATP to yield AMP and inorganic phosphate. Two isoenzymes are found in commercial preparations from *S. tuberosum*; the NADPH oxidase (nicotinamide adenine dinucleotide phosphate-oxidase) is a membrane-bound enzyme complex. It can be found in the plasma membrane as well as in the membrane of phagosome; hyaluronan synthases (HAS) are membrane-bound enzymes which use UDP-α-*N*-acetyl-D-glucosamine and UDP-α-D-glucuronate as substrates to produce the glycosaminoglycan hyaluronan at the cell surface and extrude it through the membrane into the extracellular space; glutamyl aminopeptidase is a zinc-dependent membrane-bound aminopeptidase that catalyzes the cleavage of glutamatic and aspartatic amino acid residues from the N-terminus of polypeptides.

12.2.2 Biosynthesis of cloned enzymes

Previously the basis for production of industrial enzymes was a naturally occurring microorganism producing a native enzyme with useful properties. Exploitation of the diversity of nature through efficient and laborious screening of the microsphere in various environments has yielded new products. Productivity of the natural microorganism has been boosted in some cases up to a thousand fold by mutations. Enzymes of animal or plant origin may be produced independently of the supply of animal and plant tissue. Chymosin, the active protease in calf rennin, is already being produced by fermentation. Enzymes found in minute concentrations in exotic microorganisms often difficult to grow may be produced by selecting host microorganisms which are easy to cultivate on cheap raw materials and producing a broth from which the enzyme is easy to purify. As an example, anaerobic microorganisms offer a huge untapped resource of very diverse enzymes not available today because of the difficulties in anaerobic cultivation.

The genetic engineering techniques enabling the transfer of genes between species and the modification of genes offer many benefits which are readily utilized for the production of industrial enzymes, even in large-scale fermentation processes. Productivity may be boosted by selecting an efficient gene construction. Examples are the use of multiple copies per cell of the structural gene, or the use of a strong promoter or an efficient signal sequence. Genetic engineering enables us to select host organisms and cultivation conditions that are safe to the manufacturing personnel, to the user of the product, and to the environment at large. Often an enzyme with interesting properties is found in an organism which is insufficiently characterized or is related to a toxin-producing or to an opportunistic pathogenic organism. Such uncertainty can now be avoided by transfer of the enzyme gene into a safer host.

Genetic engineering opens new era to enzymes with improved stability, activity or specificity. Through changes in the sequence of the structural gene, one or more amino acids in the enzyme molecule may be substituted by other amino acids thus changing the charges and the structure of the enzyme. By changing two amino acids in our best-selling detergent protease, Savinase, a new enzyme with greatly improved stability in the presence of bleaching agents was created. It is now marketed under the name Durazyme. Some other examples have been already mentioned in the content of Part I.

12.3 Enzyme Fermentation and Products

12.3.1 Improvement for higher producer of enzyme

Fermentation is a major process for generating enzymes in industrial purposes. It involves the selection of ideal microorganisms, like bacteria, yeasts and fungi to produce the enzymes. It's necessary to screen for a good producer and keep the strain in genetic stability. An ideal enzyme producer is never a native one, to meet an industrial purpose there is always breaking the physiological balance and pursuing only one specific enzyme being synthesized in extremely high amount. Almost all the industrial enzyme producers are mutants which undergone mutagenesis from chemical, physical and even cloning treatment. After several runs of genetic improvement, the enzyme producers gradually meet the industrial criteria: 1) being competitive for nutrient utilization; 2) having much higher productivity; 3) having higher growth rate; 4) being tolerant to poor condition; 5) being not easily contaminated; 6) having better properties for enzyme isolation and purification. The strategy and basic techniques of industrial strain improvement will be introduced in Part IV in details.

12.3.2 General process of enzyme fermentation

Both the solid and liquidized fermentation processes are practiced in enzyme industries. For liquors production, the grain particles were crashed and absorbed with water, and made into brickbat to let native microbes grow in it. The enzymes produced from the brickbat are rich enough to break down the macromolecular substrates and the resulted small molecules can be converted into alcohol and other flavor compounds which offering the liquor with unique taste and flavor. Solid fermentation can also be applied for producing enzymes used in recycling of hay or straw into useful bio-products, for the process seems much simpler and cost less.

But most of the enzyme productions adopt liquidized fermentation, for this process can be finely controlled and easily repeated (introduced in Part IV). The liquidized fermentation utilizes to a high extent alternative process modes: 1) Batch fermentation, some of the enzymes are still produced in this mode; 2) Fed-batch, through the inlets on the top of fermenter the nutrients can be further added into the tank to elongate the period of fermentation, and the productivity raised; and 3) Continuous process, the balanced volume of input and output keeps fermentation last much longer time. In the 1970's Novo Nordisk succeeded in developing a fully continuous fermentation process for the production of glucose isomerase. As far as is known, this was the first large-scale instance of continuous enzyme production. The continuous process offers considerable advantages: 1) Increased productivity, in the glucose isomerase process the productivity increased significantly because the same yield per unit of volume could be obtained at a mean residence time considerably shorter than the cycle time for the batch process; 2) Suitable limitations of nutrients, excess of easily metabolized nutrients inhibits the synthesis of many enzymes. Optimal production of glucose isomerase requires dual limitations of glucose and oxygen; this is easily maintained in the continuous process. 3) Uniform product quality, the constant milieu in the fermenter leads to a uniform composition of the broth. 4) Even and low load on utilities, such as steam, air and cooling water supply, in the evaluation of alternative process modes, the factor which really counts is the yield based on nutrient consumption. A significant productivity gain doesn't outweigh slightly reduced efficiency in nutrient use.

Today the most important task in process development is to develop control methods and process modes for the genetic engineered microorganisms. The stability and productivity of the new strains pose new challenges to fermentation control.

12.3.3 Industrial enzyme products

Some typical uses of industrial enzymes

Starch liquefaction A malt source of enzymes is used to liquefy starch at high temperatures. Starch liquefaction and saccharification of the main α-amylase starch hydrolysis product market, such as glucose and fructose. Starch is converted into high fructose corn syrup (HFCS). Because of their high sweetness, are used in the soft drink beverage industry sweeteners. The liquefaction process is used in thermal stability at high temperature α-amylase; α-amylase in starch liquefaction of the application process is already quite mature.

It is well known that malted cereal grains, such as malted wheat, oats, rye, triticale and barley, contain the enzyme, α-amylase, which is capable of hydrolyzing starch molecules to produce starch fractions that are less viscous in aqueous solutions than the cooked native starch. Thus, it is possible to prepare aqueous starch solutions containing much higher solids level with liquefied starch than is possible with cooked native starch. This is important in many applications. It is a common practice in batch starch conversion processes to add alpha-amylase to the starch before the starch slurry reaches the gelatinization temperature and then add a second quantity of the enzyme after the starch paste has been cooled to about 65℃ or below because of the instability of the enzyme at higher temperatures. The maximum temperature for malt alpha-amylase activity is generally known to be about 80℃ with the enzyme being rapidly inactivated at temperatures above 80℃.

Pulp and paper process In pulp and paper process, a great number of enzymes are employed in these fields:

Amylases are used to cleave starch molecules to reduce the viscosity, and also used for surface sizing and for starch in coatings.

Hemicellulases (xylanases) are used to cleave hemicelluloses in fiber, making the bleaching process more effective. It has been reported to be able to reduce bleaching chemicals by up to 30%. Additionally, it could improve brightness.

Lipases are used to control pitch in pulping processes. They could convert triglycerides to fatty acids which are more stable in water, won't deposit as much (Fig. 12.1).

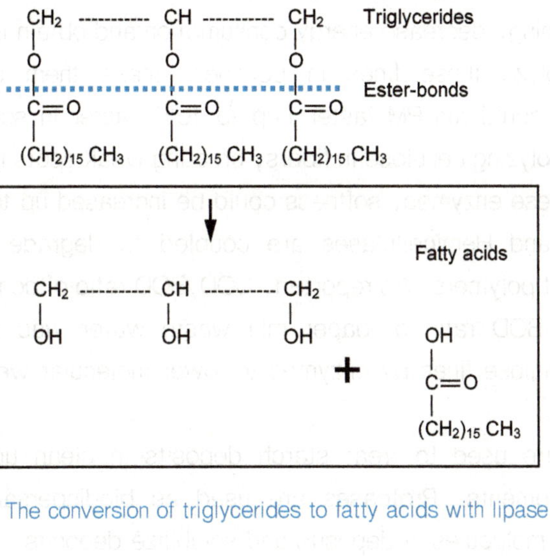

Fig. 12.1 The conversion of triglycerides to fatty acids with lipase

Cellulase and hemicellulase enzymes are used in deinking (DI) process. They hydrolyze these microfibrils, releasing the adhesives. Enzyme assisted DI is reported to remove 30%～60% more toners and to improve brightness by 4～5 points.

Esterases are used to break ester bonds in polymers used in toners and adhesives. They could improve paper cleanliness, and they are associated with less deposit, less cleaning up on polymers. These enzymes could also be used as substitutes for talc or solvent based dispersants(Fig. 12.2).

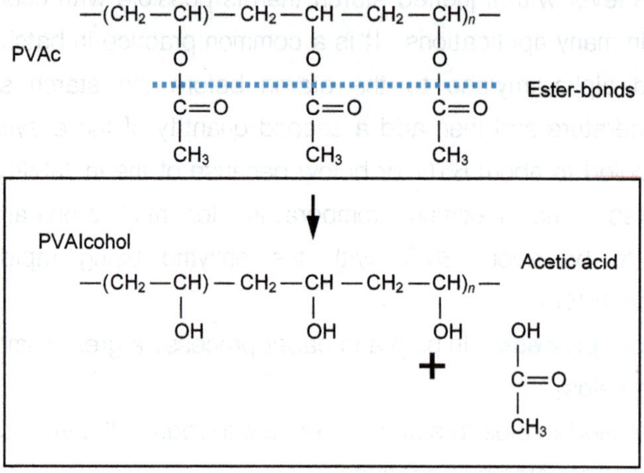

Fig. 12.2 Esterases could break ester bonds in polymers

Cellulase enzymes can be used in conjunction with or as a substitute for the caustic to swell fibers. Neutral deinking has some advantages including less COD in the white water, and lower costs; cause partial depolymerization of cellulose and swelling of fibers, more flexible fibers, and are used for the reduction of fines. Furthermore, they Improve the efficiency of refining, decrease energy consumption and obtain improved strength at higher freeness; hydrolyze these fines in polymer, make them soluble and improve PM operations, and could run PM faster (up to 10% faster in some cases); also improve softness by hydrolyzing cellulose in fibers, creating weak spots in fibers, and making fibers flexible. With these enzymes, softness could be increased up to 25%.

Cellulases and Hemicellulases are coupled to degrade cellulose and other high molecular weight polymers. As reported, COD/BOD ratio of domestic sewages was 2 : 1. However, COD/BOD ratio of paper mill waste water was up to 5 : 1. Therefore, hydrolysis of cellulose fines by enzymes to lower molecular weight sugars that are more biodegradable.

Amylases are used to treat starch deposits in clean ups of starch preparation/application equipments. Proteases are used as biodispersants or enzyme boil-outs, cleaving protein molecules in deposits and solubilize deposits

Textile process Enzymes have been found with wide application in the textile industry for improving production methods and fabric finishing. One of the oldest applications in this industry is the use of amylases to remove starch size. The warp (longitudinal) threads of fabrics are often coated with starch in order to prevent them from breaking during weaving.

Scouring is the process of cleaning fabrics by removing impurities such as waxes, pectins, hemicelluloses, and mineral salts from the native cellulosic fibers. Research has shown that pectin acts like glue between the fiber core and the waxes, but can be destroyed by an alkaline pectinase. An increase in wettability can thus be obtained.

Cellulases have become the tool for fabric finishing. Their success started in denim finishing when it was discovered that these cellulases could achieve the fashionable stonewashed look traditionally achieved through the abrasive action of pumice stones. Cellulases are also used to prevent pilling and improve the smoothness and color brightness of cotton fabrics in a process which Novozymes calls Biopolishing. In addition, cellulases could also lead to a softer handle.

Catalases are used for degrading residual hydrogen peroxide after the bleaching of cotton. Hydrogen peroxide has to be removed before dyeing. Proteases are used for wool treatment and the degumming of raw silk.

Leather process To make it soft, the leather has to be bated, which means that some of the protein fiber should be removed from the row leather. Otherwise, it will be hard-perfect for the soles of shoes but of little use for anything else. The Roman writer Pliny reported the use of pigeon droppings for this process over 2000 years ago. Thanks to the German scientist Ršhm, who developed a standardized bate in 1908, based on an extract from the pancreases of slaughtered animals. This contained trypsin, one of a mixture of enzymes found in the digestive system. Since then, all bates have been based on such kinds of proteinase preparations, though now bacterial and fungal proteinases have been used instead.

Washing powders The first enzymatic washing agent was patented by Ršhm in 1914. It was so effective that only a small quantity was required: it was sold as a spot remover. The mass-production of an alkaline protease suitable for wash conditions began in 1962. Unlike trypsin, this wasn't an animal extract but a product of microbial fermentation. This new enzyme was initially shunned by detergent manufacturers but there were exceptions. In 1963, it was incorporated into Bio-tex, which took the market by storm. Industry began to take notice of enzymes and by 1967 their widespread use in domestic detergents had become a commonplace.

The enzymes used for washing are normally produced from *Bacillus licheniformis*. They are usable at high pH and temperatures up to 60℃ and are all relatively non-specific proteases. They attack the C-terminal of carboxyl amino acids and produce small peptides which can be readily dissolved by the detergent. Currently there is considerable interest in developing better proteases for washing powders through protein engineering, particularly in engineering oxidation-resistance into the proteases.

Since the 1990's, amylases have also been added to detergents to remove stains of spaghetti, sauces, oatmeal and baby foods. In 1988 the first detergent lipase was released—the first commercial enzyme to be produced from a genetically-modified organism (GMO). Today more than 90% of detergent enzymes are made from GMOs. The detergent industry has been the largest market for industrial enzymes for over 25 years, accounting for 37% of world sales of enzymes. Apart from laundry detergents, many automatic dishwashing detergents also contain enzymes now.

Cheese making In cheese manufacture, warm milk is mixed with the enzyme rennin (rennet) [formerly extracted from (dead) calves' stomachs, but now produced from bacteria] and allowed to react for several hours. The caseinogen in the milk is uncoiled and clots to casein, which turns the milk solid. The curds (solid) are then cut with a knife and the whey (liquid) is drained away or fed to animals or as lactose fermentation. The chopped up curd is then salted and placed in a mould before squeezed to remove any trapped air (a process known as 'cheddaring'——hence Cheddar cheese). Sometimes the cheese is then dipped in brine or a solution of fungal spores to inoculate it and produce a surface "rind".

12.3.4 One-banch-one-product

A great number of enzymes have also been employed in medical uses (Tab. 12. 2)

Tab. 12.2 Medical enzymes

Enzyme	EC number	Reaction	Use
Asparaginase	3.5.1.1	L-asparaginase $H_2O \longrightarrow$ L-aspartate + NH_3	Leukaemia
Collagenase	3.4.24.3	Collagen hydrolysis	Skin ulcers
Glutaminase	3.5.1.2	L-glutamine $H_2O \longrightarrow$ L-glutamate + NH_3	Leukaemia
Hyaluronidase	3.2.1.35	Hyaluronate hydrolysis	Heart attack
Lysozyme	3.2.1.17	Bacterial cell wall hydrolysis	Antibiotic
Rhodanase	2.8.1.1	$S_2O_3^{2-} + CN^- \longrightarrow SO_3^{2-} + SCN^-$	Cyanide poisoning
Ribonuclease	3.1.26.4	RNA hydrolysis	Antiviral
β-Lactamase	3.5.2.6	Penicillin $\longrightarrow$ penicilloate	Penicillin allergy
Streptokinase	3.4.22.10	Plasminogen $\longrightarrow$ plasmin	Blood clots
Trypsin	3.4.21.4	Protein hydrolysis	Inflammation
Uricase	1.7.3.3	Urate + $O_2 \longrightarrow$ allantoin	Gout
Urokinase	3.4.21.31	Plasminogen $\longrightarrow$ plasmin	Blood clots

As biocatalysts, enzymes are expected to have a greater potential to be exploited into desirable therapeutic agents (Tab. 12. 2). Unfortunately still a number of factors reduce their potential utilization on treatment of metabolic diseases: 1) They are too large to be distributed simply within the body's cells. This is the major reason why enzymes have not yet been successfully applied to a large number of human genetic diseases. A number of methods are being developed in order to overcome this by targeting enzymes; as examples, enzymes with covalently attached external b-galactose residues are targeted at hepatocytes and enzymes covalently coupled to target-specific monoclonal antibodies are being used to avoid non-specific side-reactions. 2) Being foreign proteins to the body in general, they are antigenic and can elicit an immune response which may cause severe and life-threatening allergic reactions, particularly with continued use. It has proved possible to circumvent this problem, in some cases, by disguising the enzyme as an apparently non-proteinaceous molecule by covalent modification. Asparaginase, modified by covalent attachment of polyethylene glycol, has been shown to retain its *anti*-tumour effect whilst possessing no immunogenicity. Clearly the presence of toxins, pyrogens and other harmful materials within a therapeutic enzyme preparation are strictly forbidden. Effectively, this encourages the use of animal enzymes, in spite of their high cost, relative to those of microbial origin. 3) Their effective lifetime within the circulation may be only a matter of minutes. This has proved easier than the immunological problem to combat, disguise using covalent modification. Other methods have also been shown to be successful, particularly those involving entrapment of the enzyme within artificial liposomes, synthetic microspheres and red blood cell ghosts. However, although these methods are efficacious at extending the circulatory lifetime of the enzymes, they often cause increased immunological response and additionally may cause blood clots.

In contrast to the industrial use of enzymes, therapeutically useful enzymes are required in relatively tiny amounts but at a very high degree of purity and (generally) specificity. The favored kinetic properties of these enzymes are low K_m and high v_{max} in order to be maximally efficient even at rather low enzyme and substrate concentrations. Thus the sources of such enzymes are chosen with care to avoid any possibility of unwanted contamination by incompatible materials and to enable ready purification. The production scale of enzymes used in therapeutic and molecular biology fields would be quite varied, in the bio-companies. Normally one person or a small team is correspondent for one kind of enzyme product, thus the preparation mode was referred as "one-banch-one-product". In this way fine purified enzyme products are guaranteed in absolute high quality. The enzymatic tools used for genetic cloning are also "one-banch-one-product". The fine enzyme preparations are generally offered for sale as lyophilized pure preparations with only biocompatible buffering salts and mannitol diluent added. The costs of such enzymes may be quite high but still comparable to those of competing therapeutic agents or treatments. For example, urokinase (a serine protease, see Tab. 12. 1) is prepared from human urine (some genetically engineered preparations are being developed) and used to dissolve clot *in vivo*.

A major potential therapeutic application of enzymes is in the treatment of cancer. Asparaginase has been proved to be particularly promising for the treatment of acute lymphocytic leukemia. Its action depends upon the fact that tumor cells are deficient in aspartate-ammonia ligase activity, which restricts their ability to synthesize the normally non-essential amino acid L-asparagine. Therefore, they are forced to extract it from body fluids. The action of the asparaginase does not affect the function of normal cells which are able to synthesize enough for their own requirements, but reduce the free exogenous concentration so that induces a state of fatal starvation in the susceptible tumor cells. A 60% incidence of complete remission has been reported in a study of almost 6000 cases of acute lymphocytic leukemia. The enzyme is administered intravenously. It is only effective in reducing asparagine levels within the bloodstream, showing a half-life of about a day (in a dog). This half-life may be increased 20-fold by the use of polyethylene glycol-modified asparaginase.

Compared with industrial enzymes, the medical enzymes belong to the products of low-volume-high-value. In some cases, the engineered organisms have been cultured in flasks, but not in fermenter. However, medical uses require high purity and therefore the downstream process should be restricted rigorously.

Summary

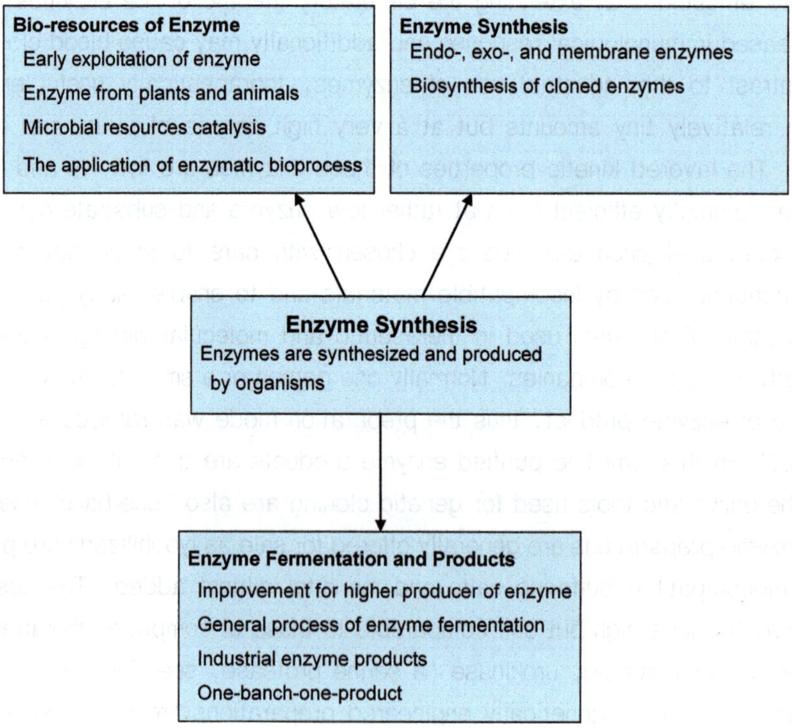

Chapter 13 Enzyme Recovery & Purification

人类掌握了酶类制备技术是生物技术发展进步的重要先决条件。酶的获取主要涉及回收与纯化,欲达到提纯目标酶的目的,通常需要根据不同酶类的性质设计回收与纯化流程。酶制备过程涉及多个步骤,有的步骤本身就是不同技术的组合。在每个步骤中,酶的回收率与纯度都不可能达到 100%,这意味着目标酶的得率必将有一定程度的流失,甚至酶的活性也会受到一定的影响。掌握酶制备的各种基本原理与技术,有的放矢地选择最合适的技术组合,尽量减少操作步骤,缩短提取周期,优化操作流程并加大过程控制,才有可能减少酶的失活,提高酶产物的回收率和纯度,使得酶生产的效益达到最大化。

13.1 Pretreatment of Crude Enzyme Sample

According to different sources, bacterial, plant or mammalian, intracellular or extracellular, the procedures of enzyme extraction are quite dependable. For the enzymes which secreted out of cell envelope would be easier to recover. Most commercial enzymes are extracellular enzymes, they are generally secreted from the microbials, which have been genetically improved, into the fermentation broth with only a few impurities, and then remove the cells and cellular pellets from the fermentation broth. In the case of solid fermentation of microbial enzymes, the first step of extraction is usually involved with the immersing of fermented solid medium with water and collection of the enzyme infusion.

For intracellular enzymes, the cell breaking procedure must be involved, and there are varieties of techniques can be used to disrupt gently the cells or bio-tissues. Since cell disruption usually leads to the release of proteolytic enzymes and causes temperature raised, the selection of an extraction technique is dependent as much upon the equipment available and the scale of operation as on the type of samples. Extraction should be performed as quickly as possible, at sub-ambient temperature, in the presence of a suitable buffer to maintain optimal pH and ionic strength, and sample should be kept in stable conditions. Samples should be clear lysate and free from any insoluble particles before running into further separation. Enzymes are very complex proteins, and their highly defined specificity as catalysts is manifested only in their native state, or says in their right structure. The original conformation of enzyme is only attained under specific conditions of temperature, pH, and ionic strength. Hence, only mild and specific methods can be used for enzyme extraction.

For isolation of intracellular enzymes, being isolated today in increasing amounts, the first step involves using physical or enzymatic means to rupture the cells. A number of methods for the disruption of cells (Tab. 13.1) are known, corresponding to the different types of cells and the problems involved in isolating intracellular enzymes. However, only few of these methods are used on an industrial scale.

Tab. 13.1 Methods for disruption of cells

Mechanical methods	Nonmechanical methods
High pressure (French-press)	Drying (freeze-drying, organic solvents)
Grinding (ball mill)	Lysis:
Ultrasound	Physical-freezing, osmotic shock
	Chemical-detergens, antibodies
	Enzymatic-lysozyme, snail enzyme

High-pressure homogenization is the most common method of cell disruption. The improved enzyme yield resulting from increased cell disruption could be counteracted by partial inactivation of enzyme caused by over heating and higher shearing forces. Therefore, efficient cooling and gentle stirring must be provided.

13.2 Removing of Impurities and Sample Concentration

13.2.1 Removing of impurities

The next step is to separate the extracellular or intracellular enzymes from crude supernatant or clear cell rupture. This step is somehow difficult to perform, because of the small size of bacterial cells and the slight differences between the density of the cell fragments and that of the enzyme molecules.

Continuous filtration is often used in industry. Large cells, e.g., yeast cells, can be removed by decantation. Today, efficient centrifuges have been developed to separate cells and cell debris in a continuous process. Tissue and organ residual can be removed with simpler centrifuges or through filters. The sedimentation rate of a typical bacterial cell with a diameter of 0.5μm is less than 1mm/h. An economical separation can be achieved only by sedimentation in a centrifugal field. The parameters of centrifuges depend on the particle size and the solids content.

The filtration rate is a complex function due to filtering area, pressure, viscosity, and resistance offered by the filter cake and the particle in medium. For a clean liquid, all these terms are constants which result in a constant flow rate for a constant pressure drop. The volume of cumulative filtrate increases linearly with time. During the filtration of suspensions, the increasing thickness of the formed filter cake and the concomitant resistance gradually decrease the flow rate. Additional difficulties may arise because of the compressibility of biological materials. In this case, the resistance resulted by the filter cake increases. Hence keeping the rate of filtration would depend on the increased pressure. But if the pressure applied exceeds a certain limit, the filter cake may collapse, and results in a total blockage of the filter.

Three modes of filtration are introduced for filtering the fluid containing enzymes: 1) **Filtration with pressure**. Plate filter and chamber filter are used to filter small volumes of solutions or to remove precipitates during preparation. The capacity to retain solid matter is limited, and the method is rather work-intensive. However, these filters are highly suitable for fine filtration of enzyme solutions. 2) **Filtration under vacuum**. This is generally the method of choice because biological materials are easily compressible. A vacuum filter is usually employed to deplete precipitates and solids in enzyme solutions quickly and efficiently. Compared with filtration with pressure, vacuum supplies a much more powerful force; therefore, the whole filtration is not time-consuming. 3) **Cross-flow Filtration**. In this mode, the input stream flows parallel to the filter area, thus preventing the accumulation of filter cake and the increased resistance to filtration. To maintain a sufficiently high filtration rate, this method consumes more energy in the form of maintaining high flux rates over the membranes. Exploitation of varieties of high quality membranes will bring this method a promising future. Filtration step removes most of the impurities from the flux, and allows the next step to concentrate the volume of the clear flux.

13.2.2 Sample concentration

The volume of flux containing enzyme is rather bulky, and the enzyme concentration is often very low. The volume of starting flux must be decreased for concentration of the enzyme. Only mild concentration can be employed which would not inactivate the enzyme, such as vacuum evaporation, precipitation, and ultra-filtration, three commonly used concentration methods are introduced individually as follows:

Vacuum evaporation Since most of the enzymes are not thermo-tolerated, that evaporate water away by heat treatment should be carefully performed. Under vacuuming, evaporators with rotating vessel or long-tube are normally used to minimize the volume of the samples by forming a thin liquid film (thin-layer evaporator, centrifugal thin-layer evaporator), under moderate temperature.

Precipitation Enzymes are peptide molecules with 3D dimensional structures folded by both hydrophilic and hydrophobic bonds. Many solvents and chemicals might interact with peptide and change its structure or the surface chemicals. Even moderate interaction changes the situation, and sometimes causes protein molecules agglomerate to form the insoluble segregates. Therefore, the solvents and chemicals used in the precipitation step must be carefully selected in order to ensure the enzymes can be precipitated from the flux without losing their activity. Precipitation is actually a simple procedure for concentrating enzymes, and the solvents routinely used could be salts, organic solvents and polymers (poly-ethylenimine and poly-ethylene glycols). Adjusting pH value of the flux to its isoionic point is a critical parameter which promotes enzyme precipitation. Precipitation is usually carried out on a small scale. Problems can arise in scaling-up of this process. The mixing time, the residence time in the reactor (which affects agglomerate formation and enzyme activity), and the shearing forces generated by stirring (which affects the aggregates formation) are all critical factors. When the starting volume is enlarged, the mixing time should be appropriately prolonged. Especially when some of organic solvents are added, protein denaturation might occur immediately. Precipitation practices have been conducted to overcome such difficulties by adopting a continuous process and using mild solvents.

Ultrafiltration A semi-permeable membrane with fine sieves permits the separation of solvent molecules from larger enzyme molecules by only allowing smaller molecules to penetrate the membrane when the osmotic pressure is exceeded. This is, as a matter of fact, the principle of all membrane separation processes (Tab. 13.2), including ultrafiltration. Under reverse osmosis, the substances with low molecular weight can be separated. Surely, the solubility and diffusion phenomenon influence the process as well. Cross-flow filtration is based solely on the sieve effect, and in the practice of isolating enzymes, cross-flow filtration is commonly used to harvest cells, whereas ultrafiltration is employed for desalting and concentration.

Tab. 13.2 The principle of membrane separation processes

Process	Application	Separation range
Cross-flow microfiltration	Concentration of bacteria, removal of cell debris	> 1000kDa
Ultrafiltration	Concentration of enzymes, dialysis	> 10kDa
Reverse osmosis	Concentration of small molecules, desalting	> 0.2kDa

13.3 Chromatographic Separation

Chromatography (**CG**) has become an essential tool in every laboratory where obtaining of pure protein is needed. Theoretic speaking, availability of various chromatography (**ion exchange, gel filtration, affinity CG, hydrophobic CG, partition CG**, and **Reverse phase CG**, etc.) with different selectivity provides a powerful combination for purifying any bio-molecule including enzymes for commercial purposes (Tab. 13.3 and Fig. 13.1). Molecules are separated according to their physical properties (size, shape, charge, sub-unit), chemical properties (covalent binding, surface chemical activity), or biological properties (bio-specific affinity).

Table 13.3 The purifying techniques related with protein properties

Protein property	Technique
Charge	Ion exchange (IEX)
Size	Gel filtration (GF)
Hydrophobicity	Hydrophobic interaction (HIC)
	Reversed phase (RPC)
Charge, ligand specificity or hydrophobicity	Expanded bed adsorption (EBA) follows the principles of AC, IEX or HIC

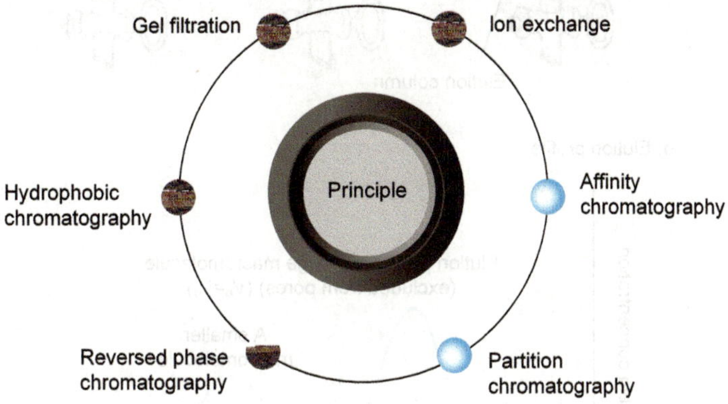

Fig. 13.1 Chromatography methods

In **gel chromatography** (also called gel filtration), the hydrophilic, cross-linked gel medium with pores of finite size are packed in the column to separate bio-molecules. The starting sample solutions must be concentrated before running the gel filtration, because the sample volume is limited to approximately; 10% of the column volume. In gel filtration, molecules are separated according to their sizes and shapes. Molecules larger than the largest pores in the gel beads, i. e., above the exclusion limit, cannot be absorbed by the gel and are eluted out of the beads first. Smaller molecules, which enter the gel beads to varying extent depending on their smaller sizes and shapes, are retarded in their passage through the column and eluted in order of decreasing molecular size. Since enzymes are various in shapes and sizes, gel chromatography helps a lot in the purification of enzymes(Fig. 13.2).

Ion-exchange chromatography is a matured separation technique based on the interaction with different surface charges of protein molecules. Enzyme molecules possess both positive and negative charges and the net charge is influenced by pH (Fig. 13.3). Thus this property has been used to separate proteins by anion exchangers (positively charged) or cation exchangers (negatively charged) (Fig. 13.3). The sample applied to the column is usually in an aqueous solution at low ionic strength, and the sample molecules will be attracted by the resin beads with opposite charges, and the elution is routinely carried out with a salt gradient due to increasing concentration. Because of the

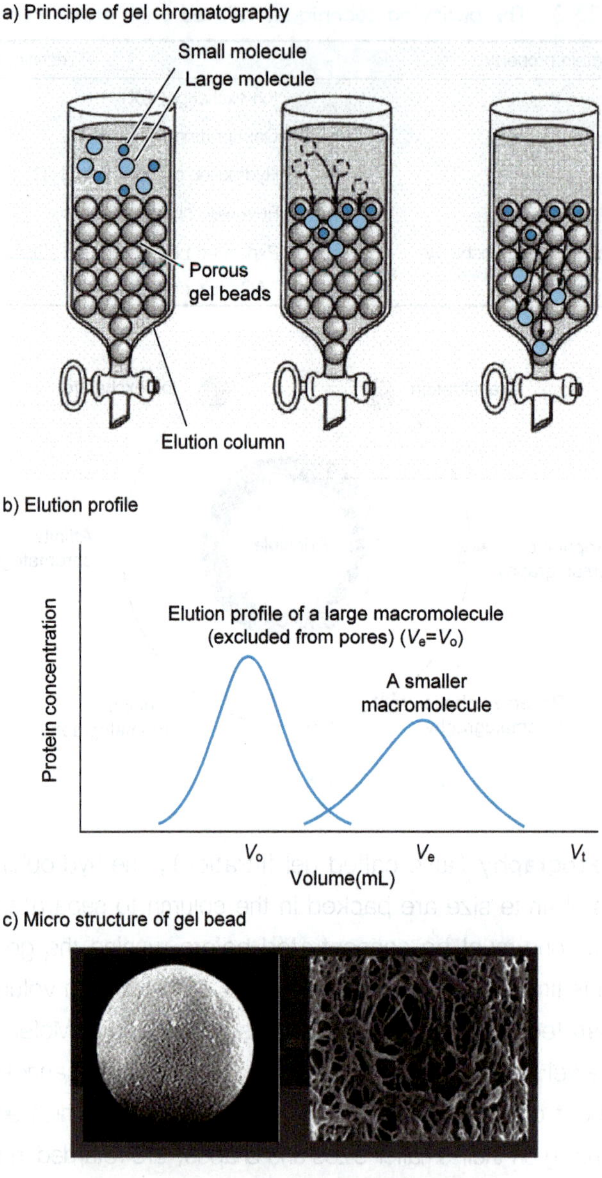

Fig. 13. 2 Gel chromatography

concentrating effects of this method, starting samples can be applied in dilute form.

For hydrophobic chromatography, media are derived from the reaction of CNBr-activated Sepharose with aminoalkanes of varying chain lengths. This method is based on the interaction of hydrophobic areas of protein molecules with hydrophobic groups on the matrix (Fig. 13. 4). Adsorption occurs when salt concentration is high, and the fractionation of bound substances is achieved by eluting a negative salt gradient. This method is ideally suited for further purification of enzymes recovered by precipitation with salts, such as ammonium sulfate.

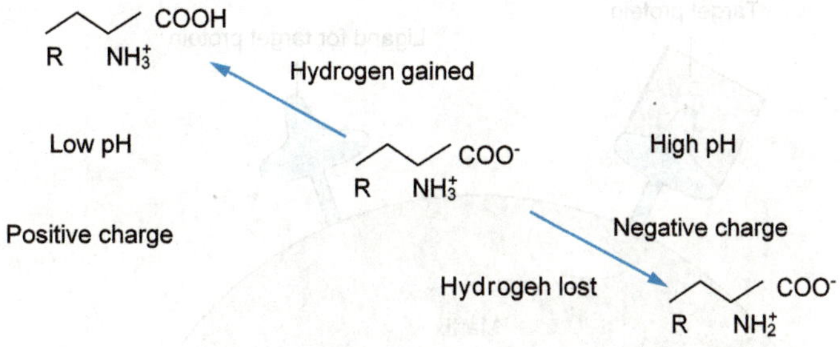

Fig. 13.3 The net charge of enzyme is influenced by pH

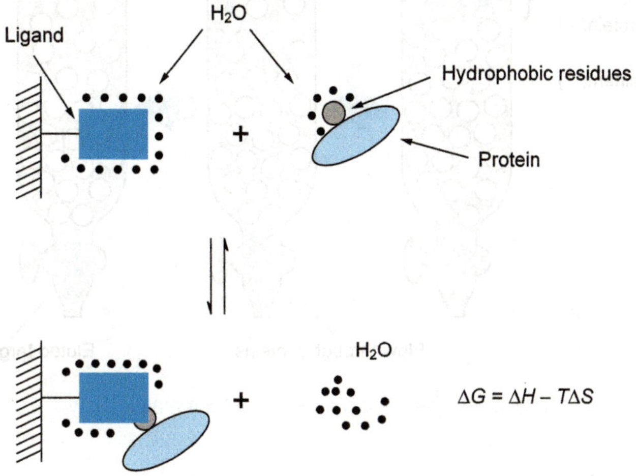

Fig. 13.4 The principle of hydrophobic chromatography

Affinity chromatography is a relatively simple procedure based on molecular recognition, and it is worthwhile to be further exploited for industrial enzyme extraction. Any given bio-molecule that one wishes to purify usually has an inherent recognition site through which it can recognize a natural or artificial molecule. If one of these recognition partners is immobilized on a poly metric carrier, it can be used to capture selectively the bio-molecule by simply passing the cell extract sample through the affinity column. The desired bio-molecule can then be eluted by changing external conditions, e. g. , pH, ionic strength, solvents, and temperature, which help the disassociation of the complex of bio-molecule and its partner; then the desired molecule will be eluted in a purified form (Fig. 13.5).

Numerous books and reviews on the theory and application of affinity chromatography have been published in recent years. Some examples of affinity purification are listed in Tab. 13.4.

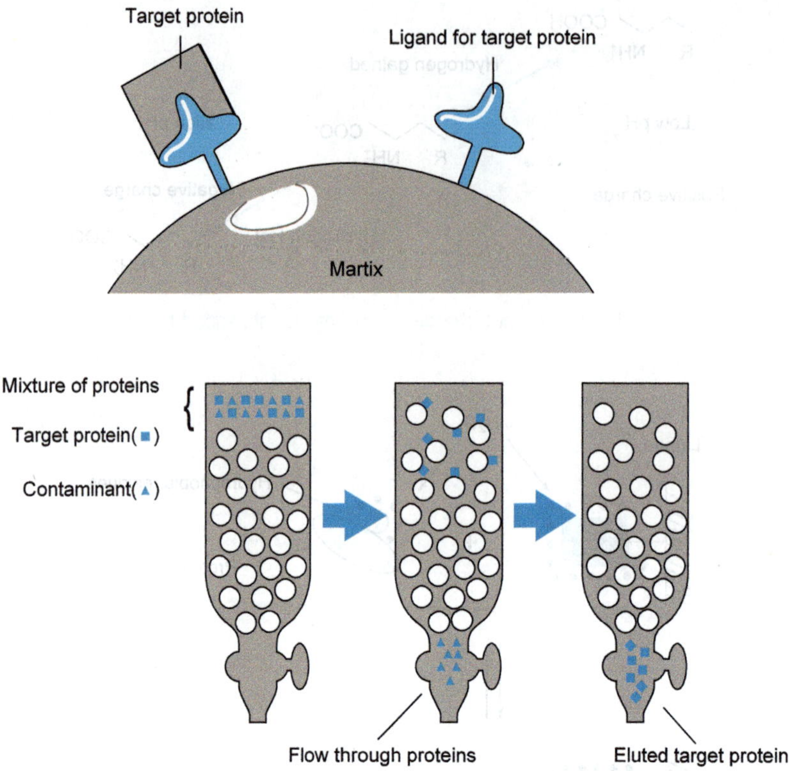

Fig. 13.5 The principle of affinity chromatography

Tab. 13.4 Bio-molecules purified by affinity chromatography

Enzymes	Other molecules
Enzymes and inhibitors	Antibodies and antigens
Regulatory enzymes	Hormone-binding proteins
Dehydrogenases	Vitamin-binding proteins
Transaminases	Receptors
	Lectins and glycoproteins
	RNA and DNA (genes)
	Bacteria, Viruses and phages
	Cells
	Genetically engineered proteins

The broad scope of the various applications of affinity chromatography has generated the development of subspecialty adaptations, many of which are now recognized from their own nomenclature as an expression of their generality and uniqueness. The techniques of affinity chromatography were summarized in Tab. 13.5.

Table 13.5 Various techniques derived from affinity chromatography (CG)

Immunoaffinity CG	Affinity density perturbation
Hydrophobic CG	Perfusion affinity CG
High performance affinity CG	Centrifuged affinity CG
Lectin affinity CG	Affinity repulsion CG
Metal-chelate affinity CG	Affinity tails CG
Covalent affinity CG	Thermphilic CG
Affinity electrophoresis	Membrane-based affinity CG
Affinity capillary electrophoresis	Weak affinity CG
Dye-ligand affinity CG	Receptor affinity CG
Affinity partitioning	Avidin-biotin immobilized system
Filter affinity transfer CG	Molecular imprinting affinity
Affinity precipitation	Library-derived affinity ligands

Among the most popular affinity-derived techniques is the immuno-affinity CG, which utilizes antibody columns to purify antigens, or antigen columns to purify antibodies. To purify some unique enzymes in natural resources, antibodies have been reported to be employed for the binding of target enzymes or for the depletion of background proteins. Other methods, such as metal-chelate affinity chromatography, apply site-directed mutagenesis to introduce various affinity tags or tails to the bio-molecule to be purified. For example, usually, a 6 His-Tag is added to the DNA sequence which encodes N-or C-terminal of target enzyme, during gene cloning process. Then in the separation procedure, via only one step of affinity chromatography with Ni-medium, most of the target enzymes could be captured (has been introduced in Part I).

Reversed-phase chromatography (**RPC**) includes any chromatographic methods that use a non-polar stationary phase. The term "reversed phase" so called has its historical background. In the 1970s, most liquid chromatography was performed on non-modified silica or alumina with a hydrophilic surface which has a stronger affinity to the hydrophilic molecules, and RPC is now also considered as "routine chromatography" for it accounts for the vast majority of analysis performed in liquid chromatography. The introduction of alkyl chains bonded covalently to the support surface reversed its polarity, therefore the elution order. When running RPC, polar molecules are eluted first while non-polar molecules are retained—so called "reversed phase" got its name based on this fact.

Expanded bed adsorption is a single pass operation in which desired proteins are purified from crude, particulate contained feed-stock without complicate separation clarification, concentration and initial purification. The expansion of the adsorbent bed creates a distance between the adsorbent particles, i. e. increased voidage (void volume fraction) in the bed, which allows for unhindered passage of cells, cell debris and other particulates during application of crude feed to the adsorption column (Fig. 13.6).

Many commercial adsorbents are available now, for example, STREAMLINE

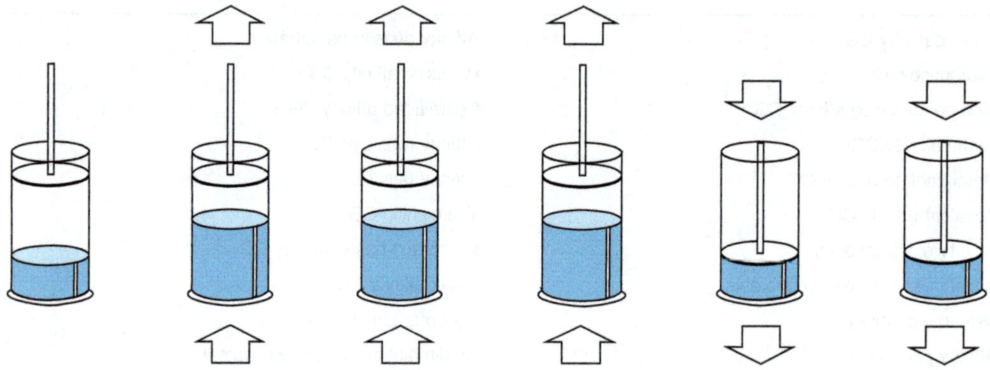

Fig. 13.6 The process of expanded bed adsorption

adsorbent is expanded and equilibrated by applying an upward liquid flow to the column. A stable fluidized bed is formed when the adsorbent particles are suspended in equilibrium due to the balance between particle sedimentation velocity and upward liquid flow velocity. The column adaptor is positioned in the upper part of the column during this phase. The crude or unclarified feed is applied to the expanded bed with the same upward flow as used during expansion and equilibration. Target proteins are bound to the adsorbent while cell particles, debris, particulates and contaminants pass through unhindered. The small portion of temporary bound solid materials, such as residual cells, debris and other particulate materials, would be washed away from the surface of expanded adsorbent by the upward liquid flow. When all weakly retained materials have been washed away from the expanded bed, the liquid flow is stopped and the adsorbent particles quickly settle down from the column. The column adaptor is then lowered to the surface of the sedimented bed. Flow is reversed and the captured proteins are eluted from the sedimented bed through a selective buffer rinsing. The target proteins in the eluate, are concentrated, clarified, partly purified, and ready for further purification by packed bed chromatography or other techniques.

After elution, the bed is regenerated by washing it with downward flow to the sedimented bed, and using buffers specific for the type of chromatographic principle applied. This regeneration step removes the more strongly bound proteins which are not easily removed during normal elution procedure.

The **simulated moving bed** (SMB) technique is a variant of high performance liquid chromatography; it is used to separate particles and/or chemical compounds that would be difficult or impossible to resolve otherwise. This increased separation is brought about by a valve-and-column arrangement that is used to lengthen the stationary phase indefinitely.

In the **moving bed technique** of preparative chromatography the feed entry and the recovery are simultaneous and continuous, but because of practical difficulties with a continuously moving bed in the **simulated moving bed technique** instead of moving the bed, the sample inlet and the analyte exit positions are moved continuously, giving the impression of a moving bed.

True moving bed chromatography (**MBC**) is only a theoretical concept. Its simulation, **SMBC** is achieved by the use of a multiplicity of columns in series and a complex valve arrangement, which provides for sample and solvent feed, and also analyte and waste takeoff at appropriate locations of any column, whereby it allows switching at regular intervals the sample entry in one direction, the solvent entry in the opposite direction, whilst changing the analyte and waste take off positions appropriately as well.

13.4 Polishing of Enzyme

"Polishing" here means refining or last step of purification. In the polishing phase of enzyme preparation, the focus is almost entirely on the high resolution of enzyme to achieve final purity. Most contaminants and impurities have already been excluded except for trace impurities such as endotoxins, nucleic acids and viruses, and closely related substances such as micro heterogeneous structural variants of the product, reagents and aggregates. It may be necessary to sacrifice the sample load or even recovery range (by peak cutting), to achieve the resolution. Recovery strategy of the final enzyme product is also a high priority and the technique must be selected to ensure the recovery in the highest yield. Product losses at this stage are more costly than those in earlier stages. Ideally the product should be recovered in buffer conditions and be ready for the next procedure, but somehow the polished enzymes are rather unstable in liquid state and thus particular enzyme stabilizers should be added to the final purification product. The technique chosen must be able to discriminate between the target protein and any remaining contaminants. The high resolution required to achieve this discrimination is not always reached by using a technique with high selectivity alone, but usually requires selection of a highly efficient media with small and uniform bead sizes. To polish the purified enzyme via molecular mass, gel chromatography and ultrafilter are the most frequently used methods, with which even the salt in solution could be removed. However, the low column capacity of gel chromatography makes it time-consuming and low recovering. Additionally, ultrafilter is considered of high loss. Therefore, usually, these protocols are practiced in small-scale extraction. Consider RPC for a polishing step provided that the target protein can withstand the running conditions. Reversed phase chromatography (RPC) separates proteins and peptides on the basis of their different hydrophobicity. RPC is a technique of high selectivity (high resolution), requiring the use of organic solvents which are suspected to have protein damage effects. However, in the polishing phase, since the majority of protein impurities have been removed, RPC is the choice for polishing, particularly when the target protein molecules are smaller and relatively stable in mild organic solvents.

13.5 Rational Formulation of Final Enzyme Products

Enzymes are sold in the form of stabilized liquid concentrates or particulate solids. The primary task of enzyme formulation is to minimize the losses of enzymatic activity during package, transportation, storage, and application. Enzymes are often exposed to humid, hot, or oxidative conditions in industrial applications, such as in detergents, textile finishing formulations, and food and beverage processing. Good formulations enhance stability by counteracting the primary factors of the deactivation especially on the catalytic-site, and the denaturation of the protein molecules. Denaturation of protein usually occurs by physical unfolding of an enzyme's tertiary molecular structure under thermal or chemical stress. Some time the slightly trace amount of proteases might cause the purified protein gradually degraded and losing its activity. Once an enzyme begins to unfold it becomes dramatically more vulnerable to deactivation and proteolysis. To minimize the unfolding of enzyme molecules, the formulations provide a protective environment so as to keep enzyme in a compact protein structure; this is done most effectively by preferential exclusion of water from the protein surface to stabilized enzyme molecules by adding water-associating compounds such as sugars, polyhydric alcohols, and lyotropic salts. The best ways to combat active site inactivation are to ensure sufficient amount of any required cofactors, e. g. , by adding protease inhibitors in the formula, and excluding the oxidative or reactive agents from the formulation.

A formulation should meet some other requirements, including preserving against microbial contamination, avoiding physical precipitation or haze formation, minimizing the formation of sensitizing dusts or aerosols, and optimizing the final enzyme products to meet the esthetic criteria such as the avoidance of abnormal color and odor. Many of these problems are best addressed by focusing as far upstream as possible, including the rational choice of raw materials in the fermentation and enzyme recovery process.

It emphasizes here, for individual new enzyme, one must make great efforts on the integration of purification techniques before obtaining the pure product of enzyme.

Summary

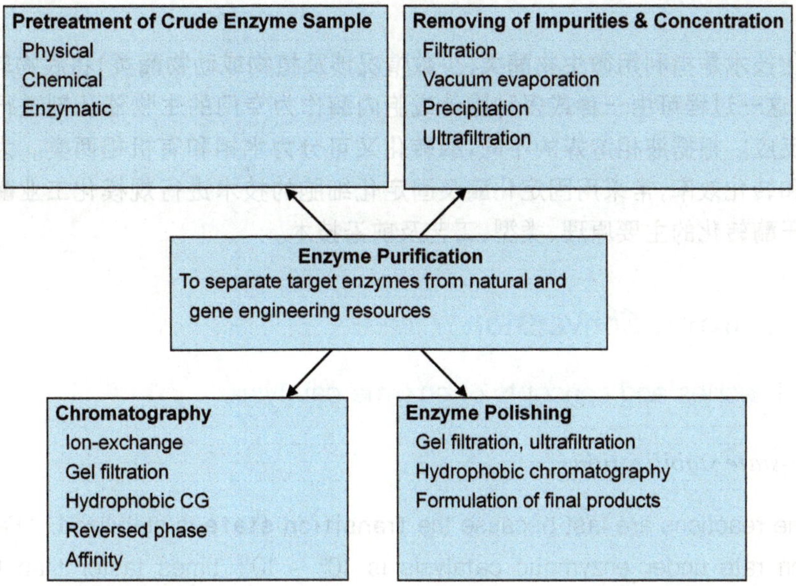

Pretreatment of Crude Enzyme Sample
Physical
Chemical
Enzymatic
Removing of Impurities & Concentration
Filtration
Vacuum evaporation
Precipitation
Ultrafiltration
Enzyme Purification
To separate target enzymes from natural and gene engineering resources
Chromatography
Ion-exchange
Gel filtration
Hydrophobic CG
Reversed phase
Affinity
Enzyme Polishing
Gel filtration, ultrafiltration
Hydrophobic chromatography
Formulation of final products

Chapter 14 Enzymatic Conversion

酶转化技术是指利用微生物酶类(少数情况涉及植物或动物酶类)将底物转化成为产物的过程,这一过程可由一种或多种胞外或胞内酶作为专门的生物催化剂进行的一种或系列化学反应。根据液相溶媒的不同,酶转化又可分为水相和有机相两类。为了提高酶的利用率和转化效率,常采用固定化酶及固定化细胞的技术进行规模化工业酶转化。本章介绍用于酶转化的主要原理、类型、工艺及前沿技术。

14.1 Enzymatic Conversion

14.1.1 Theories and concepts of enzyme catalysis

Transition-state stabilization

Enzyme reactions are fast because the **transition state** is stabilized. "Fast" means the reaction rate under enzymatic catalysis is $10^8 \sim 10^{12}$ times faster than that in the absence of an enzyme (Fig. 14.1). Thus, as indicated by the arrow sense, an enzyme manages to lower the transition-state energy from an un-catalyzed reaction to the enzyme-substrate complex ES.

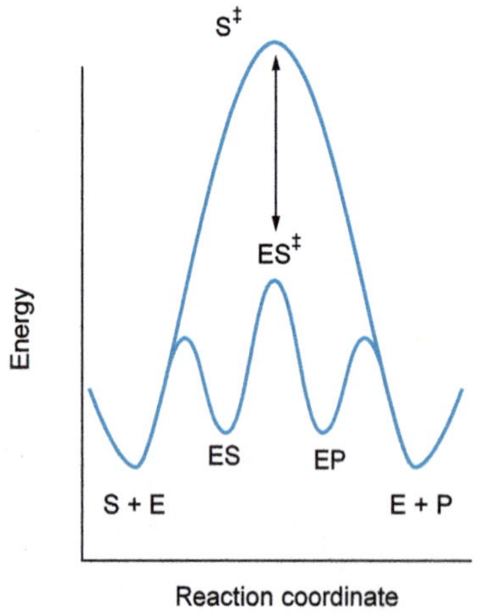

Fig. 14.1 Energy profiles showing a lowering of the transition-state energy from an un-catalyzed reaction (upper curve) to enzyme-catalyzed (lower curve) reaction

One of the best available examples of an enzyme's transition-state stabilization comes from the proteases that hydrolyze peptides (Fig. 14.2). In this hydrolysis, the amide's carbonyl oxygen is seen to become fully anionic in the tetrahedral intermediate. This implies that the oxygen assuming a partially anionic charge in the transition state leads to the intermediate. The developing anionic charge is then stabilized by hydrogen-bonding in the so-called "oxyanion hole" around the active site. What remains less clear would be the magnitude of (a) the partial negative charge on the carbonyl oxygen at the transition state and (b) the rate reduction resulted from the hydrogen-bonding within the oxyanion hole.

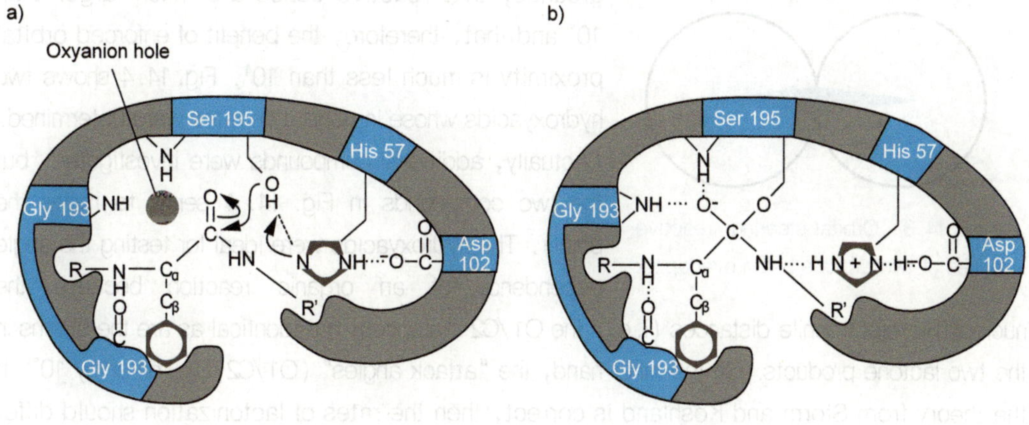

Fig. 14.2 Transition-state stabilization of a developing anionic charge on a carbonyl oxygen within the oxyanion hole of a protease

Entropy

Entropy is a popular rationalization for fast enzymatic reactions. We can witness the quote from Biochemistry by R. H. Abeles, P. A. Frey, and W. P. Jencks: "We can conclude that one effect of substrate binding at the active site of an enzyme is to increase the probability for the reaction and decrease the requirement for loss of entropy, and this can account for large rate increases in the reaction of substrates." The idea here is that bringing two reactive entities together prior to reaction normally causes a loss of entropy that impairs the reaction rate. When, however, a substrate is held tightly at an active site of an enzyme, there is no need to freeze translational and rotational motions, and thus the ensuing reaction between enzyme and substrate is accelerated. Enzyme catalyses are, according to this rationalized concept, all in the principle of entropy.

Orbital steering

Orbital steering, proposed by D. R. Storm and D. E. Koshland in 1972, was a clever explanation for enzyme catalysis. The idea is that a successful reaction occurs only after the overlap of the "reactive cones" of the two relevant orbitals. These reactive cones are represented by the dark wedges in Fig. 14.3. Simple geometric considerations show that if the cones' solid angles are only 10°, then bringing the two cones into contact results in a 10^4 fold rate increase (relative to when the reactive surfaces cover the entire spheres, and thus all random collisions are productive). If the reactive cones are indeed small, and if two such pairs of overlapped cones are present at the active site, then a 10^8 fold of catalytic activity, typical of many enzymes, can be achieved.

Orbital steering caused an immediate negative outcry. It was claimed, on theoretical

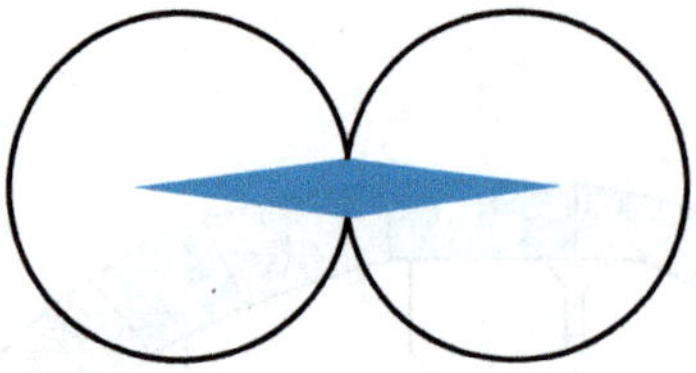
Fig. 14.3 Orbital steering: reactive cones of two orbitals being overlapped

grounds, that reactive cones are much larger than 10° and that, therefore, the benefit of enforced orbital proximity is much less than 10^4. Fig. 14.4 shows two hydroxyacids whose lactonization rates were determined. (Actually, additional compounds were investigated, but the two compounds in Fig. 14. 4 serve to make the point). The hydroxyacids were ideal for testing the angle dependency of an organic reaction because the nucleophile/electrophile distances (i. e., the O1/C2 distances) are identical as are the strains in the two lactone products. On the other hand, the "attack angles" (O1/C2/C3) differ by 10°. If the theory from Storm and Koshland is correct, then the rates of lactonization should differ by many orders of magnitude; but, on the contrary, the hydroxyacids lactonize at the same rate. The angle dependence clearly encompasses reactive cones that are much larger than 10°, thereby invalidating, at least for carbonyl additions, the orbital steering concept.

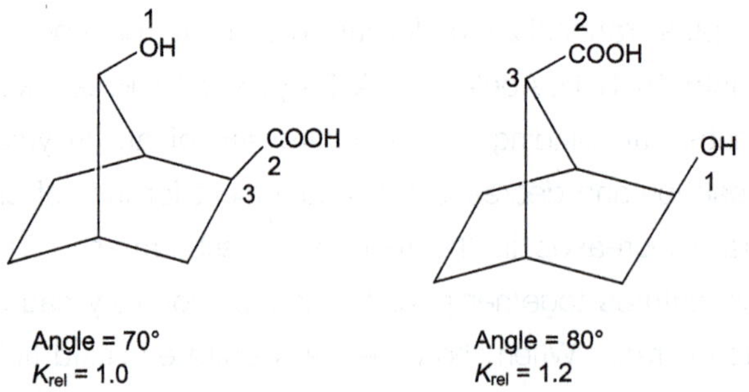

Fig. 14.4 Two hydroxyacids used to test the orbital steering concept

Proximity

Bruice put out a term "**proximity**" as the common sense parameter in 1976. It is intuitively reasonable that both intramolecular and enzymatic reactions are fast because the reactive entities are held close to each other. Since intramolecular reactions can be much faster than their intermolecular counterparts, it was quite natural to relate enzymatic reactions to intramolecular ones

14.1.2 The application of enzymatic conversion process

Applications examples in the chemical and pharmaceutical sectors

Chemical manufacturing processes are normally well established and the current overall economic advantages of biotechnological processes are rather small. New biocatalytic processes should be developed with decisive improvements, integrated into chemical and downstream processes, and performed with high rates at high substrate and product concentrations. Nevertheless, enzyme-based processes have already been quite prominent in the production of fine-chemicals and pharmaceuticals. As these product groups often include new and/or chiral compounds, it is easier for biotechnology to compete with traditional organic chemistry methods, especially if a new process is more economical and raw materials and other resources are used more efficiently. It is not surprising that only one enantiomer of a chiral drug has the desired biological activity, whereas other isomers may have no or an undesired activity. The targets of pharmaceutical compounds, such as cell-surface receptors and enzymes, are normally the chiral molecules. Because of the regulatory hurdles, the need to extend patents and the requirements to improve safety and effectiveness of their drugs, most manufacturers will increase the use of chiral molecules. The worldwide sales volume for single enantiomer drugs has already exceeded \$ 100 billion, and more than half of all new drugs reaching the market during the next five years are projected to contain optically pure active ingredients.

An important issue in the pharma sector is the large number of compounds that must be tested for biological activity to find a single promising lead. Combinatorial biocatalysis has received much attention here, as it could add a level of complexity to the diversity of existing chemical libraries or could be used to produce libraries *de novo*. An example is the use of glycosyltransferases to change the glycosylation pattern of bioactive compounds.

Biocatalytic processes

Enzyme technology has entered a phase in which new technologies, an increased understanding of fundamental molecular biology and bioinformatics are beginning to shape the discovery, development, purification, and application of biocatalysts to a much greater extent. This development is leading to new enzyme applications and increases the impact of enzyme technology in industries. Fig. 14. 5 shows the different steps in the development of a biocatalytic process and some of the issues that play a role in the different stages:

Screening stage: The biocatalysis cycle starts with the identification of a target reaction or existing industrial process that can benefit from the application of enzyme technology. The next step is the selection of a suitable biocatalyst. The characteristics of the enzyme (picked from a rather limited collection) often severely constrained the operation of the process. However, new screening protocol now makes it possible to start with the optimal parameters of the process, establish the specifications of the desired biocatalyst, and shift through the natural biodiversity for the ideal biocatalyst. In addition, the available biocatalysts can be discovered by screening functional extremophiles which

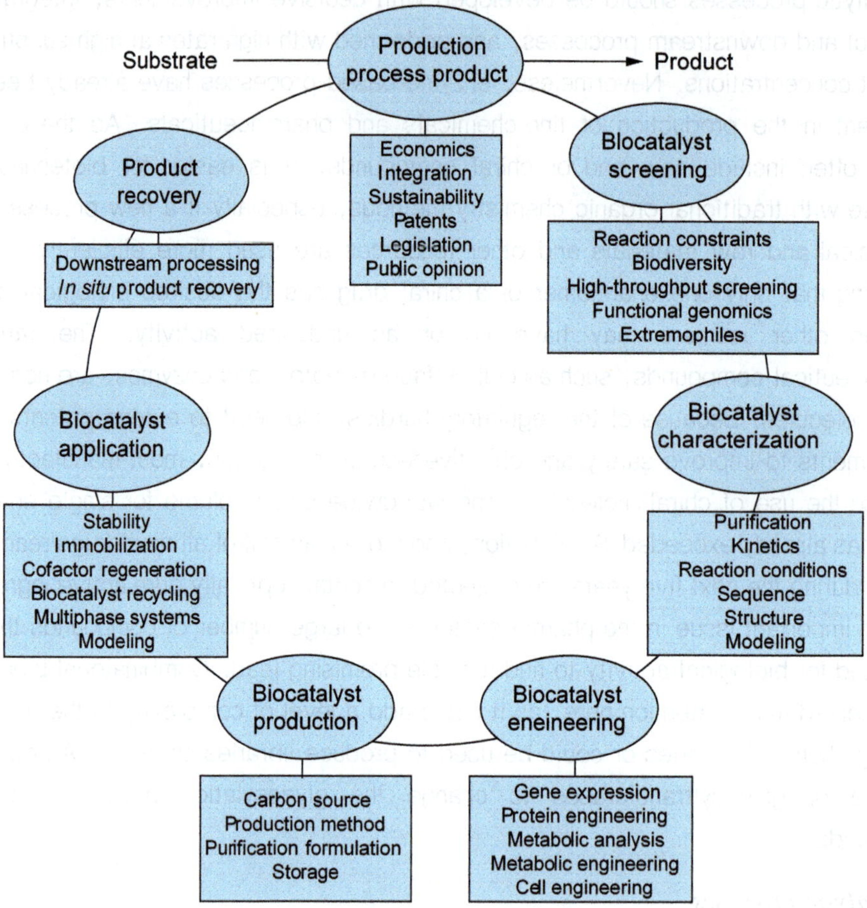

Fig. 14.5 The process cycle of biocatalysis

are tolerant to extreme operating conditions, and the physical limitations of proteins by directed evolution or gene shuffling. **Biocatalyst engineering**: Rapid developments in DNA sequencing technology make it possible that all industrially relevant organisms will be completely sequenced in the near future, and provide immediate access to useful genes provided that the link between sequence and function can be established. Within a few years, the above developments will make templates available to model the structures of most new enzymes. This will help protein engineering make specific changes that can not be selected by molecular evolution, such as surface structures for immobilization. At the same time, directed evolution can now provide us with enzymes tailored to the production process.

Artificial enzymes and chemocatalysts which are designed on the basis of knowledge gained from enzymatic catalysis are also introducing rapidly "evolving", and may well complement or replace natural enzymes in the future. Even the cofactors can be modified by chemical means; introducing new metals or modified coenzymes to an enzyme might result in new catalytic activities. On the fundamental level, sequence and structure knowledge may be combined with metabolic pathway analysis and engineering to construct strains that direct their carbon flow towards specific compounds for commercial production of chemicals. A well-known example is the biotechnological production of **indigo**. A process was brought close to commercialization by industry. Recently, the recombinant production strain was further modified to increase indigo production and to eliminate a by-product of red pigment. **Feasible application**: In some cases, enzymes are cheap and with food-grade, which can simply be added to a product stream or a product. However, many commercially available enzymes are too costly for the intended applications, even if they can be recovered and reused by immobilization. Therefore, reduced enzyme prices may dramatically increase the number of applications and even enable large-scale processes such as the production of ethanol from lignocellulosic waste. In the case of thermozymes, heterologous expression in *Escherichia coli* has an additional advantage that these proteins can be easily purified by a simple heat treatment: the host proteins precipitate while the thermozymes remain soluble. Even the formation of inclusion bodies may be turned into an advantage as protein refolding methods are further developed.

Still many enzymes are expensive but must be recovered and reused, for example, by immobilization. The material (EUPERGIT) remains one of the most popular solid supports and a recent review summarizes examples of biotransformations, including key process data, immobilization efficiency, and operational stability. The biocatalyst itself can now be changed to fit the reaction constraints of an "ideal" process, using a choice of technologies to enhance the biocatalyst characteristics. In the end, however, thermodynamics are not influenced by the properties of the enzyme. This means that enantiomeric excess and conversion often have to be improved by technical solutions, such as chemical or enzymatic racemization of one enantiomer, in situ product removal by adsorbtion and the use of biphasic systems (ionic liquids or organic solvents). In addition, it is possible to obtain superior yields by using enzyme variants that exhibit optimal activity under thermodynamically favorable conditions. Similar approaches were used in industrial processes by BASF, DSM and Lonza: choosing the acyl donor in transesterifications so that the reaction becomes irreversible; choosing the pH so that the cyanohydrin substrate racemizes; and choosing the pH so that a desired coupling reaction out-competes a hydrolysis reaction.

14.1.3 Reactors for enzymatic catalysis

An **enzyme reactor** consists of a vessel, or series of vessels, used to perform a desired conversion by enzymatic means. Enzymes are attached to solid matrix (e. g. cellulose fibers), and substrates flow into the reaction vessel and then are converted into products which then flow out, with the enzymes remaining attached to matrix. Compared to chemical reactor, enzyme reactor requires low temperature and low pressure with little consuming energy.

A number of enzyme reactors with different features are briefly introduced as follows: **Stirred Tank Reactors (STR)**, which contain all of the enzymes and substrates until the conversion is complete; and generally consists of a tank containing a stirrer, among STR: 1) **Batch Stirred Tank Reactor (BSTR)** is suitable for free enzymes and immobilized enzymes to perform catalysis; 2) **Packed Bed Reactor (PBR)** contains a settled bed of immobilized enzyme particles, and it is suitable for immobilized enzymes; 3) **Fluidized Bed Reactor (FBR)**, in which the flow of gas and/or substrate keeps the immobilized enzyme particles in a fluidized state, and it is suitable for immobilized enzymes. 4) **Bubble Column Reactor (BCR)**, the reactor filled with water and gas bubbles are injected from the bottom. It is suitable for free enzymes and immobilized enzymes. 5) **Membrane Reactors (MR)**, there are all kinds of membrane reactors which hold enzyme on the membrane film or rotate membrane sheet, and the substrate would be converted into product by flowing through the enzyme immobilized membrane. So membrane plays both support and filter functions. 6) **Recycle Reactor (RCR)**, in which a portion of the product stream is recycled and mixed with the inlet flow to the reactor, to allow the remained substrate being further converted. Recycling packed column reactor allows the reactor to operate at high fluid velocities, and allows an insoluble substrate to be completely converted.

14.2 Immobilization of Enzyme and Cells

14.2.1 Immobilization of enzyme

The principle of enzyme immobilization

Immobilized enzymes are used in organic syntheses to fully exploit the technical and economical advantages of biocatalysts based on isolated enzymes. Immobilization enables the easy separation of the enzyme catalyst from the reaction mixtures, and can lower the costs of enzymes dramatically.

Man-made usage of binding enzymes to solid materials goes back to the 1950s, when immobilized enzymes with restricted mobility, were first prepared intentionally. Immobilization was achieved by the inclusion into polymeric matrices or binding to carrier materials. Considerable efforts were also put into the cross-linking of enzymes, either by cross-linking of protein alone or with the addition of inert materials. In the course of the last decades numerous methods of immobilization have been developed on a variety of different materials. Binding to pre-fabricated carrier materials appears to have been the preferred method so far. Recently, cross-linking of enzyme crystals has also been reported to be an interesting alternative.

Immobilized enzymes are currently the object of considerable interest. This is due to the expected benefits over soluble enzymes or alternative technologies. The number of applications of immobilized enzymes is increasing steadily. Occasionally, however, experimental investigations have produced unexpected results such as a significant reduction or even an increase in activity compared with soluble enzymes. The properties of immobilized enzyme preparations are governed by the properties of both the enzyme and the carrier material. The specific interaction with the latter provides an immobilized enzyme with distinct chemical, biochemical, mechanical and kinetic properties (Fig. 14.6).

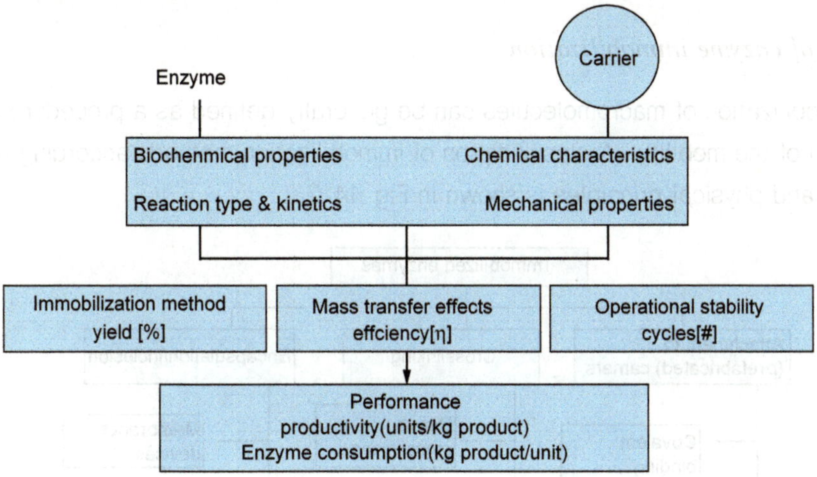

Fig. 14.6 Characteristics of immobilized enzymes

Of the numerous parameters which have to be taken into account, the most important parameters are outlined in Tab. 14. 1.

Tab. 14.1 Selected characteristic parameters of immobilized enzymes

Enzyme	**Biochemical properties** Molecular mass, prosthetic groups, functional groups on protein surface, Purity (inactivating/protective function of impurities), Kinetic parameters, Specific activity, pH, temperature profiles, Kinetic parameters for activity and inhibition, Stability against pH, temperature, solvents, contaminants, impurities
Carrier	**Chemical characteristics** Chemical basis and composition, functional groups, swelling behavior, Matrix volume of and pore size, chemical stability, mechanical properties, Mean wet particle diameter, single particle compression behavior, flow resistance (for fixed bed application), sedimentation velocity (for fluidized bed), abrasion (for stirred tanks)

Continued

Immobilized enzyme	**Immobilization method** Bound protein, yield of active enzyme, intrinsic kinetic parameters; **Mass transfer effects** Partitioning (different concentrations of solutes inside and outside the catalyst particles), external and internal (porous) diffusion; **Stability** Operational stability, storage stability; **"Performance"** Productivity, enzyme consumption

Methods of enzyme immobilization

Immobilization of macromolecules can be generally defined as a procedure leading to restriction of the mobility. A classification of immobilization methods according to different chemical and physical principles is shown in Fig. 14.7.

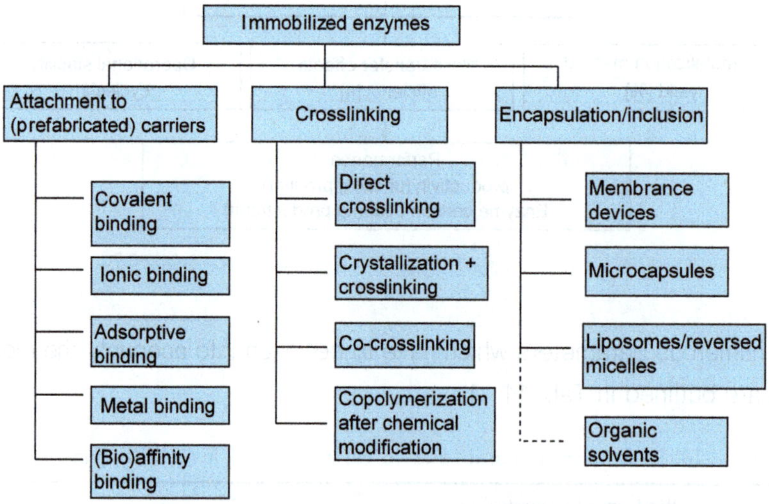

Fig. 14.7 Classification of immobilization methods

The immobilization procedure should be conducted in a way that allows the enzyme to maintain its active conformation and necessary catalytic flexibility, and catalytically essential residues of the enzyme should be undisturbed and conserved. Furthermore, by immobilization novel characteristics can be induced to make enzymes tolerant to even harsh reaction environments. Several factors effecting critically on the immobilization of enzymes onto supporting materials are introduced:

(1) A suitable activation of the enzyme for the immobilization process is performed prior to binding. This approach often suffers from significant loss of activity, because the

protein is modified by highly reactive chemical compounds that are often not strictly group specific and may alter catalytically or structurally essential residues. Also intra-and intermolecular cross-linking has to be considered.

(2) The support is modified and activated. The native enzyme is bound in a subsequent step under well-defined conditions using the natural reactivity of the molecule. This is the most prominent technique to covalently bind enzymes to carrier surfaces.

(3) A bi-or multifunctional coupling agent is used to mediate between carrier and enzyme functional groups. This can also lead to intra-and intermolecular enzyme cross-linking.

(4) The enzyme is genetically modified by DNA cloning techniques to generate an engineered protein with "(bio) specific" groups, so that it can adsorb onto special carriers by (bio) affinitive binding.

14.2.2 Immobilization of cells

Principle of cell immobilization

Cell immobilization, with a history of over 25 years, is a technique to fix cells in a suitable matrix. Cell immobilization is somehow different from the **cell entrapment**, in that immobilized cells can be entrapped but also the cells are absorbed onto support materials. In comparison with immobilized enzymes, immobilized cells used as industrial biocatalysts possess a number of advantages, especially the economical one. A substantial advantage is the elimination of expensive enzyme purification and isolation. Surveys of the preparation and utilization of immobilized cells point to a general shift of interest in both applied and basic researches, especially in the field of industrial biotransformation. Immobilized cells can be repeatedly or continuously used for the production of industrially and pharmaceutically valuable compounds including metabolites, intermediates (amino acids, peptides, sugars, etc.) and for the stereo-specific resolution of optically active compounds as well.

In comparison with immobilized enzymes, immobilized cells have provided new possibilities since they can be used as natural, water-insoluble carriers with required enzyme activities. The evident economic advantages are accentuated by the relatively simple technology of preparation of immobilized cells and by the fact that some enzymes display the highest activity when physiologically associated with cell structures, such as the plasmalemma, the cell wall, sub-cellular organelles and the cytoplasm. The ability of immobilized cells and sub-cellular organelles to retain partial or entire sequences of metabolic reactions after gentle and suitable chosen immobilization permits the use of microbial, plant or animal cells as biocatalysts in controlled synthesis of a number of industrially important compounds, such as hormones, alkaloids and enzymes.

An important factor influencing the use of immobilized cells is the selection of a suitable organism to ensure a high production of desirable compounds and high biotransformation ability or to possess a high initial specific activity of desirable enzyme (s). This requires the knowledge of regulation of synthesis and activity of the enzyme(s) in the organism. The biosynthesis and activity can then be influenced by a controlled cultivation. Another important tool is represented by genetic methods, such as strain improvement, selection and concentration of high-production cell clones or gene manipulation.

Immobilization is now a well-established technique and includes many industrial applications. Immobilized plant cells have been employed to perform biotransformation and reported to have higher production rates than freely suspended cells. For example, immobilized Capsicum cultures, treated with precursors, accumulated more quantities of biotransformed compounds than freely suspended cell cultures. Compared with plant cell immobilization, the immobilization of microorganism is less well developed in terms of large-scale application, although there are great potential to be further developed.

Immobilization models

Cell immobilization can be complemented without water-insoluble carrier. This mode (requiring no organic, inorganic, artificial or natural water insoluble carriers) is the simplest and is employed for industrial biotransformation by microbial cells (bacteria and microscopic fungi). The resulting immobilized cell preparations are either single cells with required stabilized enzyme activities, whose sedimentation rate does not substantially differ from that of native cells, or cell aggregates — particles consisting of mutually immobilized cells (covalently bound to each other), whose sedimentation properties are substantially improved by immobilization. The principles of this model are shown in Fig. 14.8.

Physical entrapment of cells into various polymer materials

Cell immobilization is mostly done by physico-mechanical entrapment of cells in different types of polymers; the immobilized particles have a variety of shapes and forms and highly favorable sedimentation properties. The easy separability from reaction media and good hydrodynamic properties of the particles facilitate the performance of the desirable biochemical processes in a continuous mode.

The entrapment of cells (Fig. 14.9) can be performed by a number of techniques according to the selection of the polymer material, its properties (e.g., solubility) and mechanism of immobilization: 1) Precipitation of cells with polymers insoluble in water and completely soluble in organic solvent (e.g., cellulose and its derivatives); 2) Entrapment of cells in gels soluble in hot water-thermoreversible gels (agar, gelatin); 3) Entrapment of cells in gels combined with hardening of gel particles by a forming ionic network (e.g.,

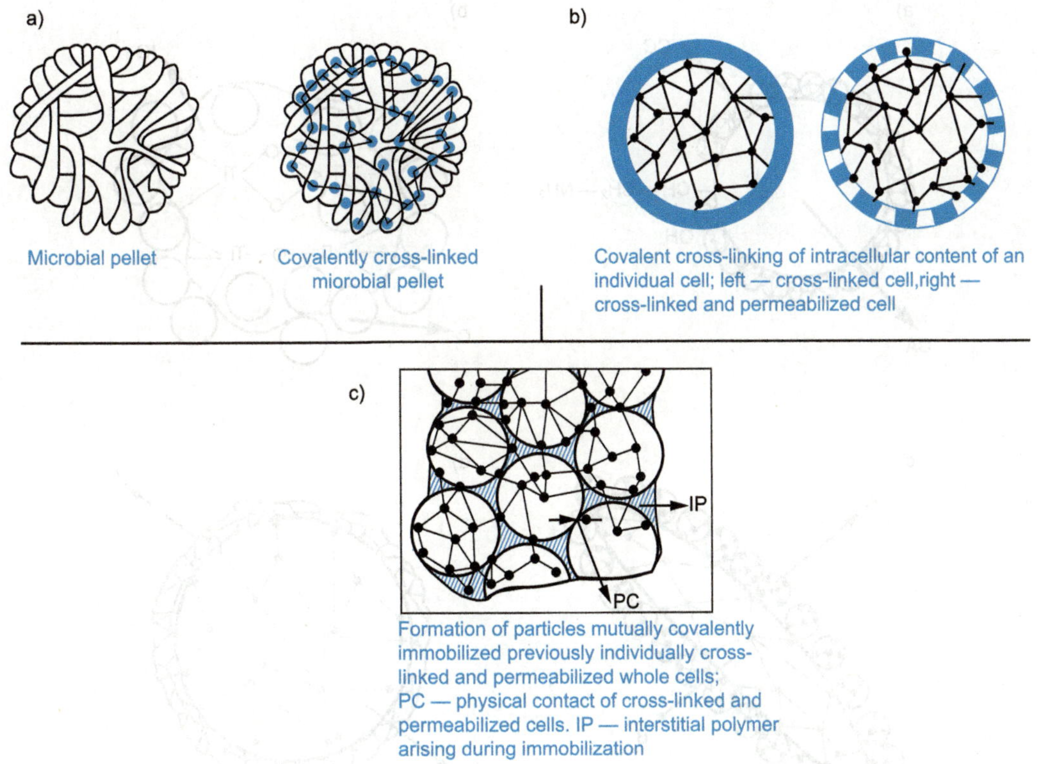

Fig. 14.8 Different mechanisms for cell immobilization

calcium alginate); 4) Entrapment of cells in a forming polymer network-copolymeration (e.g., acrylamide); 5) Entrapment of cells by dispersing them in natural materials-eomplexation (e.g., collagen); And 6) Combination of physical cell entrapment with chemical pre-or post-stabilization of cells or modification of particle surfaces with covalent crosslinking (e.g., glutaraldehyde).

Cell immobilization on particles of water-insoluble carriers

Particles, of various synthetic or natural materials, both organic and inorganic, with different size distributions, surface specificity, porosity, swelling, etc., are used as carriers in this type of immobilization. The techniques include physical sorption of cells on the carrier, physico-chemical (ionic, polar) bond of cells on a carrier possessing an ion-exchanger character, or chemical, coordination or covalent binding of cells on the carrier either directly or by means of spacers after previous chemical activation or modification of the carrier, or combinations of these techniques (e.g., physical sorption of cells on a carrier followed by covalent cross-linking)(Fig. 14.9).

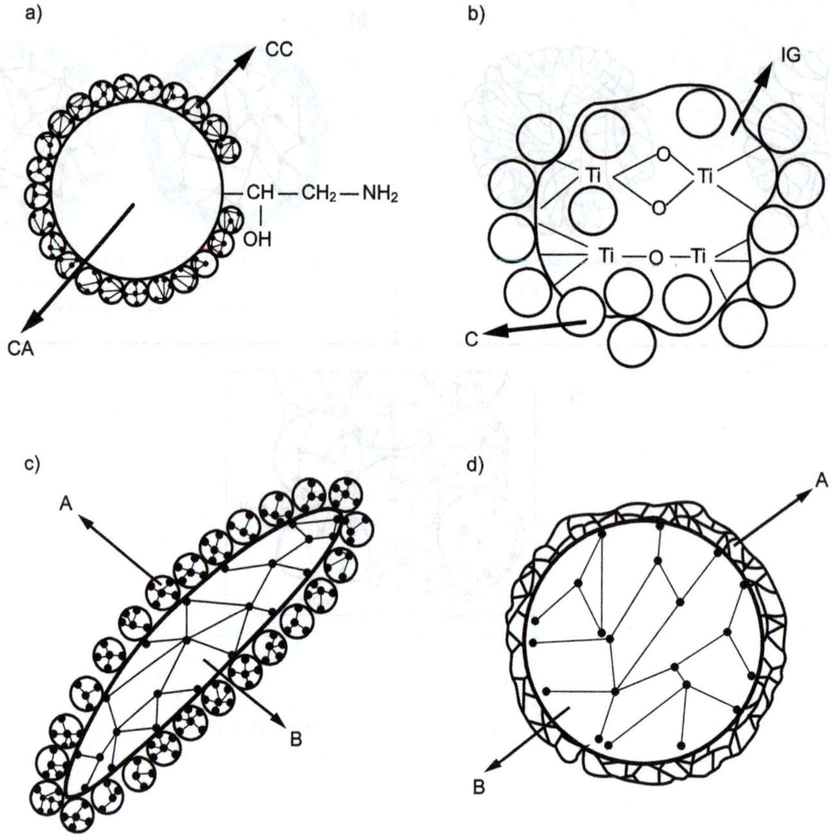

Fig. 14.9 Different mechanisms of cell immobilization

a) Physical and physico-chemical immobilization of cells on a carrier; b) Chemical immobilization of cells on a carrier and combined immobilization techniques; c) d) Immobilization of cells and enzymes on cell surfaces with microbial yells as carriers

Advantage and disadvantage of cell immobilization

In an immobilized system, the cell growth and production phases can be decoupled and controlled by chemical and physical stress conditions. This allows cells to be retained in the bioreactor for extended periods, by alternating rejuvenation/growth and secondary metabolite production cycles. Processing engineering problems may develop from the tendency of plant cells to aggregate, which can lead to blockages in pipes and openings and to the culture's rapid sedimentation, if it is not continually agitated. However the shear sensitivity of the culture means that mechanical agitation may be detrimental to cells and the culture can not be transported by using conventional pumps without significant loss of viability. Again immobilization of plant cells may be a solution to these problems and may offer a microenvironment protected from sustained shearing. The physical separation of the cells and medium in the immobilized system also makes for easy exchange of medium for the purposes of metabolic control or nutrient replenishment. The composition of the culture medium can be readily and continuously monitored by the sensors, which make the concentrations of CO_2, O_2, and sugar, etc. adjusted as required. Similarly, extracellular products can be harvested continuously by adsorption on a suitable resin, or by other means.

The main disadvantage of immobilization is that it is only of use with the cell lines which excrete the product of interest into the culture medium. Attempt to induce the release of products, which are normally retained within the cells by such techniques as permeabilization, have generally decreased cell viability to an undesirable extent, although reversible permeabilization of Catharanthus roseus using dimethyl sulfoxide (DMSO) was promising.

14.3 Enzymatic conversion in non-aqueous Phase

14.3.1 Principle of non-aqueous conversion

In nature, enzymes usually function in aqueous solution, for the conventional wisdom dictates that water is required for enzyme action; this conclusion stems from the fact that water participates (directly or indirectly) in all non-covalent interactions maintaining the native, catalytically active enzyme conformation. Hence, the removal of water should drastically distort the conformation and inactivate the enzyme; it is not surprising that most of the enzymology studies so far have used water as the catalysis medium. However, there are numerous advantages of conducting enzymatic conversions in organic solvents as opposed to water, from the biotechnological standpoints: 1) High solubility of most organic compounds in **nonaqueous** media; 2) Ability to carry out new reactions impossible in water because of kinetic or thermodynamic restrictions; 3) Greater stability of enzymes; 4) Easy recovery of product from organic solvents as compared to water; 5) The insolubility of enzymes in organic media which permits the reuse of enzyme, and thus eliminates the need for immobilization.

14.3.2 Non-aqueous mediums and applications

Aqueous-organic monophasic system, which is usually composed by water miscible organic solvents and small amount of water. The Advantages of this system would be: Shifting equilibrium towards the formation of the desirable product; increasing the solubility of the substrate; and inhibiting the side reactions and variation of the enzyme's characteristics (especially selectivity). For example, ethanol-water could be employed instead of water for isomerization of glucose to fructose. In conventional aqueous system, the yield is under 50% due to the limitation of equilibrium. For practical use, chromatography is needed to get a mixture with fructose content being above 50%.

In ethanol-water, the yield could be higher than 50% and the costly operation of chromatography could be avoided. For bioconversion of steroid, an aqueous-organic monophasic system was used to improve the solubility of the substrate.

(1) **Aqueous/organic biphasic system**, which is composed by water and water immiscible organic solvents. This system is advantageous to the In situ extraction of product and substrate to avoid the substrate and product limitations and the reverse reaction; leading to higher reaction rate, yield, enantiopurity of the product and longer lifetime of the catalyst; simplifying the product separation; and inhibiting the chemical non-selective reaction to enhance the purity of the product. For example, hexane/water biphasic system was used for asymmetric reduction of prochiral ketones to enantiopure chiral alcohols. Ethyl acetate/water was used as the reaction medium for oxynitrilase-catalyzed asymmetric transcyanation of prochiral ketones with acetone cyanohydrin for the preparation of chiral cyanohydrins, a group of compounds of great importance in the synthesis of many chiral drugs.

(2) **Microaqueous system** which is composed by water immiscible organic solvents and small amount of water which is not enough to form a real phase. This system enables many synthetic reactions such as esterification, interesterification, amidation and ammonolysis which could not proceed in aqueous phase due to equilibrium limitation (the dominant reaction would be the hydrolysis of the product rather than its formation). For example, 1-β-D-Arabinofuranosylcytosine (ara-C) could be enzymatically acylated in microaqueous system with high yield and region selectivity, while this reaction couldn't be carried out in water.

(3) **Reversed micelle system** is composed of water immiscible organic solvent, water and surfactants such as AOT (sodium di-2-ethyl-hexyl-sulphosuccinate). Be similar to the natural environment of enzyme, this system be used for basic studies on fundamental enzymology. Enzymes display "super-activity" in this system, for example, lipase could catalyze the asymmetric transesterification of racemic 2-octanol with vinyl acetate in reversed micelles with high enantioselectivity.

(4) **Supercritical fluids** means the fluids were kept at/near critical temperature and pressure; the most frequently used one is supercritical CO_2, which has been broadly employed in many enzymatic reactions based on the advantages of high diffusion coefficient, low viscosity, less mass transfer limitation, high density and solubility, easiness in modification of solvent properties, easiness in recovery of enzyme and product separation, non-toxicity and low cost of the solvent. For example, lipase could catalyze esterification, transersterification and hydrolysis reactions in supercritical CO_2, for the synthesis of esters, and kinetic resolution of racemic compounds.

(5) **Solvent free system.**

Using substrate as medium. Substrates and products are mixed as miscible liquids in this system, which has advantages of high substrate concentration; high reaction rate; operational stability of enzyme; easy product separation and avoidance of organic solvents (Not all substrates and products are miscible liquids at the reaction temperature). For example, Vinyl acetate is often used as both a solvent and an acylating agent in enzymatic esterification and transesterification reactions. Palm-oil mid-fraction (mainly 1, 3-dipalmitoyl-2-oleyl-glycerol) could be converted into cocoa butter-equivalent fat by lipase-catalyzed transesterification with palm-oil second stearin fraction (mainly 1,3-distearoyl-2-oleyl-glycerol) or 1,3-specific acyl exchange in the presence of methyl stearate. Esterase could catalyze the region-selective acylation of hydrophobilized sugars using a slight molar excess of molten fatty acids.

Eutectic mixtures. Two substrates are solids in the pure state. A liquid phase was spontaneously formed by mixing the substrates together as the mixture displays a dramatically lower melting point than its constituents. The formation of the eutectic and its melting temperature could be affected by the addition of small amounts of modifiers (water or organic solvents). This system is practiced in some cases, for example, a number of proteases (subtilisin, papain, chymotrypsin, thermolysin) retain their catalytic activity in eutectic media and high yield could be obtained in the synthesis of several bioactive and flavor peptides, meanwhile with another advantage of operational stability of the enzyme .

(6) **Ionic liquid containing system**. Ionic liquids are a group of salts, which stay in liquid state under or near room temperatures. They can be divided into hydrophobic and hydrophilic ones, and can form neat ionic liquid system (containing very small amount of water), ionic liquid/water biphasic system and ionic liquid-water monophasic system. The Advantages of this system include: Low vapor pressure, environmentally friendly (green solvent); easy to be modified in solvent properties (designed solvent); high stability; high enzymatic selectivity and stability and in some cases; high enzymatic activity.

Many kinds of biocatalysts displayed catalytic activity in ionic liquid containing systems, including hydrolase (lipase, protease etc), oxidoreductase (alcohol dehydrogenase, peroxidase), lyase (oxynitrilase) and microbial cells (yeast cells). For example, papain could catalyze the asymmetric hydrolysis of amino acid ester D, L-p-hydroxyphenylglycine methyl ester in ionic liquid 1-butyl-3-methylimidazolium tetrafluoroborate containing system with high enantioselectivity. When performed at reduced pressure, the reaction rate and the enantiopurity of the product could be further enhanced by the removal of methanol formed by the reaction system.

In summary, on-aqueous enzymatic conversion is of great potentials in the field of biochemical industry.

Summary

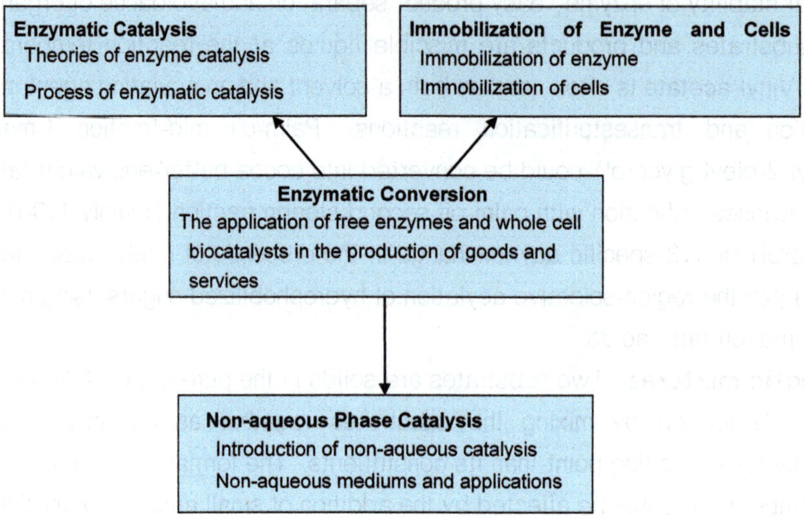
Enzymatic Catalysis
Theories of enzyme catalysis
Process of enzymatic catalysis
Immobilization of Enzyme and Cells
Immobilization of enzyme
Immobilization of cells
Enzymatic Conversion
The application of free enzymes and whole cell biocatalysts in the production of goods and services
Non-aqueous Phase Catalysis
Introduction of non-aqueous catalysis
Non-aqueous mediums and applications

Chapter 15 New Approaches of Enzyme Technology

近年来酶技术得到了长足的发展，酶的研究和应用领域得到了进一步的拓宽。酶技术进展涉及：利用基因克隆及化学修饰等手段，改造提高目标酶分子的活性和稳定性，或改变酶的催化特性等；利用酶制造各种生物复合材料，应用于工业化生产；发现了非蛋白质酶类（核酶）及其特殊的功能；以及酶抑制剂的开发和应用等。酶技术的新发展对生命科学与生物工程的进步具有较大的推动作用。

15.1 Modification on Enzymes

15.1.1 Molecular modification of enzyme by cloning

According to the well known theory of enzymology, the function of an enzyme depends on its structure, and the structure definitely depends on the sequence of the peptide molecules. Therefore, the changing of DNA sequence coding for enzyme molecules would create some opportunity of alteration of enzymatic properties. The cloning procedures introduced in Chapter 4 can be applied to manipulation of molecular modification on any particular enzyme, and some of the specific methodologies are introduced as follows.

Site-directed mutagenesis

Mutagenesis is a broad term defining as the alteration of the genetic material of an organism and allowing the changed hereditary being passed to the offspring. **Site-directed mutagenesis** is that the amino acid sequence of a given enzyme or other protein molecules may be altered by deliberately and precisely mutating the cloned gene encoding that molecule. It is a very useful technique that can be used in the study of protein function, the identification of enzymatic active sites, and the design of novel proteins. With this technique, it is possible to exchange a single amino acid in the sequence of a protein for another amino acid with different chemical properties. In this way, the function of the protein can be altered by this specific amino acid at the particular site. The basic protocol for this process was developed by M. Smith, who was awarded the Nobel Prize in Chemistry in 1993 for his fundamental contributions to the establishment of oligonucleotide-based, site-directed mutagenesis and its development for protein studies and enzyme modification.

In order to carry out site-directed mutagenesis, a DNA primer must be designed at the site of interest. The primer segment should contain the necessary nucleotide differences in order to affect the change in the protein sequence. The technique has been described in detail in Chapter 4 of the Part I.

Directed evolution

Directed evolution has now been used for over two decades as an alternative to rational design for protein engineering. Protein function, however, is complex, and modifying enzyme activity is a tall order. We can now improve existing enzyme activity, change enzyme selectivity and evolve, *de novo*, the function using directed evolution. Mimicking natural, directed evolution aims to generate a protein with new activity by screening or selecting for the desired function from a large pool of protein variants. Here we look at what can now be achieved with directed evolution experiments — improving enzyme fitness for a given application, changing substrate selectivity and even evolving function *de novo*. Directed evolution has been and continues to be employed for: 1) Improving robustness and activity of enzymes: Enzymes generally show optimal activity in aqueous solution at neutral pH, and the robustness and activity of enzymes are needed to be improved for industrial and biotechnology applications. 2) Modifying substrate selectivity: The relationship between enzyme structure and substrate selectivity is complex. Natural enzymes typically are quite specific to their intended substrates; and it is often necessary to modify or broaden their substrate selectivity for industrial or biotechnology applications. However, modifying substrate selectivity of an enzyme is more difficult than improving the activity of an enzyme in different environments. 3) Evolving activity de novo: Although natural enzymes have already catalyzed many diverse chemical transformations, including carbon-carbon bond forming reactions, hydrolysis reactions and oxidation reactions, the ultimate goal of directed evolution from both a practical and an academic perspective is to induce the de novo evolution in many ways. Several methods such as error-prone PCR and DNA shuffling have been proved as powerful techniques for directed evolution of enzymes, which have also been introduced in Part I Chapter 4.

Fusion expression

In recent years, *Escherichia coli* gene **fusion expression systems** have circumvented many of the problems inherent in the use of this bacterium for the production of recombinant proteins. These systems also provide a powerful means for identifying peptides or proteins with desired binding specificities. Gene fusion technology continues to expand with the introduction of new fusion partners, purification and detection tags, cleavage reagents and ways to display peptides on the surface of bacteria. A protein or a peptide is located either on the C-or N-terminal of the target protein, which facilitates one or several of the following characteristics: 1) **Solubility (S)**—Fusion of the N-terminus of the target protein to the C-terminus of a soluble fusion partner often improves the solubility of the target protein. 2) **Detection (D)**—Fusion of the target protein to either terminus of a short peptide (epitope tag) or protein which is recognized by an antibody (Western blot analysis) or by biophysical methods (e. g. GFP by fluorescence) facilitates the detection of the resulting protein during the expression or purification. 3) **Purification** (**P**) —Simple purification schemes have been developed for proteins used at terminus which bind specifically to affinity resins. 4) **Localization** (**L**) —Tag, is usually located on N-terminus of the target protein, which acts as an address for sending protein to a specific cellular compartment. 5) **Expression** (**E**) —Fusion of the N-terminus of the target protein to the C-terminus of a highly expressed fusion partner results in a high level expression of the target protein.

15.1.2 Chemical modification

The explosion of enzymes in commercial and synthetic applications has drawn much attention on the enhancement of enzyme function and stability. Covalent chemical modification, the original method available for altering protein properties, has now re-emerged as a powerful complementary approach to site-directed mutagenesis and directed evolution for tailoring proteins and enzymes.

Enzyme cross-link

One of the challenges for the practical implementation of protein biocatalysts is to increase stability, particularly for non-aqueous applications. In this regard, the stable effects of both inter-and intra-molecular cross-linking enzymes with bi-and poly-functional reagents have been exploited. The use of glutaraldehyde for enzyme cross-linking (Fig. 15. 1) was first recognized by Quiocho and Richards, who employed this procedure to stabilize carboxypeptidase for X-ray structure determination.

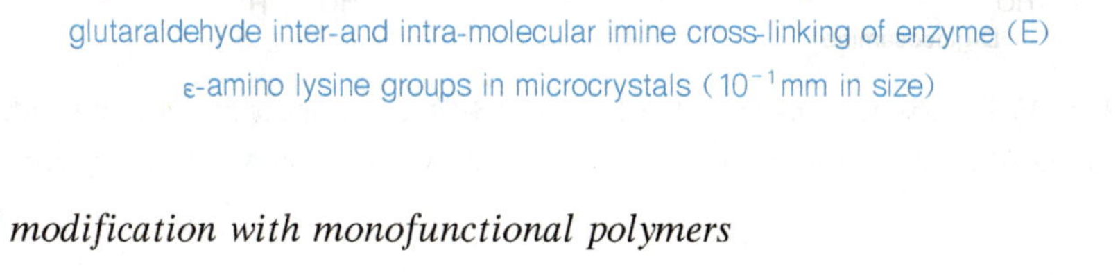

Fig. 15. 1 CLECs (cross-linked enzyme crystal) are generated from the glutaraldehyde inter-and intra-molecular imine cross-linking of enzyme (E) ε-amino lysine groups in microcrystals (10^{-1}mm in size)

Chemical modification with monofunctional polymers

In contrast to enzymatic cross-linking, chemical modification with monofunctional reagents permits the binding of specific monomeric or polymeric functionalities.

In particular, covalent modification with the amphiphilic polymer polyethylene glycol (PEG) has been employed to solubilize proteins for the applications in organic solvents, and is useful for the preparation of therapeutically active proteins with reduced antigenicity and enhancement *in vivo* stability. For example, the Candida rugosa lipase (CRL) treated with para-nitro-phenylchloroformate and cyanuric acid chloride activated PEG (Fig. 15.2) was found to exhibit enhanced stability in isooctane and the former was found to be more active than the latter.

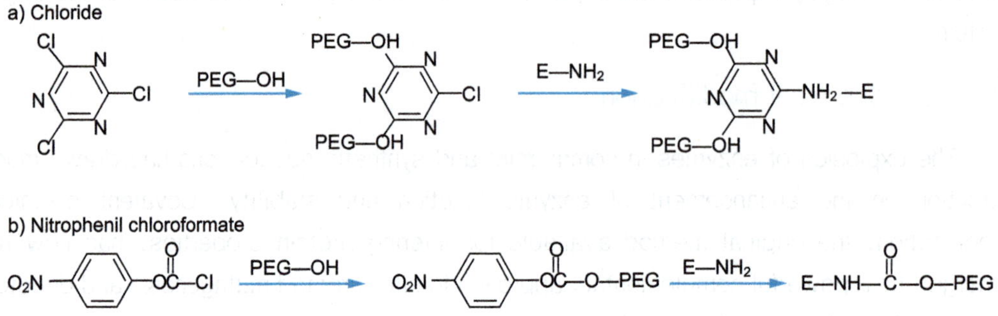

Fig. 15.2 Enzyme (E) modification with PEG with Chloride and Nitrophenol chloroformate

Chemical introduction of small moieties and atom replacement

Chemical introduction of small molecular functionalities has also proven to be beneficial. For example, non-glycosylated nuclease RNase A has been chemically coupled to D-glucosamine yielding mono-and di-glycosylated enzyme (Fig. 15.3); the Asp53 and Glu49 in RNase A were identified as the probable sites of glycosylation. The chemically monoglycosylated RNase A exhibited an 80% decrease in specific activities compared to the native enzyme but had enhanced thermal stability compared to both the unglycosylated RNase A and the natural *in vivo* glycosylated material, RNase B, which is glycosylated at Asn34 and contains five glycol-forms.

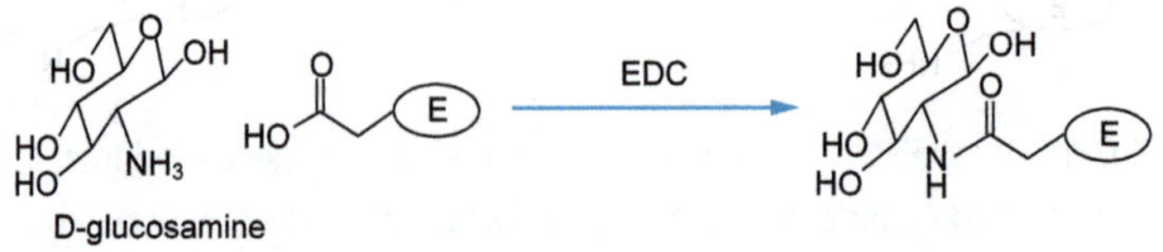

Fig. 15.3 Glycosylation using the 1-sthyl-3-(3-dimethylaminopropyl) carbodiimide hydrochloride (EDC) to mediate the coupling of carboxylate RNase A (E) to D-glucosamine

Cofactor introduction by chemical modification

The potential for chemical modification as a tool for the attachment of cofactors onto protein templates to induce new enzymatic activities was demonstrated by Kaiser. More recently, using standard solid-phase synthesis techniques, the Phe 8 residue of the C-peptide derived from RNase S was replaced with an unnatural amino acid incorporating the pyridoxamine phosphate (vitamin B_6) cofactor (Fig. 15.4). The chemically modified and reconstituted RNase S exhibited a seven-fold faster Cu(II)-assisted transamination of pyruvate to alanine for the first noncatalytic turnover of the enzyme than with the noncovalently linked cofactor.

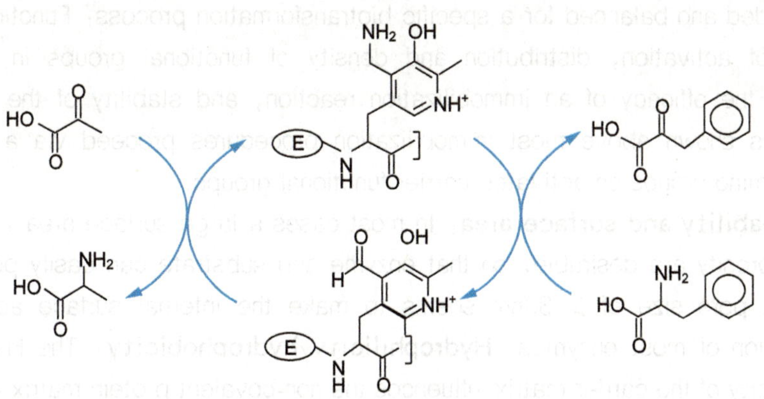

Fig. 15.4 Pyridoxamine chimera catalyzes transamination of pyruvate alanine and recycling of pyridoxal phenylalanine

15.1.3 Site-directed mutagenesis associated with chemical modification

In this method, the unnatural amino acids synthesized by chemists would be incorporated into a peptide sequence in order to modify the function of enzyme. The desirability of having a unique handle for protein modification has been recognized, as this permits the introduction of any desired molecules at a controlled position in the enzymes or proteins. Such a unique handle may be incorporated by site-directed mutagenesis and can subsequently be chemically modified to introduce an unnatural amino acid side chain. This combined site-directed mutagenesis with chemical modification approach has been applied to alter the catalytic properties of the alkaline serine protease subtilisin Badus lentus (SBL). The strategy involves the introduction of one cysteine residue at a key active-site position via site-directed mutagenesis, which is then thioalkylated with a methanethiosulfonate reagent (CH_3SO_2S-R) to give chemically modified mutant enzymes (CMMs) Methanethiosulfonate reagents, react specifically and quantitatively with sulfhydryls and thus do not react with any other protein functional groups, resulting in a homogenous product. SBL is an ideal template for this approach as it contains no natural cysteines.

15.2 Enzyme Carriers for Immobilization

15.2.1 General properties of enzyme carriers

Several enzymes have been immobilized to different kinds of carriers for environmental and industrial uses. The characteristics of the carriers have a strong influence on the performance of an immobilized enzyme. The following principles should be well comprehended and balanced for a specific biotransformation process: Functional groups: The type of activation, distribution and density of functional groups in the carriers determines the efficacy of an immobilization reaction, and stability of the carrier-fixed enzyme. As shown above most immobilization procedures proceed via a nucleophilic attack of amino groups on activated carrier functional groups.

Permeability and surface area: In most cases a large surface area ($>100m^2/g$) and high porosity are desirable, so that enzyme and substrate can easily penetrate into carriers. A pore size of $>30nm$ seems to make the internal surface accessible for immobilization of most enzymes. **Hydrophilicity/hydrophobicity**: The Hydrophilicity/hydrophobicity of the carrier matrix influences the non-covalent-protein-matrix interaction in type and strength. In addition, it can influence the adsorption, distribution and availability of the substrate and product. **Insolubility**: This is essential for a carrier to have the property of insolubility, not only for prevention of enzyme loss, but also to prevent the contamination of the product by dissolved matrix and enzyme.

Mechanical stability/rigidity: Depend on the type of reactors, if a stirred tank reactor is used, the material of the carrier should be stable against sheering force to minimize the abrasion. Fine particles (below 100~50mm) can lead to plugging of sieve plates and filters. **Size of support**: The particle size of the carrier will have an influence on filtration times from stirred tank reactors in a repeated batch mode. Furthermore, this factor is correlated to the performance in column reactors, regarding back pressure and flow rates. The size of spherical particles in the range of Φ150~300mm is preferred.

Resistance to microbial attack: During long term usage, the carriers have to be stable against microbial degradation or contamination. **Regenerability**: This property is of interest in the cases of expensive carrier materials, and recycling of the carrier is critically based on the regenerability of this material.

15.2.2 Inorganic carriers

Basically, carrier materials can be divided into those of inorganic and organic origins. Inorganic carriers (e. g. glass, silica gel, alumina, bentonite, hydroxyapatite, nickel/nickel oxide, titania, zirconia) often show good mechanical properties, thermal stability, and resistance against microbial attack and organic solvents. On the other hand, non-porous materials like metal and metal oxides only have small binding surfaces. Minerals usually display a broad distribution of pore size.

Silica gels are available under the trade names like Promaxon (Promat), Spherosil (Rhone-Poulenc) or Aerosil (Degussa). Silica compounds can be prepared with defined pore sizes and binding surfaces (controlled pore glass, CPG), but which suffer from high production costs and show limited stability under alkaline conditions. Furthermore, silica carriers are chemically inert and need activation and modification. Usually, they are treated with aminoalkyl triethoxysilanes to introduce amino groups, which can subsequently be activated for enzyme coupling reactions by a variety of different methods. The most common procedure is the reaction with dialdehydes to transform amino groups into reactive unsaturated aldehyde residues. In this way, penicillin G amidase was the first coupled to dextran (see below); the modified enzyme showing significantly increased thermostability, and this preparation was then immobilized on amino-activated silica gels.

Organopolysiloxanes constitute macroporous inorganic synthetic materials carrying organic functional groups and are commercially available (Deloxan®, Degussa AG). They combine chemical resistance of the matrix and high loading capacity. Amino functional carriers are the most interesting because they allow covalent coupling via glutardialdehyde' reaction. The straightforward reaction of the aldehyde groups with the primary amino groups of the carrier and the enzyme has been questioned. Simple Schiff bases are not that stable and the reactivity of proteins with freshly distilled glutaraldehyde is remarkably low. So it was suggested that the reaction involves the addition of amino groups to ethylenic double bonds of a, b-unsaturated oligomers present in commercial aqueous glutardialdehyde solution.

15.2.3 Organic carriers

Naturally occurring organic carriers

Natural organic polymers-such as structural proteins (keratin, collagen), globular proteins (albumin) and carbohydrates-are cheap starting materials for the production of support materials and are available in large quantities. From this group, carbohydrates are of special interest, because they will be suspected on biological safety aspects like protein matrices isolated from animal sources, and they are highly hydrophilic which provides a desirable microenvironment for many enzymes. Alginate, carrageenan, chitin or chitosan (prepared from chitin by deacetylation) are particularly useful for encapsulating microorganisms by ionotropic, respectively, acidic gelation. Enzymes have been linked to carbohydrates simply by adsorption followed by a cross-linking. Among those materials, chitosan is of importance because of its primary amino groups that are susceptible for coupling reactions. Furthermore, porous spherical chitosan particles are commercially available (Chitopearl, Fuji Spinning) allowing noncovalent or covalent attachment of enzymes. This support matrix can be easily prepared and the activation methods have been summarized. Treatment with polyethyleneimine or with hexamethylenediamine and glutardialdehyde can improve the mechanical characteristics of the biocatalyst, which is poor otherwise. However, this is often accompanied by some activities' loss or increase of diffusional limitations.

Dextran and agarose have to be cross-linked, for instance, with epichlorohydrin, to improve their mechanical characteristics and compressibility. The most widespread activation method is the cyanogen bromide method, yielding isourea and imidocarbonate functions which react with amine groups of enzymes to form N-substituted carbamates. Cyanogen bromide was more easily substituted by 1-cyano-4-dimethylaminopyridinium tetrafluoroborate (CDAP). However, it should be kept in mind that this linkage is unstable at the range of pH < 5 and > 10. A different way to activate carbohydrates for covalent coupling is the periodate cleavage that generates aldehyde functions. Reaction of the aldehyde moieties with amine groups of enzymes to Schiff bases is fast. A stable linkage can also be achieved by reduction with borohydrides. In the modification step, Sepharose was first transformed into glyoxyl-Sepharose by etherification with 2, 3-epoxy propanol, then further oxidized by periodate to yield an agarose-aldehyde gel, which allows fast coupling reactions with the amino groups.

Commercially available preactivated matrices, for example, N-hydro xysuccinimide ester or hydrazide, Affi-Gel (BioRad) and different activated Sepharose derivatives (Amersham Pharmacia Biotech) had been primarily developed for affinity chromatography. Generally speaking, those carriers are too expensive to use as biocatalyst matrix.

Cellulose is also an acceptable support and can be activated in a similar way, but binding capacity of cellulose to the enzymes is rather low as compared to agarose; however, it is inexpensive and commercially available in fibrous and granular forms. The drawbacks of these kinds of materials are their rather small sizes, which impairs the use in rapid high pressure applications.

Synthetic organic carriers

Synthetic organic polymers display the greatest variability with regard to physical and chemical characteristics. In principle, they can be adapted to the requirements of nearly any enzymatic processes. Furthermore, they are inert to microbial attack. The commercially available products are related to purely adsorptive resins, ion exchangers with a variety of different basic and acidic groups, or pre-activated supports carrying, for instance, epoxide (Eupergit®, Roehm Pharma) or azlactone groups (Emphaze™, 3M). The main synthetic polymers are polystyrene, polyacrylate, polyvinyl, polyamide, polypropylene and copolymers based on maleic anhydride and ethylene or styrene, polyaldehyde, and polypeptide structures.

Polystyrene was the first synthetic polymer used for enzyme immobilization. The binding of enzyme on polymer material occurs mainly through the adsorptive forces, and is favored at low salt conditions. However, the hydrophobicity of the matrix often leads to partial denaturation of the enzyme during the binding process and therefore the enzyme activity might be lost partially. By combining adsorption and cross-linking, penicillin G amidase was covalently coupled to a fairly hydrophilic polyacrylic ester (XAD-7), which had been pre-coated with glutaraldehyde at alkaline pH. The same enzyme was adsorbed on macroporous polymethacrylate, which was optimized with regard to monomer-co-monomer composition, and cross-linked by glutaraldehyde treatment. However, the biocatalyst obtained in the latter study was of limited operational stability and could not match the demands of industrial application.

Ionic binding on ion-exchange matrices of mixed monomers [e. g., weakly basic styrene-divinylbenzene resin (Duolite A 568)] often leads to higher immobilization yields, as compared to the adsorption on hydrophobic adsorber resins. Immobilization can indeed be improved by the above mentioned crosslinking reactions. This is of particular importance if the biocatalyst is used in aqueous systems with a high salt content or in the case of processes which liberate charged compounds, thereby increasing the ionic strength (e. g., hydrolysis of esters or amides). However, most non-covalent immobilizations on (organic) supports are of interest for non-aqueous applications. Lipases and, to a lesser extent, some of proteases are the enzymes mainly used in organic solvents. The hydrophobic polypropylene matrix (Accurel EP 100, Akzo Nobel) is one of the most frequently used supports for lipases. In this case, the immobilization seems very easy and straightforward, and the enzyme protein is bound to aqueous solutions in which the support has been suspended. Then the carrier-fixed lipase is filtered off, washed and dried. However, there are a lot of parameters, e. g., water content of carrier-fixed enzyme or additives added prior to drying of the biocatalyst, influencing the enzymatic performance in organic solvents

15.3 Enzyme Inhibitors

An enzyme inhibitor is a molecule that binds to enzymes and decreases their activity. Not all molecules that bind to enzymes are inhibitors; enzyme activators bind to enzymes and increase their enzymatic activity, while enzyme substrates bind and are converted to products in the normal catalytic cycle of the enzyme. Since blocking the activity of an enzyme plays an important role in killing a pathogen or correcting a metabolic imbalance, many drugs have been developed based on the mechanism of enzyme inhibition, as the examples of herbicides and pesticides.

The binding of an enzyme inhibitor stops a substrate from entering the active site of enzyme, and hinders the enzyme from biocatalysis. Inhibitor binding is either reversible or irreversible; reversible inhibitors bind non-covalently, and different types of inhibition are grouped depending on whether these inhibitors bind the enzyme, the enzyme-substrate complex, or both. In contrast, irreversible inhibitors usually react with the enzyme and change it chemically (e.g. via covalent bond formation), by modifying the key amino acid residues needed for enzymatic activity.

With the mechanism of enzyme inhibition, the discovery and improvement of many drug molecules had formed an active area in biochemical and pharmacological researches. A medicinal enzyme inhibitor is often judged by its specificity (its lack of binding to other proteins) and potency (its dissociation constant, which indicates the concentration needed to inhibit the enzyme). A high specificity and potency means that a drug will have few side effects and thus low toxicity.

Enzyme inhibitors also occur naturally and involved the regulation of metabolism. For example, it's a normal phenomenon that some enzymes in a metabolic pathway can be inhibited by downstream products. This type of negative feedback slows down the flux through a pathway when the products begin to build up, and is an important way to maintain homeostasis in a cell. Other cellular enzyme inhibitors are proteins that specifically bind to and inhibit an enzyme target. This mechanism helps control enzymes that may be damaging to a cell, such as proteases or nucleases; a well-known example is the ribonuclease inhibitor, which binds to ribonucleases in one of the tightest protein-protein interactions. Natural enzyme inhibitors can also be poisonous and are used as defenders against predators or as ways of killing preys.

15.3.1 Reversible inhibitors

There are four kinds of reversible enzyme inhibitors.

In competitive inhibition, the substrate and inhibitor can not bind to the enzyme at the same time, as shown in Fig. 15.5. The inhibitor have an affinity for the active site of an enzyme where the substrate also binds; substrate and inhibitor compete for accessing to the enzyme's active site. This type of inhibition can be overcome by sufficiently high concentrations of substrate (v_{max} remains constant), i.e., by out-competing the inhibitor. However, the apparent K_m will increase as it takes a higher concentration of the substrate to reach the K_m point, or half the v_{max}. Competitive inhibitors are often similar in structures to the real substrate.

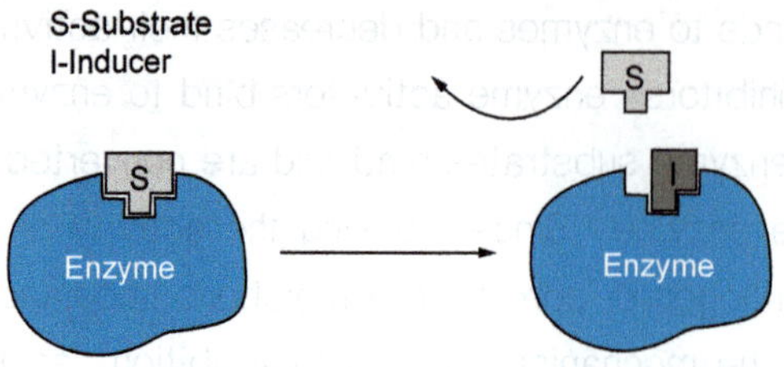

Fig. 15.5 Inhibitor replaces substrate, and bind to enzyme

In uncompetitive inhibition, the inhibitor binds only to the substrate-enzyme complex; it should not be confused with non-competitive inhibitors. This type of inhibition causes v_{max} to decrease (maximum velocity decreases as a result of removing activated complex), thus decreasing the K_m (due to better binding efficiency as a result of Le Chatelier's principle and the effective elimination of the ES complex), which indicates a higher binding affinity.

In mixed inhibition, the inhibitor can bind to the enzyme and the enzyme's substrate at the same time as well. However, the binding of the inhibitor affects the binding of the substrate, and vice versa. This type of inhibition can be weakened, but not overcome by the increasing concentrations of substrate. Although it is possible for the mixed-type inhibitors to bind to the active site, this type of inhibition generally results from an allosteric effect where the inhibitor binds to a different site on an enzyme, by changing the conformation (i. e., tertiary structure or three-dimensional shape) of the enzyme, so that the affinity of the substrate to the active site is weakened.

Non-competitive inhibition is a form of mixed inhibition where the binding of the inhibitor to the enzyme reduces its activity but does not affect the binding of substrate. As a result, the extent of inhibition depends only on the concentration of the inhibitor. v_{max} will decrease due to the inability of the reaction to proceed efficiently, but K_m will remain the same as the actual binding of the substrate, by definition, will still function properly.

Reversible inhibition can be described quantitatively in terms of the inhibitor's binding to the enzyme and to the enzyme-substrate complex, and its effects on the kinetic constants of the enzyme. In the classic Michaelis-Menten scheme below, an enzyme (E) binds to its substrate (S) to form the enzyme-substrate complex ES. Upon catalysis, this complex is broken down to release product P and free enzymes. The inhibitor (I) can bind to either E or ES with the dissociation constants K_i or K'_i, respectively (Fig. 15. 6). Each kind of inhibitor is associated with different character.

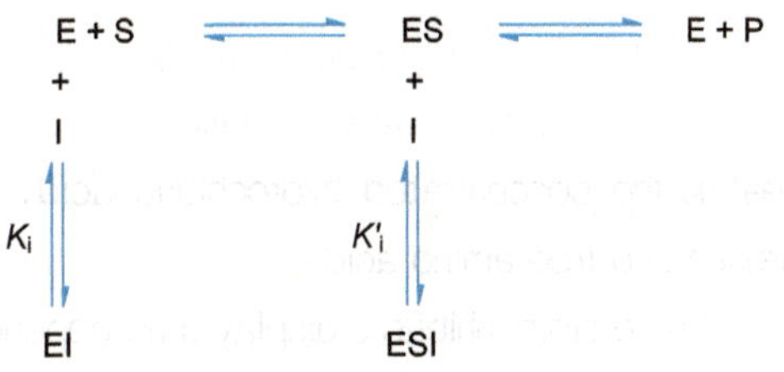

Fig. 15.6 Kinetic schemes for reversible enzyme inhibition

(1) Competitive inhibitors bind to E, but not to ES. Competitive inhibition increases K_m (i. e., the inhibitor interferes with substrate binding), but does not affect v_{max} (the inhibitor does not hamper catalysis in ES because it cannot bind to ES).

(2) Non-competitive inhibitors have identical affinities for E and ES ($K_i = K_i'$). Non-competitive inhibition does not change K_m (i. e., it does not affect substrate binding) but decreases v_{max} (i. e., inhibitor binding hampers catalysis).

(3) Mixed-type inhibitors bind to both E and ES, but their affinities for these two forms of the enzyme are different ($K_i \neq K_i'$). Thus, mixed-type inhibitors interfere with substrate binding (increase K_m) and hamper catalysis in the ES complex (decrease v_{max}).

15.3.2 Irreversible inhibition

Irreversible inhibitors usually covalently modify an enzyme, and therefore the resulted inhibition can not be reversed. Irreversible inhibitors often contain reactive functional groups such as nitrogen mustards, aldehydes, haloalkanes, alkenes, Michael acceptors, phenyl sulfonates, or fluorophosphonates. These electrophilic groups react with amino acid side chains to form the covalent adducts. The residues modified are those with side chains containing nucleophiles such as hydroxyl and sulfhydryl groups; these include the amino acids: serine (as in DFP), cysteine, threonine and tyrosine.

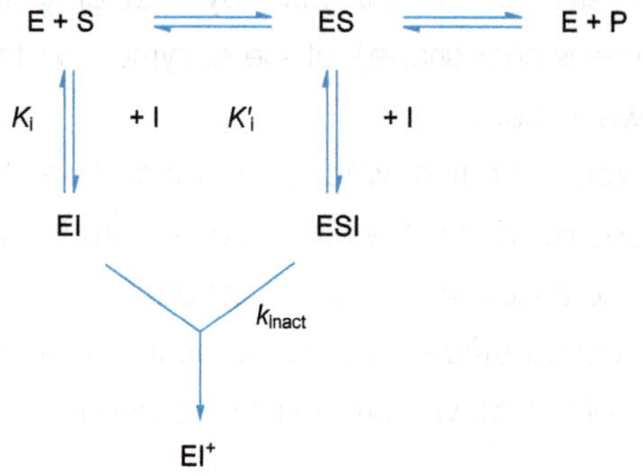

Fig. 15.7 Kinetic schemes for irreversible enzyme inhibition

Irreversible inhibition is different from irreversible enzyme inactivation. Irreversible inhibitors are generally specific for one class of enzyme and do not inactivate all proteins; they do not function by destroying protein structures but by specifically altering the active sites of their targets (Fig. 15.7). For example, extremes of pH or temperature usually cause, as non-specific effects, the denaturation of all protein structures. Similarly, some non-specific chemical treatments destroy protein structures: for example, by heating the concentrated hydrochloric acid, the peptide bonds will be hydrolyzed, and even broken into free amino acids.

Irreversible inhibitors display time-dependent inhibition and their potency therefore can not be characterized by an IC_{50} value. This is because the amount of active enzymes at a given concentration of irreversible inhibitors will be different depending on how long the inhibitors are pre-incubated with the enzymes. Instead, $k_{obs}/[I]$ values are used, where k_{obs} is the observed pseudo-first order rate of inactivation (obtained by plotting the log of % activity vs. time) and [I] is the concentration of inhibitor. The $k_{obs}/[I]$ parameter is valid as long as the inhibitor does not saturate binding with the enzyme (in which case $k_{obs} = k_{inact}$).

15.4 Ribozyme (Non-protein Enzymes)

15.4.1 Natural ribozymes

A ribozyme is a RNA molecule with a well defined tertiary structure that enables it to catalyze a chemical reaction. Ribozyme refers to ribonucleic acid enzyme. It may also be called RNA enzyme or catalytic RNA. Before the discovery of ribozymes, enzymes, which are defined as catalytic proteins, were the only known biological catalysts. In 1967, C. Woese, F. Crick, and L. Orgel were the first to suggest that RNA could act as a catalyst.

This idea was based upon the discovery that RNA can form complex secondary structures. The first ribozymes were discovered in the 1980s by Thomas R. Cech, who was studying RNA splicing in the ciliated protozoan *Tetrahymena thermophila*, and S. Altman, who was working on the bacterial RNase P complex. These ribozymes were found in the intron of an RNA transcript, which removed itself from the transcript, as well as in the RNA component of the RNase P complex, which is involved in the maturation of pre-tRNAs. In 1989, Thomas R. Cech and S. Altman won the Nobel Prize in chemistry for their "discovery of catalytic properties of RNA". The term ribozyme was first introduced by K. Kruger et al. in 1982 in a paper published in the famous periodical named "Cell".

The discovered ribozymes include: Peptidyl transferase 23S rRNA, RNase P, Group I and Group II introns, GIR1 branching ribozyme, Leadzyme—Although initially created *in vitro*, natural examples have been found: Hairpin ribozyme, Hammerhead ribozyme; HDV ribozyme, Mammalian CPEB3 ribozyme, VS ribozyme, glmS ribozyme, and CoTC ribozyme. It's believed that more ribozymes will be discovered by molecular biologists.

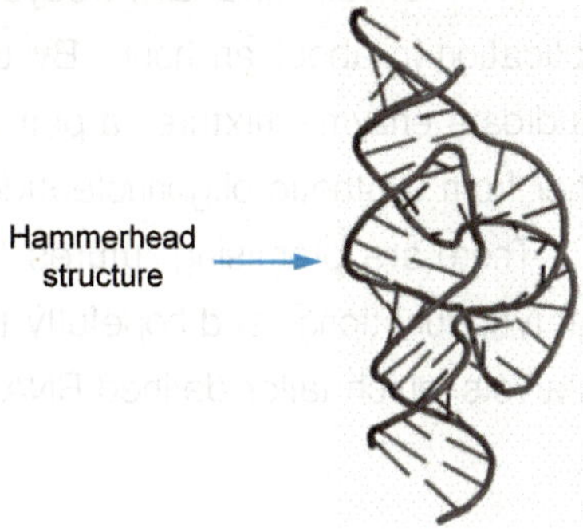

Fig. 15.8 Structure of hammerhead ribozyme

Most ribozymes contain active site that consists of one or more RNA chains, which catalyze either the hydrolysis of their own phosphodiester bonds (self-cleaving ribozymes), or other RNA molecules. Some have been found to have aminotransferase activity of the ribosome. Examples of ribozymes include the hammerhead ribozyme (Fig. 15.8), the VS ribozyme and the hairpin ribozyme.

15.4.2 Artificial ribozymes

Since the ribozymes were revealed to exist in living organisms, there have been interests in the study of new synthetic ribozymes made in the laboratory. Investigators which involving the study of life origin have produced ribozymes that are capable of catalyzing their own synthesis under very specific conditions, such as an ribozyme with RNA polymerase activity. Mutagenesis and selection have been performed resulting in isolation of improved variants of the "Round-18" polymerase ribozyme in 2001. "B6.61" is able to add up to 20 nucleotides to a primer template in 24 hours, until it is decomposed by hydrolysis of its phosphodiester bonds. The "tC19Z" ribozyme can add up to 95 nucleotides with fidelity level on 0.83% only. Tang and Breaker isolated self-cleaving RNAs by *in vitro* selection of RNAs originating from random-sequence RNAs. Some of the synthetic ribozymes that were produced had novel structures, while some were similar to the naturally occurring hammerhead ribozyme.

The techniques used to make artificial ribozymes also involve Darwinian evolution. This approach takes advantage of RNA's dual nature as both a catalyst and an informational polymer, making it easy for an investigator to produce vast populations of potential RNA catalysts using RNA polymerase activity. The ribozymes are mutated by reversedly transcribing them with reverse transcriptase into various cDNA and amplified with mutagenic PCR. The selection parameters in these experiments often differ. One approach for selecting a ligase ribozyme involves using biotin tags, which are covalently linked to the substrate. If a molecule possesses the desired RNA ligase activity, a streptavidin matrix can be used to recover the active molecules.

T. A. Lincoln and G. F. Joyce developed an RNA enzyme system capable of self replication in about an hour. By utilizing molecular competition (*in vitro* evolution) of a candidate enzyme mixture, a pair of RNA enzymes emerged, in which each synthesizes the other from synthetic oligonucleotides, without protein presented in the synthetic system.

There are promising futures for scientists to exploit more ribozymes with particular catalytic functions, and hopefully those can be applied as therapeutic agents, biosensors, catalysts which tailor defined RNA sequences, and novel tools for post-genomic studies.

Summary

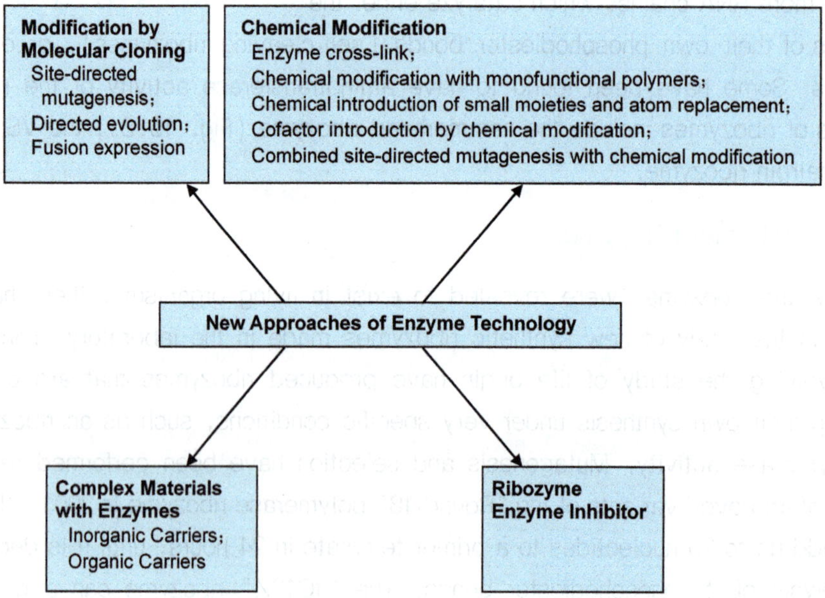

Part IV Industrial Fermentation

Chapter 16 Principles of Fermentation

发酵工程是生物技术的主要分支,发酵起源于古代,在人类生活中起到一定的作用。近代发酵工程是建立在生物化学、微生物学、分子生物学等技术与理论之上的,一般是指采用现代工程技术手段,利用微生物的某些特定功能,为人类生产有用的产品,或直接把微生物应用于工业生产过程的一种技术。

发酵工程的内容包括菌种选育、代谢调控、产品的分离提纯和生物反应器优化等方面。经过近百年的发展,发酵工程已经涵盖农产品加工、生物资源、生物能源和环境处理等诸多方面,对人类生活产生诸多影响。本章就发酵工程的基本原理、产品和应用及发酵工业的作用做相关阐述。

16.1 Fermentation Principles

It's known that the term "fermentation" is derived from the Latin verb, **fervere**, which describes the action of yeasts on the extracts of fruit or malted grain and producing bubbles of carbon dioxide to form a "boiling" appearance. However, fermentation has now come to have different meanings to biochemists and to industrial microbiologists. Commonly, it means the bio-conversion processes acted by microbial metabolism. During an industrial fermentation, not only carbon dioxide is produced, more importantly, varieties of substances such as microbial cells, enzymes, primary and secondary metabolites could also be created from the conversion of bio-resources. Meanwhile, it derives energy from the oxidation of organic substances, such as carbohydrates, and also some organic compounds might act as endogenous electron acceptors.

16.1.1 General process of fermentation

The general process of an industrial fermentation process normally involves: 1) Selection of one or multiple strains for a specific fermentation. 2) Storage of the master strain in a pure form. Sometimes the industrial strain must be mutated or genetic-engineering improved, in order to enhance fermentation productivity. 3) Preparation of an inoculation medium to grow the desired strain, and scaling up the inoculum to initiate fermentation. 4) Use of well designed fermentor facilitated with all the affiliate equipments. 5) Sterilization of the broth and all the equipment used in the process to prevent contamination of germ microorganisms. 6) Optimization of the medium ingredient to guarantee the nutrients in the fermentor meeting the requirement of microbial growth and metabolism. 7) Running the fermentation process in a precise way based upon the parameters measured and the information fed back. 8) Collection of fermentation followed by down-stream technology to separate the desired products from the other components. 9) Treatment of fermentation effluent and the solid wastes to make sure there is no negative impacts on the environment.

Generally speak, the basic principles of fermentation are to facilitate the proliferation and maintenance of desirable microorganisms in raw materials in order to produce a wide range of biochemical to meet the needs of human beings and to maintain a sustainable development of our society and environment.

16.1.2 Progress of fermentation

Fermentation has been practiced since prehistorically times. Wine brewery in China is believed to have started as early as 3000 BC. Western historians also believe that the Egyptians started to make beer 7000 years ago. The origins of a variety of indigenous fermented foods and sauces in the Orient or somewhere else are now known to have based on fermentation. The practice of enzymatic hydrolysis using surface culture or semi-solid fermentation for preparation of biocatalyst could also be dates back thousands of years.

As a matter of fact, the ancient peoples performed fermentation without any notice of the existence of microbes. Only when French microbiologist L. Pasteur proved that the microorganisms in the beef broth actually came from the preexisting life, and as matter of fact, fermentation was a biochemical reaction. Then the scientific society came to a strong belief that it is the microorganisms that are responsible for the fermentation processes. Pasteur carried out detailed investigations on beer and wine brewing; finally he came to the conclusion that living yeast cells ferment sugar into ethanol and carbon dioxide. He also explained the formation of organic acid such as lactic acid before he worked out clearly the mechanism of ethanol fermentation. Since the true story of microbes had been revealed, the exploitation of different species of industrial strains for industrial purposes has been turned into reality.

Not until the rise of chemistry, microbiology, biochemistry, and engineering in the 19th century did fermentation practices begin changing significantly. People learned how to liquefy grain starch and performed fermentation in the liquid form. Aerobic and anaerobic processes each provided different kinds of final products. The solvents such as ethanol, acetone, and butanol were produced from anaerobic fermentation; whereas, amino acids, some organic acids, and enzymes were products of aerobic processes. The observation of

A. Fleming in 1929 that staphylococcal colonies on a petri dish were inhibited by *Penicillium notatum*, which ignited the discovery of the first antibiotic named penicillin and led to a new era of antibiotic and other secondary metabolite fermentation. Fermentation of secondary metabolites seemed to be more complicated, and a big challenge was how to raise the fermentation productivity of antibiotics, in order to cure more wound soldiers and sick patients.

It is worth to mention that among the innovations in the field of fermentation during the mid 20th century, were a variety of fermentors designed specifically to meet individual requirements. The volume of fermentation tanks were gradually scaled up to as large as thousands of cubic kilo-liter (Fig. 16.1). The affiliated equipments and pipe lines were also redesigned to prevent contamination. Since microbes are capable of a wide range of metabolic reactions by using various sources of nutrients, it encourages people to run industrial fermentation with raw materials of rather low cost, such as molasses, cellulose, or waste products from other resources for production of commercially useful biomass, enzymes, and biochemicals.

Fig. 16.1 Modern fermentor

Since 20th century, biotechnology has been considered as one of the most predominant innovations, which has deeply affected human lives and government's strategies. Concerns about, in practice, large scale of the bio-productions are usually based upon fermentation process, and the technology is still in its highlight to attract more and more researchers and entrepreneurs. Today fermentation industries are also facing severe challenges, such as the challenge of rapid change of product spectrum and factory location, as well as dealing with pollution problem and an increase in labor cost, which compels large scale bioprocessing factories relocate away from high labor cost regions. With daily increased species of Genetic Modified Organisms (GMO) had been constructed to make them industrial strains, for example, the successful deciphering of the human genome with its myriad of new therapeutic targets, the advanced genetic expression systems for producing novel proteins and other bio-functional factors that is consequently constructed, and also the related fermentation processes which is further optimized. The advent of this latter range of products and processes has revolutionized the practice of clinical medicine and many areas of fundamental research, and has also significantly increased the need of skilled talents in fermentation industries.

16.2 Fermentation Products and Application

There are different ways to group the fermentation products into certain category, and in this book we classify them as follows:

Biomass or single cell proteins For certain commercial purpose, the microbial cells are cultivated in fermentor then harvested as end products, those cells are referred to as biomass. The term single cell protein usually means the microbial protein synthesized in a large scale by unicellular species, such as yeast or bacterial cells, for adding on feed stuffs to increase their protein content. For a long time, it was known that under anaerobic condition the yeast species, *Saccharomyces* ferment sugar into ethanol, but by aeration it grows fast in the sugar broth. The yeast cells harvested from fermentation might be referred to as primitive "biomass". Beside of yeast, there are other microbes employed in SCP production, including fungi (*Aspergillus oryzae*, *Sclerotium rolfsii*, *Polyporus* and *Trichoderma*), bacteria (*Rhodopseudomonas capsulata*), and algae (*Chlorella* and *Spirulina*). Typical yields of 43% to 56%, with protein contents of 44% to 60%.

In the 1960s, researchers at British Petroleum Corp. developed what they called "proteins-from-oil process", a technology for producing single cell protein from yeast fed by waxy n-paraffins from oil refineries. Initial research work was done by Alfred Champagnat at BP's Lavera Oil Refinery in France. The term SCP was coined in 1966 by C. L. Wilson at MIT. The "food from oil" idea became quite popular by the 1970s, with Champagnat being awarded the UNESCO Science Prize in 1976, and paraffin-fed yeast facilities being built in a number of countries. The primary use of the product was as poultry and cattle feed.

The technology of single cell proteins developed when microbes ferment waste materials (including wood, straw, cannery and food processing wastes, residues from alcohol production, hydrocarbon, or human and animal excreta). The problems with extracting single cell proteins from the wastes were dilution and cost. The cells are found in very low concentrations, usually less than 5% in the medium. Engineers have developed many techniques to concentrate the biomass, including centrifugation, flotation, precipitation, coagulation and filtration, or the use of semi-permeable membranes. The single cell protein needs to be dehydrated to approximately 10% moisture content and/or acidified to aid in storage and prevent spoilage. To increase the concentration of biomass and to dehydrate the moisture to adequate levels would be expensive; the methods are not always suitable for small-scale operation. It is economically prudent to feed the product locally and shortly after it is produced. Large-scale production of microbial biomass has many advantages over the traditional methods for producing proteins for food or feed.

Enzymes In the cases of other commercial purposes, the end products of fermentation would be some catalytic proteins, as we already know from the Part II, which perform originally the bioconversions for its host cells, but might be applied to provide varieties of specific chemical effects. As we all known, there are more than hundreds of enzyme products commercially used today.

Primary metabolites Theoretically speaking, every individual metabolite occurred in the life cycles of a microbial cell might be a potential product, since we can find its usage and also work out the metabolic pathways to make them accumulated through fermentation. The metabolites, with rather smaller molecular weight, playing vital roles to its host cells, are known as primary metabolites. Ethanol has many applications as a raw material, solvent and fuel and is utilized in large quantities in the chemical, pharmaceutical and food industries. Worldwide, four million tons of industrial ethanol is produced annually, 80% by fermentation. The increasing trend of ethanol production in the United States since 1980 has been noticed from 0.18 billions of gallons, and gradually to a total production capacity of 4.89 billions of gallons in 2006.

Yeasts are the preferred organisms for industrial-scale ethanol production. When certain species of yeast metabolize sugar, they produce ethanol and carbon dioxide. Different species can be used, depending on the composition of the raw material utilized. *S. cerevisiae* has been the major organism particularly suitable for fermentation of **hexoses**, and *Kluyveromyces fragilis* or *Candida* sp. can utilize lactose or **pentoses**, respectively. Yeast convert hexoses to ethanol and carbon dioxide by **glycolysis**. The chemical equation below summarizes the conversion:

$$C_6H_{12}O_6 \longrightarrow 2\ CH_3CH_2OH + 2\ CO_2 \qquad (16.1)$$

The theoretical ethanol yield over glucose is 0.51g/g, and the growth yield over glucose is 0.12g/g. Ethanol production is triggered by anaerobic conditions and this process is carried out at around 35～40℃. Trace amounts of oxygen (0.05～0.1mmHg①) are required by yeast for lipid biosynthesis and maintenance of cellular processes. The ethanol feed solution should be balanced in terms of nitrogen, phosphorus, minerals and some trace elements.

Other primary metabolites, such as acetone, butanol and organic acids are also the products from fermentation.

---

① 1mmHg = 1.33322×10^2 Pa.

Secondary metabolites Those biochemical substance occur only in certain phase of the microbial life cycle, and seem to have no vital roles but only some special bio-functions in the host life, are named as secondary metabolites, for example, the antibiotics. Antibiotics are among the most frequently prescribed medications in modern medicine. Antibiotics cure disease by killing or injuring bacteria. The first antibiotic was penicillin, discovered accidentally from a mold culture. Today, over 100 different antibiotics are available to doctors to cure minor discomforts as well as life-threatening infections. The importance of antibiotics is multifaceted: they combat human diseases and also help increase food supplies; with sales exceeding 1 billion dollars annually, their commercial impact is large; finally, biochemical engineering is, in many ways, the offspring of the development of antibiotic fermentations.

Currently more than 6000 substances are known to have antibiotic activity. The majority of well exploited antibiotics come from only limited types of drugs. These are the main classes of antibiotics:

- Penicillins such as penicillin.
- Cephalosporins such as cephalexin (Keflex).
- Macrolides such as erythromycin (E-Mycin), clarithromycin (Biaxin), and azithromycin (Zithromax).
- Fluoroquinolones such as ciprofloxacin (Cipro), levofloxacin (Levaquin), and ofloxacin (Floxin).
- Sulfonamides such as co-trimoxazole (Bactrim) and trimethoprim (Proloprim).
- Tetracyclines such as tetracycline (Sumycin, Panmycin) and doxycycline (Vibramycin).
- Aminoglycosides such as gentamicin (Garamycin) and tobramycin (Tobrex).

Fermented foods Even in the 21st century, fermented foods are still the major diets to common people. Without bakery, beer, wine, sauce, vinegar and other flavor foods, one can imagine his life will be no "taste" at all.

Together with drying and salting, fermentation is one of the oldest methods of food preservation. Its importance in modern-day life is underlined by the wide spectrum of fermented foods marketed both in developing and industrialized countries, not only for the benefit of preservation and safety, but also for their highly appreciated sensory attributes. Fermentation is a chemical change brought about by yeast, bacteria or mold. This process has been used for centuries by people in order to make and preserve certain types of food. Moreover, fermentation is not only a way to preserve certain foods, but also add nutrient value to the fermented foods. Fermented vegetables contain more vitamin C, and fermented milk products have ample amounts of B vitamins. The bioavailability of these vitamins also increases with fermentation. And one of the most important aspects is lacto-fermentation.

What is it? Lacto-fermentation happens when the starches and sugars in vegetables and fruit convert to lactic acid by a friendly lactic-acid producing bacteria. This produces not only a tangy, delicious product, but also preserves them. Lacto-fermented fruits and vegetables are of great health benefits and importance, for example, when vegetables are lacto-fermented, vitamins increase. In addition, it is more digestible and people get a plethora of good bacteria of lacto-fermented vegetable.

Lacto-fermentation processes have survived throughout the centuries because of the following benefits of this technology:

• It serves as a household technology for improving food safety.

• It serves as a low-cost method of food preservation.

• It contributes to the improvement of the nutritional value and digestibility of food raw materials.

• It contributes to the solution to control diarrhea diseases in African children.

• In some sense, "probiotics" is referred to as the industrial cultivated micro-flora for daily uses, which have been perceived to be beneficial to support human health and prevent diseases. Lacto-bacterial probiotics is now becoming a very popular health care product.

16.3 Benefits from the Fermentation Industry

Fermentation can be spontaneous or induced by specifically added microorganisms. It is a good example of traditional wisdom. The benefits of fermentation have been recognized since last century, for example:

• The acid (lactic and acetic acid) produced in fermentation helps to preserve the food.

• Fermentation improves the protein content and adds vitamins and minerals; improves the absorption of important nutrients, particularly iron and zinc.

• Fermentation can utilize a broad spectrum of raw materials as carbon sources, which include even waste products. Thus they help in the removal of pollutants also.

• Strains with high yield and good composition can be selected or produce relatively easily.

• Fermentation can produce higher purity or chiral purity product than chemical engineering.

Except of making routine bioproducts, the modern fermentation also focuses on the issues of energy shortage and environmental pollution, which would have negative impacts on sustainable development of our society.

Production of biofuel as cleaner energy The contributions of fermentation to sustainable society should also be mentioned. Except solar energy, hydraulic and wind power, the fossil energy causes sever green house effect. According to the recent report of the Intergovernmental Panel on Climate Change (IPCC), warming of the world's climate system is unequivocal and is very likely due to the observed increases in anthropogenic greenhouse gas concentrations. It's urgent to develop much cleaner energy for daily use.

Ethanol is a clean-burning fuel source made from renewable bio-resources such as corn and grain sorghum. Ethanol has been added to ordinary unleaded gasoline in order to increase octane, reduce exhaust emissions and meet clean air standards. Ethanol made biologically by fermentation from a variety of biomass sources is widely recognized as a unique transportation fuel with powerful economic, environmental and strategic attributes. For example, the ethanol industry now produces more than 10 billion gallons of fuel ethanol in USA, representing 7 percent of the gasoline supply, and 70 percent of all gasoline sold contains some ethanol. Ethanol contributes more than $45 billion to the U. S. Gross domestic product annually, creating more than 238,000 jobs, and contributing $12 billion to consumers through lower transportation fuel prices. During 2007, 6.5 billion gallons of domestically produced ethanol displaced 228 million barrels of imported oil.

Except ethanol, there is another renewable energy, algae oil. Although there is still a great difficulty in large scale production, it is the final renewable solution for vehicle fuel (Biofuel).

Bioremediation for protection of environment Bioremediation is defined as any process that uses microorganisms, fungi, green plants or their enzymes to return the natural environment altered by contaminants to its original condition. Bioremediation may be employed to break down some specific soil contaminants, such as degradation of chlorinated hydrocarbons by bacteria. An example of a more general approach is the cleanup of oil spills by the addition of nitrate and/or sulfate fertilisers to facilitate the decomposition of crude oil by indigenous or exogenous bacteria.

There are a number of efficiency advantages of bioremediation, which can be employed in areas that are inaccessible without excavation. For example, hydrocarbon spills (specifically, petrol spills) or certain chlorinated solvents may contaminate groundwater, introduce the appropriate electron acceptor or electron donor amendment, as appropriate, and may significantly reduce contaminant concentrations after a long time allowing for acclimation. This is typically much less expensive than excavation followed by disposal elsewhere, incineration or other ex situ treatment strategies, and this reduces or eliminates the need for "pump and treat", a common practice at sites where hydrocarbons have contaminated clean groundwater.

It is generally assumed that one of the major concerns of bioremediation with microbes, particularly in maritime areas, is that the microorganisms must be able to resist a variety of adverse conditions that are far removed from the ideal conditions of the laboratory. Whatever the case, and regardless of whether the organisms are genetically modified or not, microbial bioremediation is limited by factors such as mass transfer (low contaminant bio-availability), aeration, nutrient level at contaminant sites (might require bio-stimulation techniques) and problems with the thermal conditions. All of these issues should be addressed on a case-by-case basis to obtain the maximum microbial growth to metabolize the toxicants.

Over the last few years, a number of companies have been established to develop and commercialize biodegradation technologies. Existence of such companies now has become economically justifiable, because of burgeoning costs of traditional treatment protocols, increasing public resistance to such traditional technologies, as well as increasingly stringent regulatory requirements. The interest of commercial businesses in utilizing microorganisms to decomposing and detoxify effluents, soils, etc. is reflected in "bioremediation", which becomes a common buzzword in waste treatment. Companies specializing in bioremediation (or, as it was known several years ago, in biodegradation technologies) will need to develop a viable integration of microbiology and the systematic engineering. As an example of a bioremediation company, Envirogen (NJ) has developed recombinant PCB (polychlorinated biphenyl)-degrading microorganisms with improved stability and survivability in mixed populations of soil organisms. The same company also has developed a naturally occurring bacterium that degrades **trichloroethylene** (TCE) in the presence of toluene, a toxic organic solvent killing many other microorganisms. Microorganisms have also been successfully applied during the removal of the Exon Valdez oil spill. Use of microbes for bioremediation is not limited to detoxification of organic compounds. In many cases, selected microbes can also be used to treat the pollution of heavy metals (such as selenium) into a much less soluble elemental form and in much less toxic level.

Summary

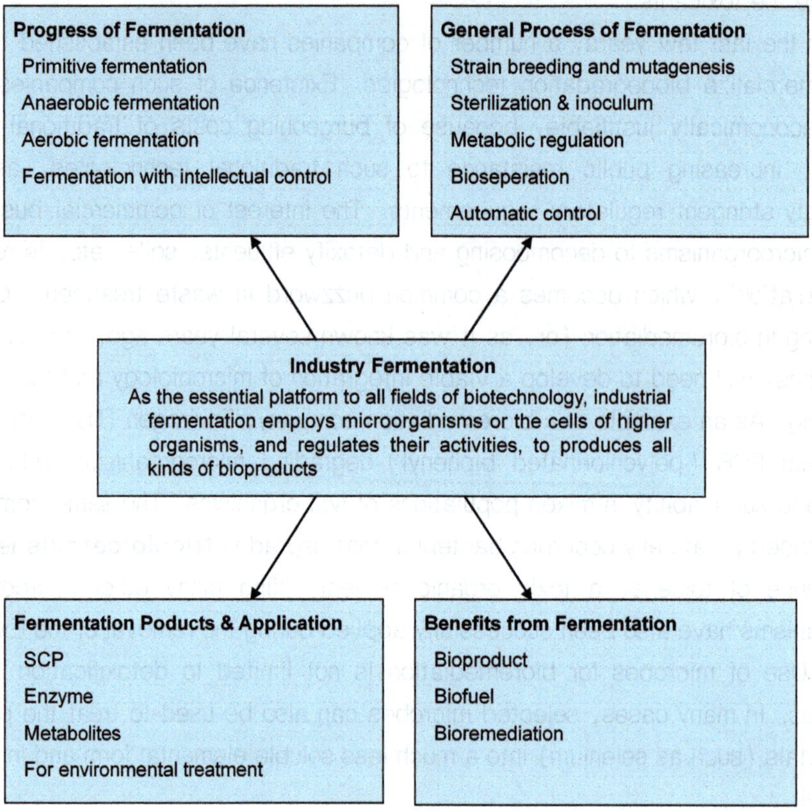
Progress of Fermentation
Primitive fermentation
Anaerobic fermentation
Aerobic fermentation
Fermentation with intellectual control
General Process of Fermentation
Strain breeding and mutagenesis
Sterilization & inoculum
Metabolic regulation
Bioseperation
Automatic control
Industry Fermentation
As the essential platform to all fields of biotechnology, industrial fermentation employs microorganisms or the cells of higher organisms, and regulates their activities, to produces all kinds of bioproducts
Fermentation Poducts & Application
SCP
Enzyme
Metabolites
For environmental treatment
Benefits from Fermentation
Bioproduct
Biofuel
Bioremediation

Chapter 17 Industrial Microorganisms

工业发酵生产主要依靠几类重要的微生物，如细菌、放线菌、酵母、霉菌，还有其他一些不断得到开发利用的新菌属，工业发酵所使用的菌株一般称为“工业菌种”。认识工业微生物的自然来源、遗传背景及生理生化特征，了解工业菌种的筛选及育种改良等基本方法，掌握其生长规律和扩增培养方法，这是发酵上游技术的核心内容。

17.1 Types and Characters of Industrial Microorganisms

17.1.1 General characters

Since the fermentative activity of microorganisms has been defined in the 18th century, the fermentation processes of alcohols, amino acids, enzymes, organic acids, vitamins, antibiotics, natural polymers such as polysaccharides, fermented food or feed, and other products have been well established. Microbes acting as "initial driving force" and "production workers" in fermentation were broadly recognized.

Industrial microorganisms are very diverse, including bacteria, actinomycetes, fungi, and micro-algae. Ever since these microorganisms were first identified, microbiologists have attempted to bring order to the way they are named and classified according to morphology and so on. But the range of morphological features that is useful in the differentiation of bacteria is fairly limited compared to animals and plants, so other characteristics have also been employed. These features include metabolic properties, pathogenicity, nutritional requirements, staining reactions and antigenic properties. Recent advances in the development of techniques for cultivation of animal and plant cells have enabled these more complex cells to be used also in industrial fermentation process.

17.1.2 Bioproducts from industrial strains

Industry microorganisms are used in brewing, winemaking, baking, pickling and other product making processes. There must be different ways to group the fermentation products into certain category; Tab. 17.1 shows a wide range of products that can be produced by it, which include SCP, enzyme and protein, primary metabolite, secondary metabolite, bioenergy and so on. Some procedure has single pure microorganism. Those microbes also take an important place in most higher-order multicellular organisms as symbionts.

Tab. 17.1 Examples of different strains and the fermentation products

Class of compounds	Products	Industrial strains
Alcohol	Ethanol	*S. cerevisiae*
	Glycerol	*S. cerevisiae*
	Citric acid	*A. niger*
Organic acids	Acetic acid	*Acetobacter*
	Malic acid	*S. cerevisiae*
Amino acids	L-lysine	*C. glutamicum*
	L-phenylalanine	*E. coli*
	L-tyrosine	*C. glutamicum*
	Riboflavin	*B. subtilis*
Antibiotics	Penicillin	*P. chrysogenum*
	Erythromycin A	*S. erythraea*
	Rifampicin	*S. mediterranei*
Enzymes	Lipase	*A. niger*
	Human proteinase	*A. niger*
	Amino peptidase	*A. oryzae*
	α-amylase	*A. oryzae*
	Subtilisin	*Bacillus* sp.
Other	SCP	*S. cerevisiae*; *Aspergillus oryzae*; *Chlorella* sp.
	Vitamins	*Blakeslea trispora*
	Poly-hydroxy-butyrate	*Ralstonia eutropha*

17.1.3 Typical industrial microbes

The most commonly involved industrial microbes are briefly introduced as follows:

Bacteria

Bacteria belong to a large group of single-celled, **prokaryotic microorganisms**, which are vital in recycling nutrients, with many steps in metabolic pathway. Depending on their different genetic background, physiological and biochemical characteristics, bacteria are quite diverted on nutrient recycling. For example, ***Acidithiobacillus ferrooxidans*** utilizes the mineral components as nutrient, which might be used in bioleaching, and the species of **soybean *rhizobia*** have special capability of nitrogen fixation, which intake the nitrogen from the atmosphere and utilize it to synthesize *N*-containing compounds. But for most of the industrial bacterial species, the nutrient recycling involves the assimilation of organic compounds and inorganic elements. The nutrients absorbed by bacterial cells are finally converted into varieties of end products, including all kinds of organic metabolites, water and carbon dioxide. There are some important industrial bacteria which are still active in modern fermentation:

Bacilli are a Gram-positive and rod-shaped bacterium easily found in soil and have the ability to form a tough, protective endospore, allowing the organism to tolerate extreme environmental conditions. Unlike several other well-known species, *B. subtilis* has historically been classified as an **obligate aerobe**, though recent research has demonstrated that it is not necessarily correct. *B. subtilis* is used as soil inoculants in horticulture and agriculture. Enzymes produced by *B. subtilis* and *B. licheniformis* are widely used as biocatalyst in starch processing and as additives in laundry detergent.

Other uses of this genus include: 1) It was popular worldwide to use *B. subtilis*, before the introduction of antibiotics, to aid treatment of gastrointestinal and urinary tract diseases. It is still widely used in Western Europe and the Middle East as an alternative medicine; 2) *B. thuringiensis* has been used to produce the biopesticide-parasporal crystal protein for replacement of certain chemical pesticides; although its biosafety effects are increasingly concerned; 3) It has been used to decompose the explosive materials into harmless compounds of nitrogen, carbon dioxide, and water; 4) It is used for biosynthesis of hyaluronic acid, which is useful for human joint lubrication.

Escherichia coli (*E. coli*)

Escherichia coli are all known a Gram-negative, rod-shaped bacterium that is commonly found in the lower intestine of warm-blooded organisms (Fig. 17.1). Most *E. coli* strains are harmless, but some serotypes can cause serious food poisoning in humans, and are often responsible for the product recalls. The bacterium can also grow easily and inexpensively in a laboratory setting. The genome of *E. coli* has been intensively investigated. It's the most popular prokaryotic model organism, and an important species in the fields of biotechnology and microbiology, where it has served as the host organism for the majority of up-to-date work with recombinant DNA.

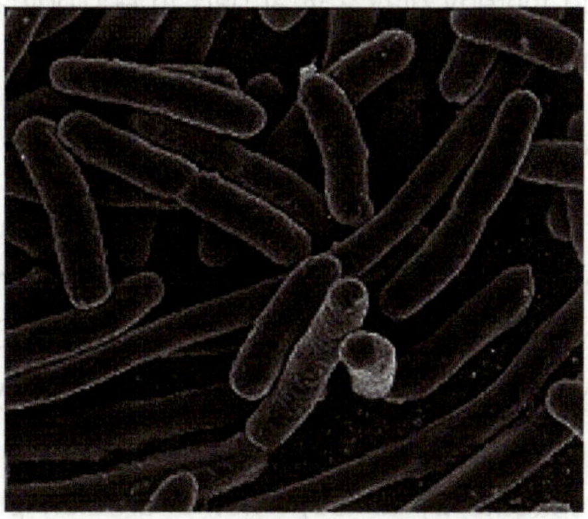

Fig. 17.1 Scanning electron micrograph of *Escherichia coli*

Actinomycetes

Actinomycetes (Fig. 17.2), terrestrial or aquatic, are the Gram-positive prokaryotes with high G + C ratio in their genomes. Many actinomycetes form branching filaments,

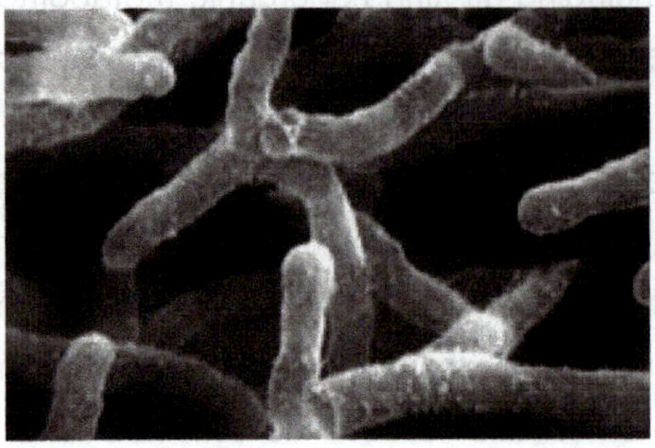

Fig. 17.2 Scanning electron micrograph of *Actinomycetes israelii*

which somewhat resemble the unrelated fungi, among which they were originally classified under the older name actinomycetes. Most actinomycetes species are aerobic, but still there are few, such as *Actinomycetes israelii*, grow under anaerobic conditions. Some of the actinomycete species even produce external spores.

Actinomycetes are well known as secondary metabolite producers and hence of high pharmacological and commercial interests. In 1940 Dr. Selman Waksman (Nobel Laureate) reported that the soil microbes he was studying produced actinomycin which kills germ bacteria. Since then, hundreds of naturally occurring antibiotics have been discovered in those terrestrial microorganisms, especially from the genus named *Streptomyces*. Now, Actinomycetes produce more than half of the world's antibiotics, which are consequently invaluable in the medical field. By degrading starch, protein, cellulose and chitin, or interfere cellular metabolism, Actinomycetes are not only inhibiting the bacteria existing in soil, some of their species are also the important decomposers, and thereby play a vital part in ecological maintaining by exerting organic matter turnover and carbon recycling.

Yeast

As eukaryotic microorganisms, yeasts are unicellular, although some species may have multicellular forms (Fig. 17.3). The sizes of yeasts vary greatly depending on the species, typically measured 3~4μm in diameter, although some yeast can reach 40μm in mycelial-like form. Most of the yeasts reproduce asexually by mitosis, and many propagate by an asymmetric division process called budding. Many species of yeasts are facultative anaerobes, which need oxygen for growth and avoid oxygen for ethanol fermentation. Under aerobic period we can see quite a lot mitochondrial appeared in the yeast cytoplasm, very active for working on respiratory electron transfer; and under anaerobic condition, one can hardly see those cell organs.

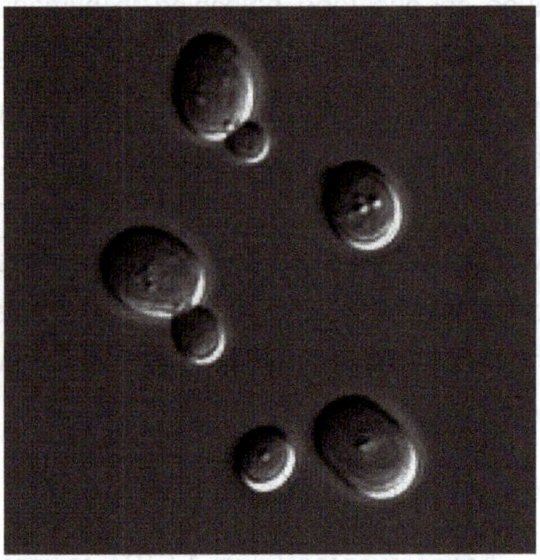

Fig. 17.3 *Saccharomyces cerevisiae*

From prehistoric period till now, yeasts

have been playing an important role in human life. The yeast species *Saccharomyces cerevisiae* has been used in baking and in fermenting alcoholic beverages for thousands of years. In the 20th century, Yeast *Candida* has been used to produce glycerol, and the fusion strains from parental *Schizosaccharomyces* and *Saccharomyces* can use xylose and glucose to produce ethanol. Along with the development of molecular biology, yeast became the new model organism and has been thoroughly researched, from which people gathered information about the biology of the eukaryotic cells. In the field of modern biotechnology, yeasts have been exploited into different GMO strains, which might accept particular human genes in order to produce human hormones or some bio-active peptides. In biofuel industry, yeasts have been used to ferment the raw bio-resources as well as produce fuel-alcohol. Some of yeast species even use paraffins as the only carbon substance for growing and producing secondary metabolites.

Mold

Unlike plants, which contain cellulose in their cell walls, species of mold utilize a combination of chitin and glucans to build their cell wall. Their lifestyles and cycles can also be very dissimilar. Molds (or moulds; see spelling differences) are fungi that grow in the form of multicellular filaments called hyphae. A connected network of these tubular branching hyphae has multiple, genetically identical nuclei and is considered to be a single organism. There are major propagation forms, either reproduce through hyphae growth or germinate from the small spores which contain the same genomes as in the cells. Mold spores can be asexual (the products of mitosis) or sexual (the products of meiosis); Many species are able to reproduce in both forms. Mold spores may remain airborne indefinitely, cling to clothing or fur, and be able to survive under extreme temperature and pressure. Some of the mould species, for example, *Aspergillus*, *Penicillium* (Fig. 17.4), and *Mucor*, *Trichoderma* are excellent producers of many primary and secondary metabolites,

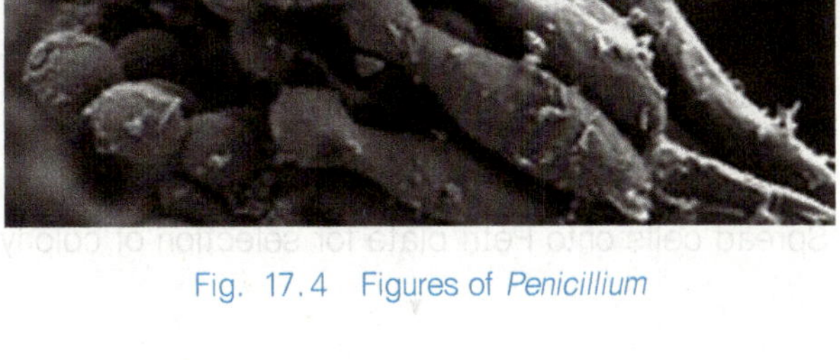

Fig. 17.4 Figures of *Penicillium*

e. g. , organic acids, antibiotics, and monacolins, which are valuable fermentation products to human life. The first scientific paper from Chinese scholars being submitted by the Journal *Science* was about the discovery of new Mucor specie selected from fermented soy bean curb. Industrial *Aspergillus* strain is capable of converting as much as 95% of the available carbohydrates into primary metabolites, such as citric acid, and *Penicillium* is actually the first microbial being found to produce antibiotic, penicillin. In addition, moulds are also able to secrete a wide range of enzymes, such as glucoamylase, pectinase, cellulose and chitinase, representing a rich source for the production of biocatalysts used in food, detergent, textile and even chemical industries.

Other industrial cell lines

Along with the progress of biotechnology, not only the microorganisms could be used to ferment valuable products, but some of plant and animal cells have also been exploited for commercial purpose, especially in production of novel biomedicines and vaccines (Introduced in Part II).

17.2 Breeding of Industrial Strains

Breeding of industrial strains is most important, which ensures the fermentation productivity as higher as possible. Two of the traditional breeding techniques are introduced as follows:

17.2.1 Separation and selection

It's necessary to separate the microbial culture into isolated colonies for selection of superior individual strains. Usually the industrial microbial cells are cultivated into log phase, and spread on Petri plate, in which the growth medium is specifically designed, to let individual cell grow into colony, then pick up the one with good quality, the procedure is called strain selection.

The following flowchart illustrates the general steps of selection for industrial stains:

Background research

↓

Design of selection scheme

↓

Sampling from native resource

↓

Duplication under suitable condition

↓

Spread cells onto Petri plate for selection of colony

↓

Pick up positive colony, and assay

↓

Establish industrial culture condition

↓

Domesticate for getting better progeny

↓

Toxicity test

17.2.2 Mutagenesis

Native species are obtained through natural selection and separation of industrial strains, such as the present industrial use of yeast in beer and wine brewing and so on. Native microbial strains in their living conditions may undergo random mutations or natural genetic variation, producing excellent long-term production traits formed after the accumulation. They are called wild-type. Mutagenesis is referred to as the special procedure designed for altering the genotype of the target strains, which aims on different purposes. The selected wild-type cells are treated with mutagenic factors either chemically on DNA bases or by radiation hits physically on the genome, which forces the genetic traits of the starting strain thus changed, resulting the genotype be passed on to their offspring. The badly hurt individual cells will not survive under the defined growth condition, so that there is an opportunity for people to pick up ideal mutants for industrial fermentation.

Mutation is the fountainhead of genetic variation. Although cell hybridization might induce genomic recombination, inducing artificial mutagenesis is an important means of accelerating mutation with the rate about thousands times greater than that of natural mutation. At present, most of the various production strains of microbes used in the international and domestic fermentation industries are mutated strains. In addition, so called *in vitro* mutagenesis derived form gene cloning technique is now enriching the pool of industrial strains, and actually a lot of industrial strains are derived from this method.

17.2.3 Chromosomal recombination

Chromosomal recombination used to be another choice for breeding of microorganism, but not often used recently. Recombination can occur between similar molecules of DNA, as in homologous recombination, dissimilar molecules or as in non-homologous end joining. Homologous recombination is a type of genetic recombination in which nucleotide sequences are exchanged between two similar or identical sequences of DNA. The phenomenon occurs when there is a need for cells to repair harmful breaks that occur on both strands of DNA accurately. Homologous recombination also produces new combinations of DNA sequences during meiosis. Homologous recombination is also used in horizontal gene transfer to exchange genetic materials between different strains and species of bacteria and viruses.

On the other way, recombination can occur between DNA sequences that contain no sequence homology. This is referred to as **nonhomologous recombination** or nonhomologous end joining, such as **bacterial conjugation**, which is the transfer of genetic materials between bacterial cells by directing cell-to-cell contact or by a bridge-like connection between two cells. It can occur either in the same species, or occasionally between different species. Bacterial conjugation is often incorrectly regarded as the bacterial equivalent of sexual reproduction or mating since it involves the exchange of genetic material. During conjugation, the donor cell provides a conjugative or movable genetic element that often being a plasmid or transposon. Most conjugative plasmids have systems ensuring that the recipient cell does not already contain a similar element. Many *E. coli* mutants had been constructed through chromosomal recombination during the last century, and some of the knowledge on molecular genetics has also been provided.

17.2.4 Genetic engineering

Genetic engineering of industrial strain is the direct manipulation of the microbial genetic material in a way that does not occur under natural conditions. It involves series of gene cloning techniques that are mentioned in Part I of this book (not be repeated here).

17.3 Industrial Microbial Growth and Culture Media

The increase of cellular constituents leading to a rise in cell number is called microbial growth. It's critical to prepare the growth medium which contains all kinds of the nutrients needed in suitable amount. Other factors effecting the microfloral growth should be optimized to ensure the cells grow well and reproduce in a normal way. Without carefully designed growth recipe and optimal growth condition, one could not assure a good fermentation.

17.3.1 Bacterial growth

Microbes reproduce in form of binary fission which is commonly occurred in all prokaryotes, some protozoa, and some organelles even within eukaryotic types. A dividing bacterial cell, on average, must have precisely as twice as much of everything found in a newborn cell. How does a bacterial cell ensure that all cell components are duplicated between divisions, and how does a cell ensure that it does not divide prior to duplicating all of its components? These questions have been investigated by studying the biosynthesis of cell components during the division cycle. Cell cytoplasm increases uniformly and exponentially during the division cycle. At division, the components of the cell are segregated to two daughter cells (Fig. 17. 5). The genome segregates stochastically but the cytoplasm segregates randomly, and the peptidoglycan segregates in a manner consistent with the fixed location of the synthesized peptidoglycan (main component of cell wall). The key "events" during the division cycle are the initiation and termination of DNA replication and the initiation and termination of pole formation. There appears to be no other cell-cycle-specific events or syntheses during the division cycle.

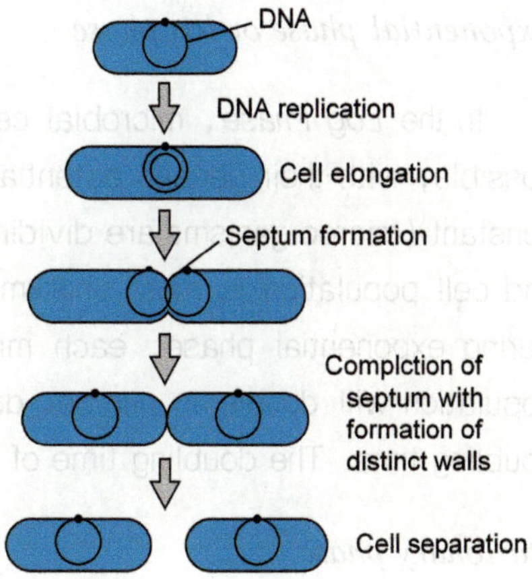

Fig. 17.5 Bacterial division by binary fission

Microbial growth is usually studied by following the change in population number. Population growth is illustrated by the Growth Curve of a microbial culture (Fig. 17.6). Growth of microbial strains in form of binary fission can be plotted as the logarithm of the number of viable cells verses the incubation time. The growth curve of bacteria usually shows four phases of growing states:

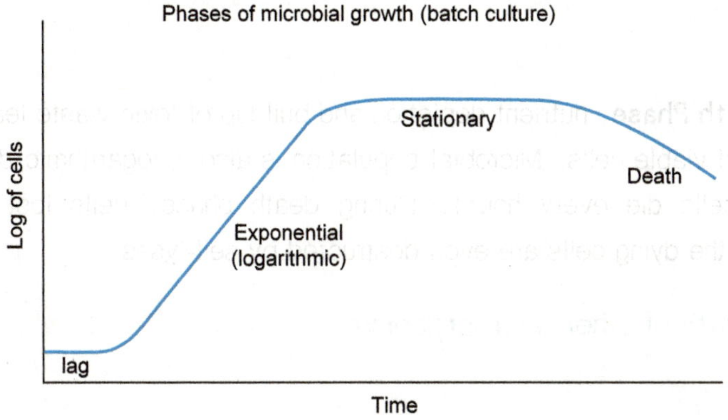

Fig. 17.6 Microbial growth curve

Lag phase

When microorganisms are inoculated into fresh growth medium, there is usually no immediate increase in cell numbers. A short period, called Lag Phase is needed for microbial cells to adapt to the new environment, where cells resuscitate eventually and are ready to replicate DNA.

Exponential phase or log phase

In the *Log Phase*, microbial cells are growing and dividing at the maximal rate as possible, with their genetic potential and rich nutrients in medium. Rate of cell growth is constant (microorganisms are dividing and doubling in logarithmic form) during this phase and cell population is most uniform in terms of chemical and physiological properties. During exponential phase, each microorganism is dividing at constant intervals; thus population will double in number during a specific length of time - generation time or doubling time. The doubling time of *E. coli* is only 20 minutes.

Stationary phase

In the *Stationary Phase*, the increase of cell population ceases and the growth curve becomes horizontal. Total number of viable microorganisms remains constant or balanced between cell division and cell death. The population may cease to divide though remaining metabolically active. The reasons why microorganisms enter stationary phase are as flows: depletion of essential nutrients; insufficient oxygen supply; feedback inhibition from the end-product which might be toxic to the cells; reaching of high cell density etc.

Death phase

In the **Death Phase**, nutrient depletion and buildup of toxic waste lead to the decline in the number of viable cells. Microbial population is also in logarithmic decline (constant proportion of cells die every hour). During death phase, cells lost their metabolic activities, and the dying cells are even destructed by self-lyses.

17.3.2 Growth of other microorganisms

Yeast growth

Many species of yeasts may undergo both vegetative and sexual reproduction. Budding is the most common mode of vegetative growth in yeasts, and multilateral budding is a typical reproductive characteristic of yeasts. Yeast buds are initiated when mother cells attain a critical cell size at a time coinciding with the onset of DNA synthesis. This is followed by localized weakening of the cell wall and this, together with tension exerted by turgor pressure, allows extrusion of cytoplasm into an area bounded by new cell wall material. The regulation of cell wall synthetic enzymes and the transport of specific bud plasma membrane receptors are key steps in the emergence of a bud. Chitin forms a ring at the junction between the mother cell and the newly emerging bud, resulting the generation of a daughter cell. After cell separation, this ring will be retained at the surface of the mother cell, form the so called bud scar (Fig. 17.7), and a birth scar at the surface of the daughter.

In the mode of sexual reproduction, the yeast ascospores of different sexual genders might mate to form a new progeny. Mitotic division is initiated when yeast attains a critical cell size, and it is stimulated by exogenous or endogenous signals during interphase. A prerequisite is that DNA synthesis guarantees proper duplication of the genetic material and that all substrates critical to anabolic pathways are available. In yeast cells, the mitotic phase can be subdivided into a number of steps that are characterized by particular states of orienting the spindle pole bodies, sorting and movement of the duplicated chromosomes, and finally, cell separation after cytokinesis. While all processes pertinent to the cell cycle have been highly conserved among ascomycetous fungi and *S. cerevisiae* in particular, exhibition of several peculiarities with respect to cell division and cytokinesis are put on.

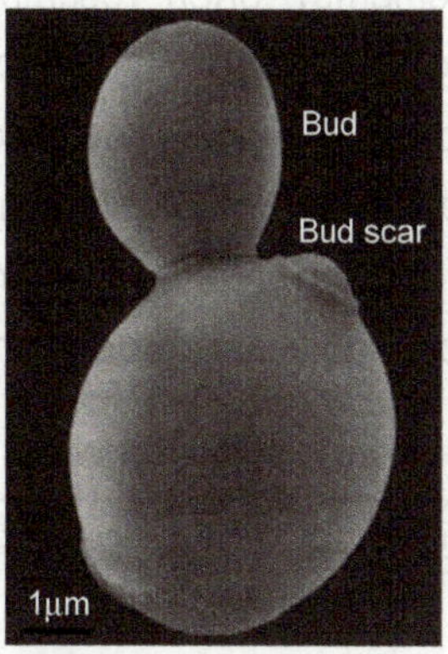

Fig 17.7 Scanning electron micrograph of budding yeast

Mold growth

Very few molds are composed of a single cell that is reproduced via binary fission, while others can be enormous, taking up entire fields. Most of the common molds are reproduced from spores. Mature fungi release these spores and they subsequently begin to divide and eventually form a structure called hyphae, which is thread-like root that made of haploid cells. When hyphae meet two different spores, they intertwine and form one cell with two distinct nuclei, and eventually sprout the fungus progeny.

Fungi reproduction is very complex and many fungi using more than one form of reproduction models mainly depend on the given condition in their life cycle. So called anamorphic stage means that the molds are undergoing the asexual reproduction by dropping a variety of spores with distinct hyphae that eventually grow into a new fungus. This process is much faster than sexual reproduction and results in clonally environments.

17.3.3 Culture media

Defined and complex media

Defined media contain glucose, $(NH_4)_2SO_4$, KH_2PO_4, $MgCl_2$ and other trace elements dissolved in water, and the concentration of each component would be defined. Complex media contain natural ingredients, such as yeast extracts, peptone, molasses, and corn steep liquor. The chemical compositions of the ingredients are not known exactly, but they usually provide the essential growth factors, such as vitamins, hormones, and trace elements, often resulting in higher cell yields, compared to the chemically defined medium. Often, complex media are less expensive than defined media. The primary advantage of defined media is that the results are more reproducible and the operator can easily control the fermentation. Further, the recovery and purification of fermentation product would be easier and cost-lower. The examples of typical defined media and complex media are summarized in Tab. 17.2 and Tab. 17.3.

Tab. 17.2 Example of typical defined media

(In the mode of batch fermentation) Constituent	Purpose	Concn. (g/L)
Basic composition		
Glucose	C, energy	30
KH_2PO_4	K, P	1.5
$MgSO_4$	Mg, S	0.6
$CaCl_2$	Ca	0.05
$Fe_2(SO_4)_3$	Fe	15×10^{-4}
$ZnSO_4 \cdot 7H_2O$	Zn	6×10^{-4}
$CuSO_4 \cdot 5H_2O$	Cu	6×10^{-4}
$MnSO_4 \cdot H_2O$	Mn	6×10^{-4}
Additional ingredient: A		
$(NH_4)_2HPO_4$	N	6
$(NH_4)H_2PO_4$	N	5
Additional ingredient: B		
$C_6H_5Na_3O_2 \cdot 2H_2O$	Chelator	4
Additional ingredient: C		
Na_2HPO_4	buffer	20
KH_2PO_4	buffer	10

Tab. 17.3 Example of complex media in penicillin fermentation

(In the mode of fed-batch)	
Glucose or molasses	10% of total
Corn steep liquor	1%~5% of total
Phenylacetic acid	0.5%~0.8% of total
Antifoam	0.5% of total
pH 6.5~7.5	

One may already notice that any microbe requires water, (oxygen), energy source, carbon, nitrogen source and inorganic elements for growth. It can be summarized in the overall fermentation system as follows: Carbon source + nitrogen source + energy + O_2 + other growth nutrients ⟶ Biomass + Product + byproducts + CO_2 + H_2O + heat.

Carbon source

Starch is the most important carbon source used in fermentation processes. It may be used in the form of crushed cereal grains such as corn, rice, wheat, or smashed roots of potato and cassava. Cellulose present in wood is also a potential carbohydrate source if pretreatment on wood fiber will be easier and cost-lower. Sucrose is available for use in fermentation processes normally in syrup form from molasses, a by-product of sugar refinery, or from cane juice. Lactose presents in whey at a concentration of 4%～5% and whole whey or deproteinized whey is used as a cheap source of carbohydrate in some alcohol production processes. Glucose is usually prepared by direct enzymatic conversion of starch for use in fermentation media. In some instances more expensive refined glucose, as syrup or in crystalline form, is used for production of "small volume and high value" pharmaceutical products.

Nitrogen

Nitrogen is also very essential to microbes, since nucleotide, protein and other biochemicals contain N element. Most industrial microbes can utilize both sources of nitrogen, inorganic or organic. The routine sources of inorganic nitrogen for fermentation are ammonia, nitrates, and urea. When ammonium sulphate is used as inorganic nitrogen source, it will create acid conditions as the ammonium ion is utilized and the free acid will be liberated. On the other hand, ammonia gas and nitrates will normally cause an alkaline drift as they are metabolized.

Organic nitrogen presents in cereals, root crops and protein rich hydrolysate. The complex substrates not only contain nitrogen, but also vitamins, growth factors and minerals, which have a key influence on the fermentation process. Organic nitrogen may also be supplied by amino acid, protein or urea. Pure amino acids are only used in very special cases in fermentation processes usually as precursors or for particular amino acid auxotrophic strains, and those amino acid that defect genotype might have an absolute requirement for corresponding amino acids. In many instances, cell grows faster with a rich supply of organic nitrogen. Commonly used natural carbon and nitrogen sources are listed in Tab. 17.4.

Water

Without water, any living activities of a life will not occur. Water is definitely the key component in all kinds of microbial growth, and it is needed in many of the ancillary services, such as sterilization and cooling system. Clean water of consistent composition is therefore required in large quantities from reliable permanent sources. Some of the factors which need to be considered might include pH, dissolved solids and the impurity in water. Pretreatment of water will be a critical step in a specific fermentation procedure.

Tab. 17.4 Commonly used natural carbon and nitrogen sources

Carbon sources	Nitrogen sources
Starch	Soya meal
Cane juice	Yeast extract
Molasses	Beef extract
Whey	Dried blood
n-Alkanes	Corn steep liquor
Oil	Fish solubles
Celluloid material	Cotton or rape seed extract
Domestic sewage	Distillers solubles

Energy

Energy for growth comes from either the oxidation of medium components or from the sun light. Most industrial micro-organisms are chemo-organotrophs. Therefore the commonest sources of energy will be the carbon sources such as carbohydrates, lipids and proteins. Some of the micro-organisms can also utilize methane, methanol, even paraffin as carbon and energy sources. Existent of chloroplast in algae cells makes them have similar metabolic pathway as plants. By utilization of light, CO_2, and other chemical elements from sewage, algae might play an important role to environmental protection, if one establishes some rational processes to make the blue algae itself a valuable bioresource.

Vitamins

Vitamins are bioactive substances and some of them are co-enzymes or metabolic precursors. Many of the natural carbon and nitrogen sources contain all or some of the required vitamins. When there is a vitamin deficiency, one should be able to eliminate the deficiency by careful blending of organic materials. It is important to remember that if only one vitamin is required, it may be sometimes more economical to add the pure vitamin.

Oxygen

Oxygen is a very important component for microbial growth and metabolism, and its availability can be extremely important in controlling growth rate and metabolite production. The medium may influence the oxygen availability in a number of ways including the following: 1) Fast metabolism; 2) Medium with solid; 3) Rheological property; 4) Antifoams added. In aerobic fermentation, the air compressor is the key facility to provide the aeration to the medium and ensure there is enough dissolved oxygen in the medium to support cell growth and metabolite production. The aseptic air filter keeps germ and dust particle away from the air inlet, and prevent contamination.

Precursors and metabolic regulators

Some components in a medium help to regulate the production of a specific metabolite rather than to support cell growth. Such additives include precursors, inhibitors and inducers, all of which may be used to manipulate the metabolism in the progress of fermentation.

17.4 Inoculation of Industrial Strains

17.4.1 Inoculation criteria

The development of the inoculum, which goes through various stages before the inoculation off the production fermentor, has a definite effect on the subsequent performance of the process. In some cases, a bad preparation of inoculum may ruin the fermentation and result in complete loss of productivity. Hence, it is essential that the culture used to inoculate fermentation meets the following criteria:

(1) In a healthy, active state, thus minimizing the lag phase in the fermentation.

(2) To provide an inoculum of optimum size.

(3) Keep in a suitable morphological form.

(4) Retain its productivity, without degeneration.

(5) Extremely free of contamination.

In laboratory research, it is also imperative, when attempting to optimize a process, that a reproducible and reliable baseline condition is achieved. Without such a baseline, any increases in yield/productivity are difficult to assess.

The simplest inoculation procedure uses vegetative seed stages on similar medium all the way through the stages from shake flask to production fermentor. This implies vigorous growth throughout the stage. The media should not become exhausted or highly unbalanced prior to the transfer to the next stage. The desirable amount of inoculum is at least 5% of the next volume. In a laboratory fermentation, the inoculum stages rarely exceed two stages, whereas as many as six may be required for certain large scale fermentation.

When shake flasks are used to prepare inocula for laboratory fermentors, it is very difficult to obtain online information regarding the physiological state of the micro-organisms contained within the culture. Dry cell weight, viable counts etc., are of little use when the delay time involves in such analysis. Without such information, it is not easy to predict which part of the growth curve the inoculum is in prior to the inoculation of the production fermenter. The difficulty in identifying the physiological state of the industrial inoculum is made worse when using complex media (containing sparingly soluble components) where even visual estimation of growth is not possible. Sometimes the experience experts are needed to monitor the process.

17.4.2 Inocula for bacterial process

Inocula for aerobic and anaerobic processes are rather different. The main objective of developing inoculation procedure for bacterial fermentations is to produce an active inoculum which will ensure as short a lag phase as possible in subsequent culture. A long lag phase is disadvantageous in that not only it cause time wasted but also the medium is consumed in maintaining a valve culture prior to growth. An example of inoculation for an aerobic bacteria process being introduced is related to *Corynebacterium glutamicum*. It is of immense industrial significance and is one of the biotechnologically most important bacterial species with an annual production of more than two million tons of amino acids, mainly L-glutamate and L-lysine. In the glutamate fermentation, *Corynebacterium glutamicum* cultures with urea and carbohydrates from sugar beets, sugar cane, tapioca or molasses, excretes amino acids into the culture broth from where L-glutamate is synthesized. The formula of media should be precisely made to ensure the fermentation providing as much glutamate as possible.

Another example of anaerobic fermentation is provided by the *Clostridial* acetone-butanol fermentation. Although this process is now only in operation in South Africa, it was a very important development in the fermentation industry during the First World War, and may even be an economic process in the future.

17.4.3 Preparation of actinomycetes inoculum

The vast majority of industrially important actinomycetes will produce asexual spores, and it is common practice to use sporulation medium first for preparing as many spores as possible and scribe them from the slant to make a spore suspension. Little information is available on the morphological mechanisms of actinomycetes when growing in the submerged culture, but it appears that forming of pellet does play an important role in further fermentation. Lawton et al. have demonstrated that a wide range of pellet morphological types were found amongst the *streptomycetes*, and that the morphological form of some species was influenced by the concentration of spores in the inoculum, the medium composition and the shear forces operating during culture. However, Stark and Smith observed that the agar used to solidify the medium might have an inhibitory effect on the growth and/or sporulation of the organism and that this detrimental action might be reduced by washing the agar to remove toxic substances. The examples of sporulation media for some industrial streptomycetes are given in Tab. 17.5.

Tab. 17.5 Example media for the sporulation of streptomycetes

Organism	Product	Medium		Reference
S. aureofaciens	Tetracycline	Beef extract	2.0%	Di Marco and Pennella(1959)
		Asparagine	0.05%	
		Glucose	1.0%	
		K_2PO_4	0.05%	
		Agar	1.3%	
S. aureofaciens	Tetracycline*	Malt extract(Difco)	1.0%	William et al. (1974)
		Yeast extract(Difco)	0.4%	
		Glucose	0.4%	
S. erythruys	Erythromycin*	Beefextract(Difco)	0.1%	Williams et al. (1974)
		Yeast extract(Difco)	0.1%	
		Casamino acids(Difco)	0.2%	
		Glucose	0.2%	
S. melanochronogenes	Actinomycin*	l-asparagine	0.1%	Williams et al. (1974)
		Beef extract(Difco)	0.2%	
		Glucose	1.0%	
		KH_2PO_4	0.025%	
S. vinaceus	Viomycin*	Corn-steep liquor	1.0%	Williams et al. (1974)
		Starch	1.0%	
		$(NH_4)_2SO_4$	0.3%	
		NaCl	0.3%	
		$CaCO_3$	0.3%	

17.4.4 Yeast and fungi inoculation

The commercial production of "bakers" yeast involves the development of an inoculum through several stages. For mass production of yeast cell, every inoculation stages should be kept under aseptic condition. The development of inoculum for ethanol production by yeast *Sacchromyces* involves five stages, the first two being aseptic while the remaining stages are carried out in open vessels, gradually scaled up, since the contaminants might have weaker growth tendency in the alcohol containing medium. The culture media of yeast are various, for example, there are YPD and YPAD medium, Hartwell's complete (HC) medium, Synthetic complete (SC) medium, and Yeast Nitrogen Base (YNB) etc. The fitness of different media to a particular yeast strain should be carefully studied before using.

The preparation of inocula for fungal fermentations is probably more involved than that for bacterial and yeast processes. The majority of industrially important fungi is capable of asexual sporulation, so it is common practice to use a spore suspension and let it germinates and being used as the seed to cultivate fungal mycelia for next stage of inoculum development.

As shown in below, there are many formula under different purposes, and some of the typical culture media of fungi include: Water Agar (WA)—— for isolating fungi from surface-sterilized substrates; Antibiotic Agar (AA)—— for isolating fungi from substrates not readily surface-sterilized, or cleaning up a culture contaminated with bacteria; Acidified Cornmeal Agar (ACMA)——for isolating fungi from substrates with acidity inhibition on bacteria and the medium supports the growth of a wide range of fungi; Cornmeal Agar (CMA) —— Provides a good balance of mycelial growth and sporulation; Potato Carrot Agar (PCA) ——considered a relatively weak medium somewhat comparable to CMA; Malt Agar (MA) ——Lacks peptone, and is useful for culturing many *Ascomycota*; the sporulation in some species is inhibited by peptone; Malt Extract Agar (MEA) ——Growth medium for soil or wood fungi, such as, basidiomycetes, etc, also an all-purpose type of medium; Potato Dextrose Agar (PDA)——A relatively rich medium for growing a wide range of fungi; Potato Dextrose-Yeast Extract Agar (PDYA) ——For growing cultures derived from mushrooms.

17.5 Preservation of Industrial Strains

17.5.1 Preservation of industrial strains

Industrial microorganisms must be kept pure and with excellent traits for the used in the production. So the preservation of those strains is important. The culture must be pure without contamination during the preservation process, i. e. at least 70% to 80% of the cells must survive during storage, and the cells must remain genetically stable. Varieties of preservation methods can be employed, but for particular strains no single method is applicable to all situations.

The spores of mold can be mixed with clean sand particles and dehydrated into a dried state; bacterial cells might be kept in defatted milk powder and be protected by protein molecules, and lyophilization became very popular for preservation. In many cases, microbial cultures are preserved at low temperatures, in order to suspend growth processes. For short periods, most of the strains are kept at refrigerator temperature (around 4℃), but for longer-term storage, the cells are suspended in protective medium and are kept at −80℃ (under deep freezing) even for several years, this guarantees the maximum genetic stability of the stored strains.

17.5.2 Revitalization of industrial organism

Whatever the method used to preserve microbial strains is, the cells are placed under stress (particularly during lyophilization). Consequently, the stored cells are not suitable to be used directly. They must be revitalized first or rejuvenated by seeding in nonselective medium, i. e. a medium that ensures maximum growth. After this step has been performed, it is possible to inoculate the cells in selective media to allow the industrial traits recovered. Likewise, we should remember that some microorganisms will tolerate a given preservation technique better than others and that special precautions may be necessary for the preservation of certain strains. As is mentioned at the beginning of this chapter, there is no single method that is appropriate for the preservation of all microorganisms, but it is not difficult to identify the most appropriate method in each case. By the way, the cold stored cultures should be thawed rapidly (37℃ water bath), since slow thawing causes **recrystallization**. Lyophilized or dried cultures must be rehydrated for a few minutes in an appropriate liquid medium. In both cases the reconstituted cells must be inoculated into appropriate culture media as soon as possible.

Summary

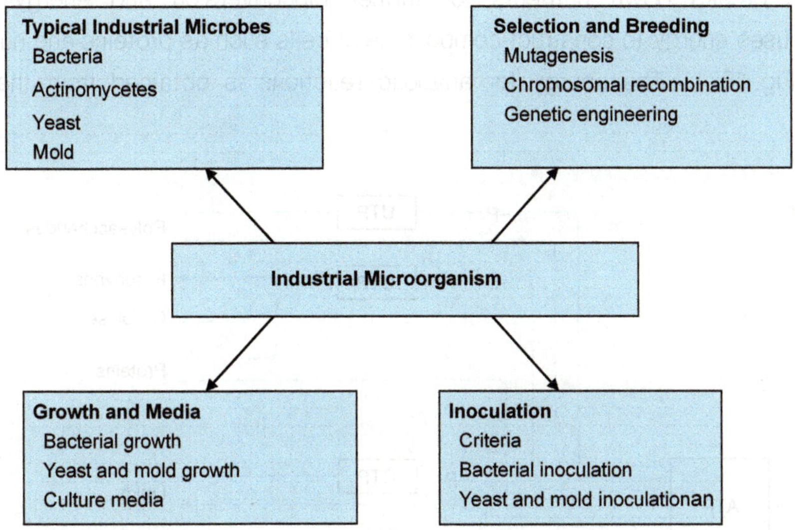

Chapter 18 Microbial Metabolism and Fermentation Kinetics

微生物代谢是工业发酵的核心，是生化反应工程学的基础理论，直接指导着生物技术实践。本章介绍微生物代谢的基本途径；次级代谢；呼吸与能量代谢；讨论了通过对酶的合成与活力的调控进行理性的工业发酵。在此基础上对微生物发酵过程化学计量学进行描述；最后就微生物反应和代谢动力学，以及批次发酵和连续发酵过程的基本动力学进行了阐述。

18.1 Metabolism Related to Basic Pathway

Metabolism is defined as the sum of all biochemical reactions occurring within a living organism, which can be divided into two categories: Catabolism and anabolism. Catabolism breaks down nutrients for further bioconversion and energy obtaining; anabolism uses energy to construct components of cells such as proteins and nucleic acids shown in Fig. 18.1. The energy for anabolic reactions is obtained from the catabolic reactions.

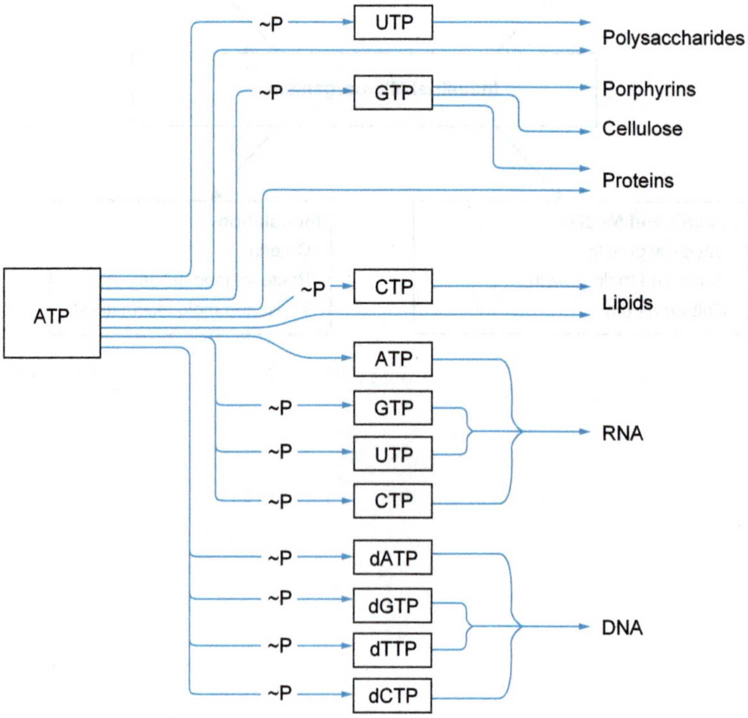

Fig. 18.1 Macromolecular anabolism related to ATP

18.1.1 Carbon catabolism

Carbohydrates are the most important substrates for fermentation, although some of the microbial species can also utilize hydrocarbon, carbon dioxide, amino acids or other compounds. It's notable that single sugar glucose is the widely employed carbohydrate for fermentation. There are different glucose fermentation pathways, and which one can be regulated and what kind of end products would be produced depend on the individual microorganism employed. The main concerns in this section will be the observation of basic metabolic pathways and the related intermediates which are important to fermentation industries.

EMP cycle

As the essential pathway for glucose catabolism, the Embden-Meyerhof-parnas pathway (also known as EMP pathway, seen in Fig. 18.2) involves ten enzyme-catalyzed steps which start with glucose and end with **pyruvate**. Each step in EMP pathway is simple and involves the reactions of isomerization, ring splitting, or transfer of small group

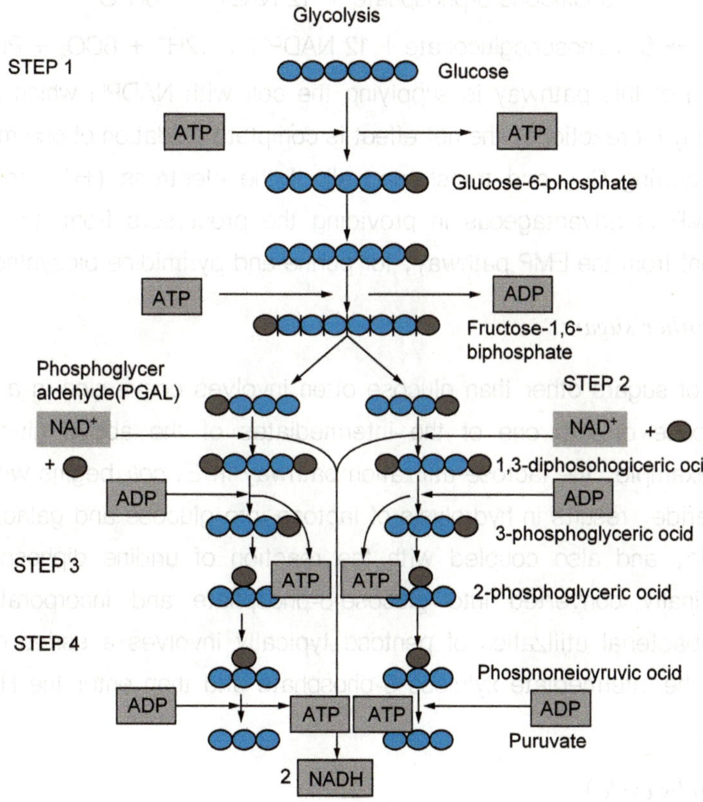

Fig. 18.2 EMP pathway (Glycolysis pathway)

such as H_2O, hydrogen or phosphate to certain compounds. The critical fact is that two moles of pyruvate are produced per mole of glucose through the EMP pathway. The overall stoichiometry of EMP pathway is

$$C_6H_{12}O_6 + 2\ NAD^+ + 2\ ADP + 2\ H_3PO_4 \longrightarrow 2\ NADH + 2\ C_3H_4O_3 + 2\ ATP + 2\ H_2O + 2\ H^+ \quad (18.1)$$

The energy storage accomplished by this or other substrate catabolic pathways is called substrate-level phosphrylation.

Another key function of EMP pathway is provision of carbon skeletons for cellular biosynthetic reactions. Not only the pyruvate, but also the intermediates in the pathway can be used as further substrates for biosynthesis of various products. Most of microbes, animals and plants employ EMP as the core pathway for their metabolism; biotechnologists should be acknowledged with the details in EMP pathway.

Pentose phosphate cycle

Another glucose catabolism named pentose phosphate cycle or hexose monophosphate pathway (HMP) begins with the oxidizing of glucose phosphate

$$6\ \text{Glucose 6-phosphate} + 12\ NADP^+ + 6H_2O \longrightarrow 5\ \text{6-phosphogluconate} + 12\ NADPH + 12H^+ + 6CO_2 + Pi \quad (18.2)$$

A major function of this pathway is supplying the cell with NADPH which in turn carries electrons to many bioreactions; the net effect is complete oxidation of one mole of glucose 6-phosphate liberating CO_2 and transferring all of the electrons (H^+) to NADP. More importantly, HMP is advantageous in providing the precursors from its intermediates, which are absent from the EMP pathway, for purine and pyrimidine biosynthesis.

Utilization of other sugars

Utilization of sugars other than glucose often involves conversion in a few reactions either into glucose or into one of the intermediates of the above glucose catabolic pathway. For example, the lactose utilization pathway in *E. coli* begins with assimilation of this disaccharide, results in hydrolysis of lactose into glucose and galactose. Through cyclic, catalytic, and also coupled with the reaction of uridine diphosphate (UDP), galactose is finally converted into glucose-6-phosphate and incorporated into EMP pathway. The bacterial utilization of pentose typically involves a series of conversions which produce the intermediate xylulose 5-phosphate and then enter the HMP for further metabolism.

TCA cycle (Krebs cycle)

Krebs cycle was named after professor Krebs, for he has described the Citric acid cycle — also known as the tricarboxylic acid cycle (TCA cycle) in a well accepted pattern (Fig. 18.3). The Krebs cycle— is a series of enzyme-catalyzed chemical reactions, which is of central importance in all living cells that use oxygen as part of microbial respiration. Throughout the entire cycle, acetyl-CoA changes into citrate, isocitrate, α-ketoglutarate, succinyl-CoA, succinate, fumarate, malate, and finally, oxaloacetate, which are very important intermediates for other pathway or they could be regulated to secrete out of the cells as fermentation products. The mechanism of TCA cycle is the key of the organic acid fermentation. Also TCA cycle is critically important to amino acid synthesis.

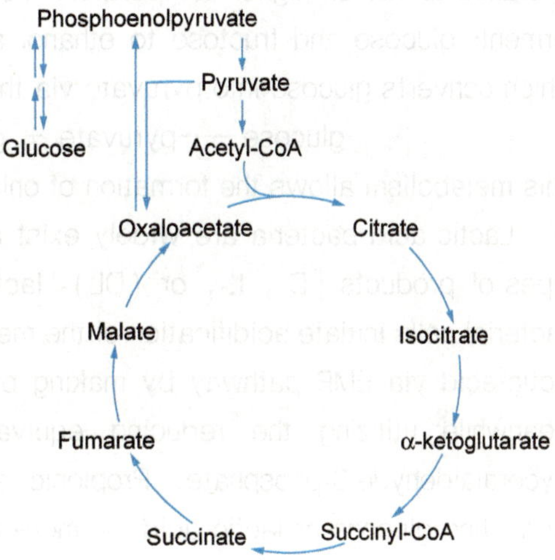

Fig. 18.3 TCA cycle

Anaerobic metabolic pathway and related fermentation products

The section is organized according to the fact that many small molecular products are metabolited under anaerobic reaction, such as methane, ethanol, lactic acid, acetone, propionic acid, butyric acid and butanol etc, which encompass the parts of traditional fermentation products. For example, Methanogens play a key role in the carbon cycling by reduce CO_2 to methane and also fixing one-carbon substrates to the cellular carbon for bacterial growth. The application of methanogens is receiving increasing attention to the production of biofuels and SCP from waste disposal, and also to making novel biochemicals such as coenzymes and special metabolites. The rather diverse methanogenic species have common capacity of coupling oxidation of hydrogen with the concomitant reduction of carbon dioxide:

$$CO_2 + 4H_2 \longrightarrow CH_4 + 2H_2O \quad (18.3)$$

These organisms apparently lack cytochromes, Calvin cycle, pentose phosphate pathway and portions of TCA cycle (Fig. 18.3). Regeneration of ATP might involve a membrane-bound ATPase coupled to a transmembrane proton gradient.

Ethanol produced by yeasts under anaerobic conditions is the basis of alcohol industries. The yeast Saccharomyces represents the bulk of starch and molasses-based processes. Through the EMP (Embden-Meyerhof-Parnas) pathway, glucose is finally converted into ethanol and CO_2 under several enzymatic reactions. The conversion of xylose to ethanol by yeast has been a practical project when considering of lignocellulose hydrolysate utilization. Bacterial ethanol attracts producers for their rapid growth and tolerance to rather higher temperature (37℃) than yeast *Saccharomyces*. *Z. mobilis* ferments glucose and fructose to ethanol and CO_2 by a different pathway for example, which converts glucose into pyruvate via the ED (Entner-Doudoroff) pathway:

$$\text{glucose} \longrightarrow \text{pyruvate} + \text{glyceralaldehyde-3-phosphate} \quad (18.4)$$

This metabolism allows the formation of only 1mol of ATP per mole from glucose.

Lactic acid bacteria are widely exist and can be grouped depending on the optical types of products [D-, L-, or (DL)- lactic acids]. The homofermentative lactic acid bacterial cells initiate acidification of the medium, and convert 1mol of glucose into 2mol of lactic acid via EMP pathway by making pyruvate first, then reduce it into lactic acid, meanwhile utilizing the reducing equivalents generated in the EMP oxidation of glyceralaldehyde-3-phosphate. Propionic acid is formed by some anaerobic bacteria fermenting glucose or lactic acid. A more complex series of reaction by which lactate is converted to propionate relates succinate-propionate pathway which involves a carbon fixation step in clostridium species.

Acetone formation is observed primarily in *Clostridium acetobutylicum*, which is responsible for commercial fermentation of acetone-butanol, and the acetone is produced via:

$$\text{Acetoacetate} \longrightarrow \text{acetone} + CO_2 \quad (18.5)$$

And butanol is converted from butyrate through the involvement of acetoacetate CoA: butyrate-acetate-transferase and two dehydrogenase reactions. Although the related anaerobic acetone-butanol fermentation is no longer to be performed in China, but the commercial processed of acetone-butanol fermentation is still being carryied out at large scale in African country.

18.1.2 Basic anabolism

Anabolism uses energy and monomeric building blocks to synthesize cell components by joining the mono-units such as deoxynucleotides, nucleotides, amino acids, saccharides and lipids to form deoxyribonucleic acids, ribonucleic acids, proteins, polysaccharides and complex lipids respectively. In the case of synthesis of peptide molecules, 20 different amino acids are involved. Living cells assimilate nitrogen by incorporating it into glutamic acid first, and the reaction between ammonia and α-ketoglutaric acid (one of the TCA intermediates). Amino acids synthesis can be grouped based on different precursors from basic pathways, as shown in Tab. 18. 1. All the branched pathways are investigated, the fermentation engineers are now able to regulate the biosynthesis of each amino acid and produce them in bulky volume.

Tab. 18.1 Precursors of amino acid synthesis

Precursors	Amino acids	Group
α-ketoglutarate	Glutamate Glutamine Arginine Proline	Glutamate
Oxalacetate	Aspartate Asparagine Threonine Methionine Lysine (in part) Isoleucine (in part)	Aspartate
Pyruvate	Alanine Valine Leucine (in part)	Pyruvate
Phosphoenolpyruvate + erythrose-4-phosphate	Phenylalanine	Aromatic
3-phosphoglycerate	Serine Glycine Cysteine	Serine
Phosphoribosyl Pyrophosphate + ATP	Histidine	Histidine

In the case of nucleic acid synthesis, the ribose as one of the building blocks of nucleotide is converted through HMP, and several amino acids (serine, glycine, glutamine and aspartate etc), phosphate and CO_2 are also the substrates for purine and pyrimidine ribonucleotides (or deoxyribonucleotides) synthesis (Fig. 18.4).

Now based on the advanced biotechnology, all the nucleotide monomers and DNA or RNA molecules with different sequences can be synthesized for commercial or laboratory uses. The synthesis of informational biopolymers is a considerably molecular biological process and will not be discussed in this section.

18.1.3 Secondary metabolite synthesis

Most of the organisms at cellular level, including microbes, are able to synthesize secondary metabolites when the cells and their environment are faced to special conditions. Many of secondary metabolites are produced by microbial fermentation or botanic cell cultivation. Normally, these compounds are not synthesized in substantial quantity during the log phase of cell growing. More and more secondary metabolites have

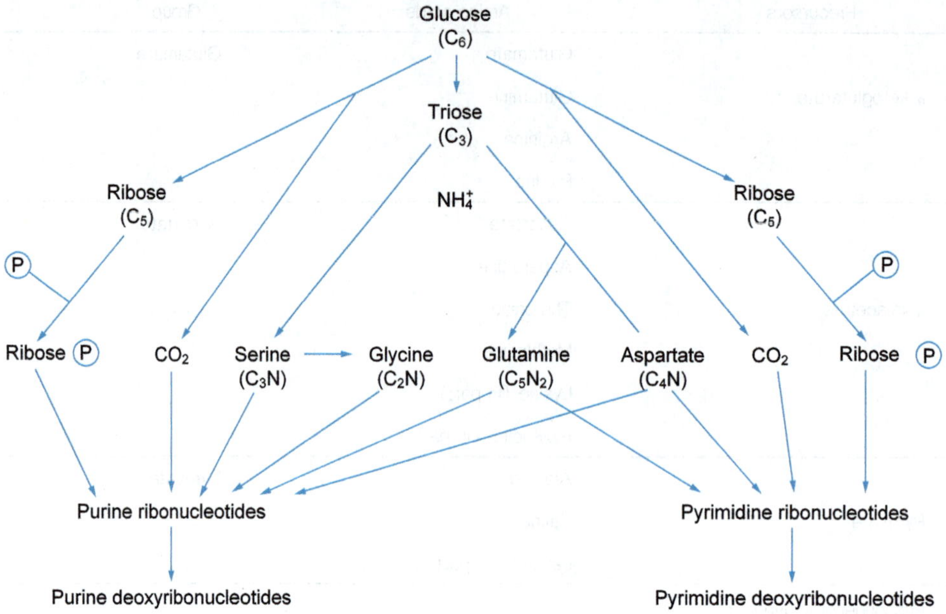

Fig. 18.4 Flowchart of nucleotides basic anabolism

been isolated and many of them are found to have very complicate chemical structure and sophisticate biological functions. Most of the antibiotics (thousands have been found) belong to secondary metabolites which are playing critical roles on human health. Even single microbial specie might produce nearly hundreds of different antibiotics, as in the case of *Streptomyces griseus*. It's clear that many primary metabolites are the raw materials for secondary metabolite synthesis; the intermediates in EMP, HEP, TCA pathways and also amino acids or nucleotides are often employed by individual secondary metabolic routes to form more complicate compounds.

Also the energy from the cellular metabolism are usually provided by the catabolic reactions and quite often involved with ATP, which is considered as one of the most important modulators for secondary metabolism. That's why the phosphate concentration of fermentation medium must be kept in an optimal level to favorite the production of secondary metabolites.

18.1.4 Energy metabolism

Based on energy metabolism, microorganisms can be divided into **light autotrophic**, **chemoautotroph** and **chemoheterotrophs**: The energy in sunlight is captured by the light autotrophic organisms, such as, plants, **cyanobacteria**, **purple bacteria**, **green sulfur bacteria** and some **protists**. This process is often coupled to the conversion of carbon dioxide into organic compounds, as part of photosynthesis. However, the energy capture and carbon fixation systems can be operated separately in prokaryotes, as purple bacteria and green sulfur bacteria can use sunlight as a source of energy, while the carbon metabolism can be switched between carbon dioxide utilization and the fermentation of organic compounds.

Some of the bacteria or protozoa are chemoautotrophic, which obtain their nourishment from the oxidation of inorganic chemical compounds as opposed to photosynthesis. In addition to deriving energy from chemical reactions, chemoautotroph synthesizes all necessary organic compounds from carbon dioxide. Other inorganic energy sources, such as hydrogen sulfide, elemental sulfur, ferrous iron, molecular hydrogen, and ammonia can also be used by chemoautotroph, which fall into several groups, for example, **nitrobacteria**, **hydrogen-bacteria** and ***acidithiobacillus***. Those microbial species are important for energy recycling and bioleaching.

Most industrial microorganisms are chemoheterotrophs; by secretion of the digestive enzymes into their surroundings and decomposition of macromolecules into small units which would be further metabolited, those organisms grow quickly and are able to obtain large amount of energy from very efficient oxidative **phosphorylation**, which relates to the cell respiration in details.

Respiration

Respiration means "breath", but also is known as "oxidative metabolism" in the field of biochemistry. It's one of the key activities for a microbe to gain useful energy. There are aerobic and anaerobic respirations with totally different mechanisms.

Aerobic respiration Aerobic respiration is usually illustrated as three processes: the EMP cycle, the Krebs cycle, and the Electron Transport Chain which produces ATP through chemiosmotic phosphorylation. The overall energy conversion is as follows:

$$C_6H_{12}O_6 + 6O_2 \longrightarrow 6CO_2 + 6H_2O + \text{energy (ATP)} \qquad (18.6)$$

The net energy gained from one cycle is 3 NADH, 1 FADH2, and 1 ATP. Thus, the total amount of energy yield from one whole glucose molecule (2 pyruvate molecules) is 8 NADH, 2 $FADH_2$, and 2 ATP. The equation for Krebs cycle:

$$2CH_3COCOOH + 8NAD^+ + 2FAD^+ + 2ADP + 2Pi + 4H_2O$$
$$\longrightarrow 6CO_2 + 8NADH + 2FADH_2 + 2ATP \qquad (18.7)$$

The energy in these molecules will be used to synthesize ATP through the electron transport system. If oxygen is present, therefore each TCA cycle yields ATP of 38 molecules as energy supply for other bioreactions (Tab. 18.2).

Tab. 18.2 Energy resulted from Krebs cycle

Krebs cycle	Substrate phosphorylation	Oxidative phosphorylation
Glucose ⟶ 2Pyruvate		2 * 3ATP
2Pyruvate ⟶ 2Acetyl-CoA	2ATP	2 * 3ATP
2Acetyl-CoA ⟶ $2CO_2 + 4H_2O$	2ATP	2 * 12ATP

Electron transport chain Electron transport chain is an oxidative reaction that affects the transfer of electrons through a series of carriers. This is also called the respiratory electron chain, which is the final stage of aerobic respiration. An electron transport chain (ETC) couples a reaction between an electron donor (such as NADH) and an electron acceptor (such as O_2) to the transfer of H^+ ions across a membrane through a set of mediating biochemical reactions. These H^+ ions are used to produce **adenosine triphosphate** (ATP), the main energy intermediate in living organisms, as they move back across the membrane.

Electron transport chains are used for extracting energy from sunlight (photosynthesis) and from redox reactions such as the oxidation of sugars. Mitochondria existing in the eukaryotic organisms play important roles as the platform for the electron transfer in the oxidative reaction. The overall electron transport chain:

$$\begin{array}{c} \text{NADH} \rightarrow \text{Complex I} \rightarrow \text{Q} \rightarrow \text{Complex III} \rightarrow \text{cytochrome c} \rightarrow \text{Complex IV} \rightarrow O_2 \\ \uparrow \\ \text{Complex II} \end{array}$$

Anaerobic respiration The term anaerobic respiration in biology means a way for a life form to generate usable energy beyond the oxygen. So in brief it involves chemical oxidation under anaerobic condition. Anaerobic respiration starts by breaking down the molecules of glucose and producing **pyruvic acid**. This acid then converts into smaller compounds and produces ATP, which is used by the cells as energy supply. The anaerobic respiration is the oldest model of cellular respiration. This type of respiration, primarily works by fermentation, is also known as glycolysis. In this process, the pyruvate is converted to lactate in a process called lactic acid fermentation: pyruvate + NADH + H^+ ⟶ lactate + NAD^+. This process occurs in lacto-bacteria involved in making yogurt (produced lactic acid causes the milk to curdle). Some organisms convert NADH back to NAD^+ in a process called ethanol fermentation. In this process, the pyruvate is converted first to acetaldehyde and carbon dioxide, then from acetaldehyde to ethanol. Yeast cells grow quickly under aerobic condition but produce etnanol by anaerobic fermentation (Fig. 18.5). One might observe the phenomenon of the disappearing of yeast mitochondria when the *Saccharomyces* cells shift from aerobic to anaerobic fermentation.

18.2 Metabolic Regulation and Control

Metabolic regulation allows microorganisms to respond to signals and interact actively

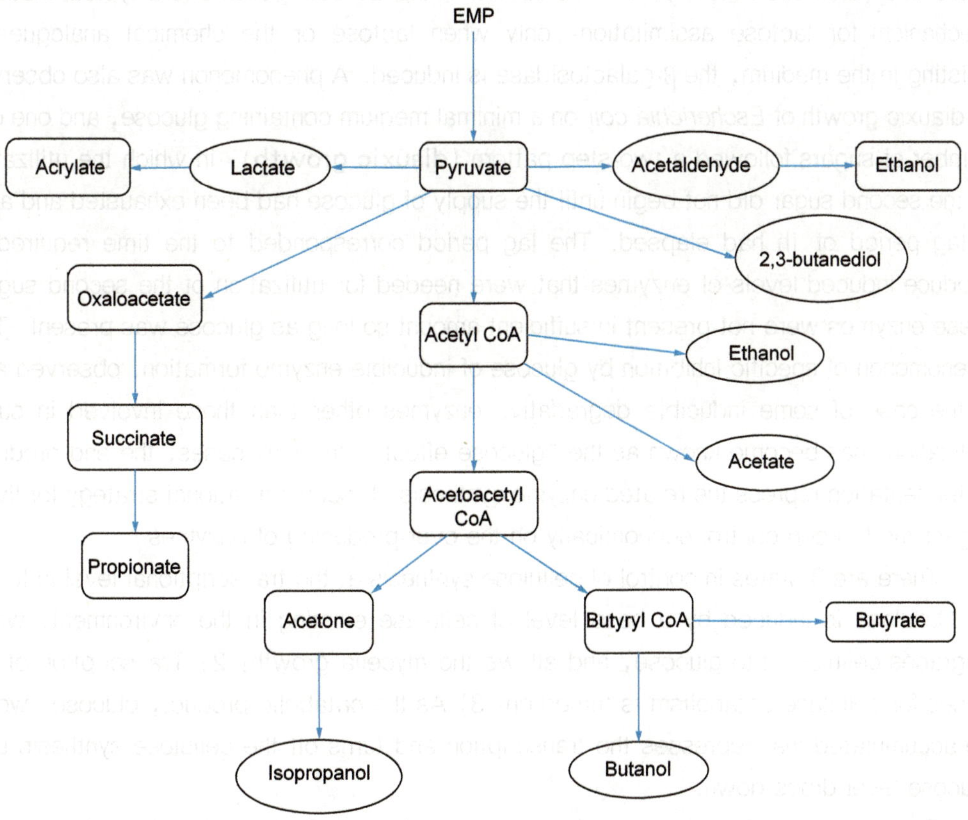

Fig. 18.5 End products of fermentation from pyruvate

with the changed environments. There are multiple levels of metabolic regulation. Extrinsic control mainly involves the control of fermentation conditions, such as pH, dissolved oxygen and temperature. In intrinsic regulation, the metabolic pathway self-regulates to respond to changes in the levels of substrates or products, for example the regulation of enzyme activity. Since most of the metabolic activities are catalyzed by enzymes, the timing of enzyme synthesis and the amount of enzyme produced by cells are also critical to fermentation process. In this section, the control of enzyme synthesis and the regulation of enzymatic activity are the main discussed. Usually, the targets of the self-regulation are the key enzymes, for example, through the mechanisms of induction or repression, the enzyme synthesis can be either turned up or closed down, and through the substrate activation or product inhibition the enzyme activity can be magnified or reduced. The self-regulation patterns in living organisms are now clearly revealed, and most of them have been utilized in the state of arts.

18.2.1 Example of control on enzyme synthesis

In Fermentation process, some of the enzymes are induced while some others are repressed (also see Part III). The lac operon in the *E. coli* genome is a typical inducing mechanism for lactose assimilation; only when lactose or the chemical analogues is existing in the medium, the β-galactosidase is induced. A phenomenon was also observed in diauxic growth of *Escherichia coli* on a minimal medium containing glucose, and one of a number of sugars followed a two-step pattern (**diauxic growth**), in which the utilization of the second sugar did not begin until the supply of glucose had been exhausted and after a lag period of 1h had elapsed. The lag period corresponded to the time required to produce induced levels of enzymes that were needed for utilization of the second sugar; these enzymes were not present in sufficient amount so long as glucose was present. This phenomenon of specific inhibition by glucose of inducible enzyme formation, observed also in the case of some inducible degradative enzymes other than those involved in sugar utilization, has become known as the "glucose effect". In many cases, the end products of fermentation repress the related enzyme synthesis. It seems a rational strategy for living organisms to keep control economically on the over-producing of enzymes.

There are 3 states in control of cellulose synthesis at the transcriptional level in fungi: 1) Cellulase is induced by a base level of cellulase existing in the environment, which degrades cellulose into glucose, and allows the mycelia growth; 2) Transcription of the genes for cellulase catabolism is turned on; 3) As the catabolic product, glucose, would be accumulated that represses the transcription and turns off the cellulose synthesis until glucose level drops down.

For production of enzymes by fermentation, it's critical to revise the native control mechanism of enzyme synthesis. Usually, the biotechnologists breed the industrial strains with effective mutation for relieving the enzyme induction or repression on genetic level. Gene cloning as the revolutionary technique allows almost any kind of enzymes being produced beyond the primitive cellular control, which pushes industrial enzyme production into the state of arts.

18.2.2 Regulation of enzyme activity

For biosynthesis of primary and secondary metabolites, more concerns would be focused on the regulation of enzyme activities which directly affects the productivity of metabolites. Obviously, the native regulation mechanisms are related to keeping enzyme activity within the microbiological efficiency and well-coordination. For the same reason, cells are not allowed being over-produced in any metabolites. If fermentation engineers would like to achieve a higher production of a specific biochemical, he or she must be acknowledged with the regulation mechanisms, especially the precursor activation and feedback inhibition on enzyme activities.

Precursor activation

The accumulation of a substance within a cell may specifically activate an enzyme that sets in motion a sequence of reactions for which that substance is the initial substrate. This reduces the concentration of the initial substrate. In precursor activation, the regulator molecule increases the affinity of the enzyme in series for its substrate. Actually this may happen in several different ways, 1) Directly - by altering the enzyme itself; 2) Indirectly - through a signaling cascade, such as when the substrate binds a kinase, and the kinase phosphorylates the enzyme, increasing its activity; 3) Gene regulating the substrate binds some sorts of gene regulation elements, which cause an increase in the amount of enzyme within the cell.

Feedback inhibition

Feedback inhibition is a way of regulation for enzymatic system activity exerted via initial enzymes of the system. At least four different types of feedback mechanisms are known that would affect the fermentation productivity. For example, most of the pathways of amino acid biosynthesis are regulated by feedback inhibition.

(1) Sequential feedback inhibition (Fig. 18.6). The common steps are inhibited by the product before the branch, and the first enzyme of each branch is inhibited by the branch product. High levels of F and H inhibit enzyme e_4 and e_6, respectively → D will accumulate →the pathway is inactivated if both F and H are high.

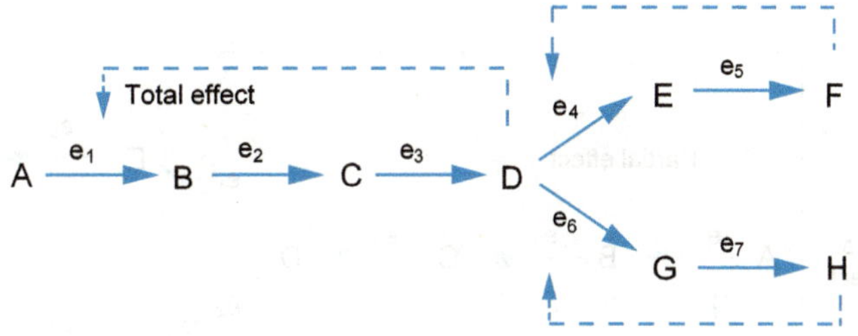

Fig. 18.6 Sequential feedback

(2) Isoenzyme feedback inhibition (Fig. 18.7). Different proteins which catalyze the same reaction are called **isoenzymes**. The catalyzation of isoenzymes can be located either in different places of one cell or in different tissues. It is said, isoenzymes have tissue specificity. A number of separate enzymes initially carry out the same conversion, each of which is sensitive to inhibition by a different end product.

(3) Concerted feedback inhibition (Fig. 18.8). More than one of the end products or all end products must be present in excess to repress the first enzyme.

(4)Cumulative feedback inhibition (Fig. 18.9). A single allosteric enzyme may have

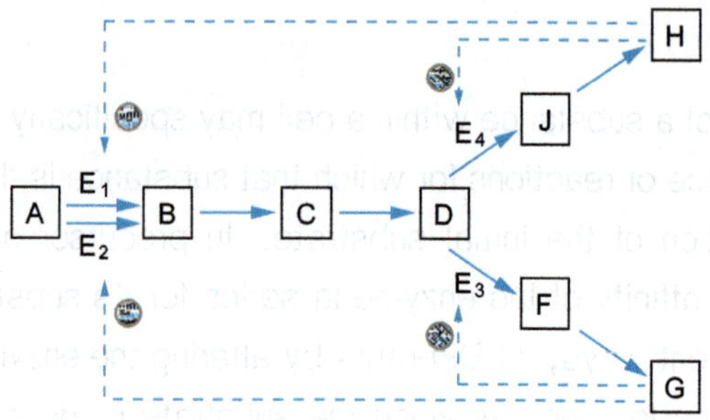

Fig. 18.7 Isoenzyme feedback

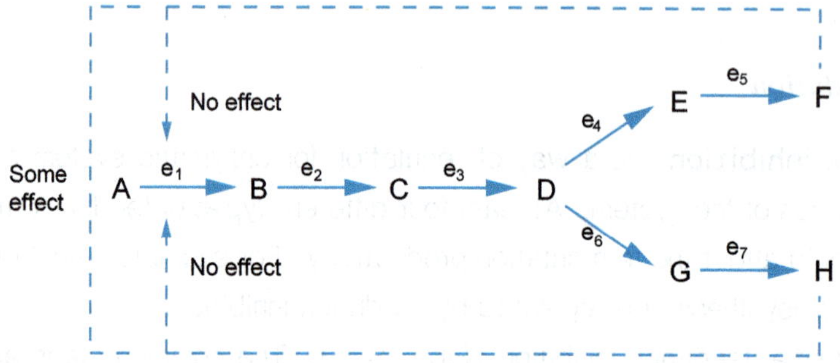

Fig. 18.8 Concerted feedback

effector sites for several end products of a pathway. Each effector causes only partial inhibition and the full inhibition is the result of a cumulative effect.

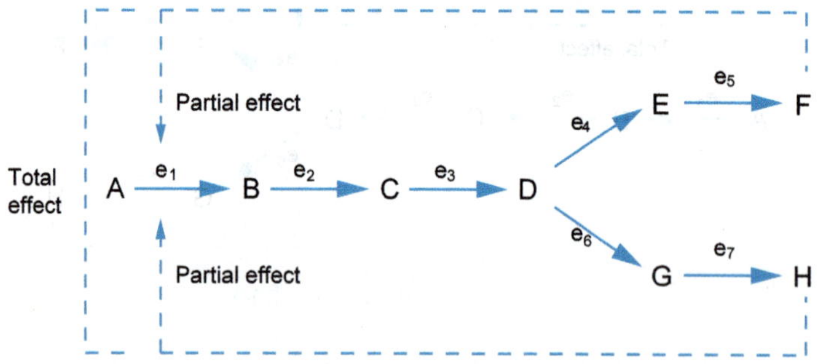

Fig. 18.9 Cumulative feedback

In production of metabolites by fermentation, learning of the related metabolic pathway and the regulation pattern on the enzyme activities would be helpful to the process performance, for example, the amino acid higher producers can be selected by anti-analogues screen system after mutation. The accumulation of branched end products usually means that the related inhibition has been removed.

18.2.3 Other important regulation

Energy charge

Many reactions in primary metabolism are controlled by the energy status of the cell. One index of the energy status is the energy charge, which is proportional to the mole fraction of ATP plus half the mole fraction of ADP, given that ATP contains two anhydride bonds whereas ADP contains one. Hence the energy charge is defined as:

$$\text{Energy charge} = \frac{[\text{ATP}] + \frac{1}{2}[\text{ADP}]}{[\text{ATP}] + [\text{ADP}] + [\text{AMP}]}$$

Regulation of ribosome synthesis

The formation of functional ribosomes is a highly complex process requiring the interplay of a large number of molecular reactions. Ribosome formation in eukaryotic cells involves the participation of three different RNA polymerases for the production of four different rRNAs and about 80 different **ribosomal proteins** (r-proteins), most of which present in a single copy per ribosome. Although there are various stages in the production of the different ribosomal constituents, and they assemble into active ribosomes, which occur in different cellular compartments, normally growing cells do not contain free pools in most of the ribosomal components. Therefore, the expression of the various ribosomal genes must be subject to tight coordinate control to ensure the production of equimolar amounts of the different rRNA and r-protein constituents. In addition, cells can adjust to the expression of the whole set of ribosomal genes in a concerted fashion in response to variations of environmental conditions, such as changes in nutritional conditions or temperature etc., which alter the demand for protein biosynthetic capacity.

18.2.4 Regulation examples of secondary metabolism

Secondary metabolism (also called special metabolism) is a term for pathways and small molecule products of metabolism that are not absolutely required for the survival of the organism. It is a complex process. Examples of the products include antibiotics and pigments. Secondary metabolites are produced by microbes, plants, fungi and animals. Streptomycin is a typical secondary metabolite, and also an important antibiotics produced by Streptomyces bacteria.

Streptomycin is a low-molecular-weight natural product. It is produced only by a limited number of closely related organisms. Production ability is easily lost from natural mutation. **Streptomycin** can be synthesized by washed mycelial suspensions. The antibiotic contains three moieties: streptidine, streptose, and *N*-methyl-L-glucosamine.

All three parts arise from glucose.

Derepression of enzymes responsible for streptomycin production after growth appears to be the regulatory mechanism which triggers antibiotic synthesis. Enzymes shown to be formed only during the idiophase include the amidino transferase and streptidine kinase. The well known phosphate inhibition of streptomycin formation is due to repression or inhibition, or both, of the numerous phosphatases involved in the biosynthesis of the antibiotic, for example, the feedback inhibition may also play a role in streptomycin formation in view of the complete inhibition of streptidine kinase by streptomycin.

18.3 Stoichiometry of Fermentation Process

For many fermentation process calculations, we wish to compare potential substrates in terms of cell mass yield, product yield, or evolution of heat. Also, we may need to know how close to its thermodynamic limit a system is operating. Although the cell is complex, the **stoichiometry** of conversion of substrates into products and cellular materials is often represented by a simple pseudo-chemical equation. Hence, we will discuss how these equations can be written and how useful estimates of key yield coefficients can be made.

18.3.1 Definition of growth rate

As we know, the growth rate of microorganism is defined as:

$$r = \frac{dx}{dt} \tag{18.8}$$

On the basis of r, we defined the special growth rate:

$$\mu = \frac{1}{x}\frac{dx}{dt} \tag{18.9}$$

The special substrate uptake rate and the special generation rate are:

$$q_s = \frac{1}{x} \cdot \frac{ds}{dt}, q_p = \frac{1}{x}\frac{dp}{dt} \tag{18.10}$$

Where,

x = Biomass concentration, mg/L;

t = time, s^{-1};

s = substrate concentration, mg/L;

p = products concentration, mg/L.

The rates of biomass production and product formation are:

$$\frac{dX}{dt} = \frac{dX}{ds}\frac{ds}{dt} = -Y_{\frac{X}{S}} \cdot \frac{dS}{dt} \quad \text{and} \quad \frac{dP}{dt} = -Y_{\frac{P}{S}} \cdot \frac{dS}{dt} \tag{18.11}$$

Where the yields of biomass and product are:

$$Y_{\frac{X}{S}} = -\frac{\Delta X}{\Delta S} \quad \text{and} \quad Y_{\frac{P}{S}} = -\frac{\Delta P}{\Delta S} \tag{18.12}$$

For instance the theoretical yield of ethanol fermentation based on fermentation of glucose results in two moles of ethanol:

$$\frac{2\text{mol EtOH}}{1\text{mol } C_6H_{12}O_6} = \frac{2 \times 46}{180} = 0.511 \frac{\text{g EtOH}}{\text{g } C_6H_{16}O_6} \tag{18.13}$$

The average biomass concentration is defined as the product of yield of biomass and change of substrate concentrations in inlet and outlet streams. The biomass balance is:

$$\bar{X} = Y_{\frac{X}{S}}(-\Delta S) = Y_{\frac{X}{S}}(S_i - S_o) \tag{18.14}$$

Where,

S_i = The end substrate concentration, mg/L;

S_o = The initial substrate concentration, mg/L.

18.3.2 Stoichiometric calculations

Elemental balances

A material balance on biological reactions can easily be written when the compositions of substrates, products, and cellular material are known. Electron balances are required in addition to elemental balances to determine the stoichiometric coefficients in bioreactions. Accurate determination of the composition of cellular material is still a major problem. A cellular composition on average can be represented as $CH_{1.8}O_{1.5}N_{0.2}$. One mole of biological material is defined as the amount containing 1 gram atom of carbon, such as $CH_\alpha O_\beta N_\gamma$.

Consider the following simplified biological conversion, in which no extra-cellular product other than H_2O and CO_2 are produced.

$$C_wH_xO_yN_z + aO_2 + bH_gO_hN_i \longrightarrow c\, CH_\alpha O_\beta N_\gamma + dCO_2 + eH_2O \tag{18.15}$$

Where $C_wH_wO_yN_z$ represents 1mole of carbohydrate, and $CH_\alpha O_\beta N_\gamma$ stands for 1mole of cellular material, the simple elemental balances on C, H, O, and N yield the following equations:

C balance: $w = c + d$ H balance: $x + bg = c\alpha + 2de$

O balance: $y + 2a + bh = c\beta + 2d + e$ N balance: $z + bi = c\gamma$

The material balance in a cell can be described as:

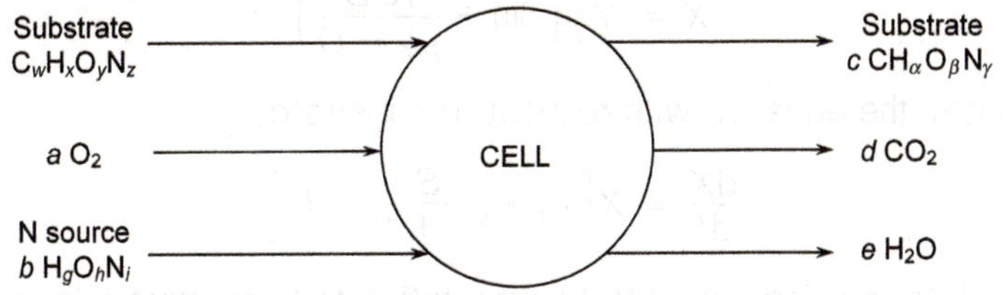

Notice that we have five unknown coefficients (a, b, c, d and e) but four equations. This means we need an additional equation to solve a system of five equations with five unknowns. An important and measurable parameter in a living system is the **respiration quotient** (RQ), which is defined as moles of CO_2 produced per mole of O_2 uptake.

The respiratory quotient (RQ) is

$$\text{Respiratory quotient (RQ)} = \frac{\text{Moles of } CO_2 \text{ Produced}}{\text{Moles of } O_2 \text{ Produced}} = \frac{d}{a} \tag{18.16}$$

When RQ is given, the stoichiometric coefficients can be solved simultaneously.

18.3.3 Material balance

For cell recycling, studying **chemostat** efficiency has a practical mean. During the process, the material balance in terms of substrate is often discussed.

The substrate balance is given based on following equation:

$$-\frac{dS}{dt} = \left(\frac{F}{V}\right)(S_{in} - S_{out}) - \left(\frac{\mu}{Y_{\frac{X}{S}}}\right)X - \left(\frac{q_p}{Y_{\frac{P}{S}}}\right)X - mX \tag{18.17}$$

In general, $\mu >> m$, so we can therefore neglect the last term.

For steady-state no product is formed which means there are no changes in substrate and product concentrations:

$$-\frac{dS}{dt} = 0 \quad \text{and} \quad \left(\frac{q_p}{Y_{\frac{P}{S}}}\right) = 0 \tag{18.18}$$

Also, $F/V = D$ then $\mu = D$

$$0 = D(S_{in} - S_{out}) - \frac{\mu X}{Y_{\frac{X}{S}}} \tag{18.19}$$

Rearranging the above equation:

$$\bar{X} = Y_{\frac{X}{S}}(S_{in} - S_{out}) \tag{18.20}$$

For steady-state condition:

$$D = \mu_m \frac{S}{K_S + S} \Rightarrow S_{out} = \frac{K_s D}{\mu_m - D} \tag{18.21}$$

Where, K_s = Saturation or Monod constant, g/L.

So, we can get:

$$\bar{X} = Y_{\frac{X}{S}}\left(\text{Sin} - \frac{K_s D}{\mu_m - D}\right) \tag{18.22}$$

For unsteady state, the equation with respect to substrate:

$$\frac{dX}{dt} = X\left[\mu_m \cdot \frac{S}{K_S + S} - D\right] \tag{18.23}$$

It's a routine work for running chemostat fermentation to keep materials well balanced.

18.3.4 Metabolic energy stoichiometry

Cell use chemical energy quite efficiently but, like any real process, some of the energy in the substrates is released as heat. Metabolic heat generation dictates cooling requirements for bioreactors which harbor cells. Cell heat production is primarily the result of energy and growth metabolism. It is necessary to consider it and have an energy calculation.

For instant, we can get it below, when $C_6H_{12}O_6$ is used as nutrient source, and supposed would be burnt completely:

$$C_6H_{12}O_6 + 6O_2 \longrightarrow 6H_2O + 6CO_2 + 2871\text{kJ} \qquad (18.24)$$

If the Metabolic product is ethanol or lactic acid, then the heat generated would be:

$$C_2H_5O + 3O_2 \longrightarrow 3H_2O + 3CO_2 + 1368\text{kJ} \qquad (18.25)$$

$$CH_5CHOHCOOH + 3O_2 \longrightarrow 3H_2O + 3CO_2 + 1337\text{kJ} \qquad (18.26)$$

Or the heat generated in ethanol fermentation and lactic acid fermentation is 136kJ and 197kJ per 1mole glucose.

We define Y_{kJ} analogous to the using of the heat generated during the fermentation, and ΔH_c and ΔH_a are the heats of combustion (kilocalories per gram) of substrate and of cell mass material, respectively we can write:

$$Y_{kJ} = \frac{\Delta X}{(-\Delta H_a)(\Delta X) + (-\Delta H_c)} \qquad (18.27)$$

This arises from approximate energy balances over two pathways for aerobic growth. Provided that the predominant oxidant is oxygen, the heat generation ΔH_c per gram of substrate completely oxidized minus $\Delta X \Delta H_a$, the heat obtained by combustion of cells grown from the same amount of substrate, will reasonably approximate the heat generation per gram of substrate consumed in the fermentation which produces cells, H_2O and CO_2.

And we know:

$$-\Delta H_c = (-\Delta H_s)(-\Delta[S]) - \sum(-\Delta H_p)(\Delta[P]) \qquad (18.28)$$

The ΔH_p is the enthalpy of the products, so we can get:

$$Y_{kJ} = \frac{\Delta X}{(-\Delta H_a)(\Delta X) - (-\Delta H_s)(-\Delta[S]) - \sum(-\Delta H_p)(\Delta[P])}$$

$$= \frac{Y_{\frac{x}{s}}}{(-\Delta H_a)Y_{\frac{x}{s}} + (-\Delta H_s) - \sum(-\Delta H_p)Y_{\frac{p}{s}}} \qquad (18.29)$$

As we have known the example that citric acid is manufactured by submerged cultures of *Aspergillus niger* in a batch reactor operated at 30℃ for 2 days fermentation.

We can figure out by Stoichiometric calculation, that 2500kg glucose and 860kg oxygen were consumed to produced 1500kg of citric acid, 500kg of biomass and other by-products; power input to the system by mechanical agitation was about 15kW; and approximately 100kg of water was evaporated during the 2 day operation.

18.4 Fermentation Kinetics

The purpose of fermentation modeling is to sort out the mass of data which results from a practical fermentation experiment, and to express the results in a concise form which is intelligible to colleagues and those who wish to make use of the results. The term "modeling" is used frequently in engineering and is more popular in science which simply means something we want to understand, describe, predict, systemized, computerized and control. The components of modeling include the control volume, variables, parameters and the equations. Assumptions are also very important and indispensable components of modeling. It means that we need the ideal reactor for kinetics measurement.

18.4.1 Kinetics of cell growth

In our attempts to model and understand cell population kinetics, we shall use unstructured models in which only cell mass or number concentrations will be employed to characterize the growth phases. Therefore, this section will introduce the commonly used explicit mathematical expressions for the volumetric rates of microbial activity. These basic forms of the kinetic rate expressions will be used in the latter sections.

Explicit rate expressions for microbial activity

Unicellular organisms divides by binary fission; each cell grows to its full size, replicates its genetic material, and then divides into two identical daughter cells. By identical means, two cells divide into four and four into eight and so on, leading to an exponential increase in cell numbers:

$$1 \rightarrow 2 \rightarrow 4 \rightarrow 8 \rightarrow 2^n \qquad (18.30)$$

Cell growth

A variety of methods are available to predict cell growth by direct or indirect measurements. Cell dry weight, cell optical density, cell turbidity, cell respiration, metabolic rate and metabolites are quite suitable for analyzing cell growth, substrate utilization and product formation. The rate of cell growth is described in this section. Cell metabolism is made up of hundreds of sequential, branched and parallel biological reactions that are normally catalyzed by enzymes.

Normally the specific growth rate function $\mu(S)$ is simply abbreviated as μ, and it has the dimensions h^{-1}. Using the definition of the volumetric and specific rates for growth, from Equation 18.31 we have

$$r_x = \mu x_V \tag{18.31}$$

The relationship between the specific growth rate and limiting substrate concentration proposed by Monod states that:

$$\mu = \mu_{max} \frac{S}{(K_S + S)} \tag{18.32}$$

Where r_x is the volumetric rate of cell growth, kg cells m^{-3}/h; μ_{max} is the maximum specific growth rate, h^{-1}; S is limiting substrate concentration, kg substrate m^{-3}; K_S is the saturation constant, kg substrate m^{-3}; x_v is the viable cell concentration, kg cells m^{-3}. When the substrate concentration is not limited, that is when $S >> K_S$ numerically, K_S can be ignored in Equation 18.32 and then S cancels each other, the specific growth rate approaches μ_{max} and the growth rate becomes independent of S and only proportional to the cell concentration.

Doubling time

Doubling time, t_d, is defined as the time required for the biomass to be doubled and assuming a constant maximum value of specific growth rate this will be:

$$t_d = \frac{\ln 2}{\mu_{max}} = \frac{0.693}{\mu_{max}} \tag{18.33}$$

Maximum specific growth rate

The rate of increase in microbial biomass concentration when all the above requirements are provided and if cell death can be ignored, will be proportional to the viable microbial biomass concentration, i.e.:

$$r_x = \frac{dx}{dt} = \mu x_V \tag{18.34}$$

where μ is described as the specific growth rate. When the specific growth rate is the maximum specific growth rate (i.e, $\mu = \mu_{max}$), which occurs during the exponential phase of microbial growth, the relationship between the variation biomass concentration and time may be deduced as follows:

$$\mu = \mu_{max} \tag{18.35}$$

18.4.2 Kinetics of batch fermentation

The batch culture is a simple, well-controlled vessel in which the concentration of nutrients, cells and products vary with time as the growth of the microorganism proceeds.

Since the concentrations change with time, a batch culture is an unsteady state operation. And a bioreactor can be a simple open container with very limited control or regulation such as the temperature, or a very sophisticated vessel with strict containment and computer monitoring and control, e. g. , bio-productions involving genetically modified organisms. Central to the whole process is the bioreactor (which will be discussed in Chapter 19). In general, the batch operations are preferred to reduce the risk of contamination, reduce financial loses due to spoilt cultures, and provide better containment and accumulate products so that inherently low product concentrations can be thus increased, which then help with downstream processing.

Kinetic of balanced growth

Balance equations are written for every extensive property of interest in each control region. For the batch culture, with the assumptions listed above, we shall take the bulk liquid as the control region: 1) Extensive properties are those that are additive over the whole of a system; 2) Each balance equation should be linearly independent of the others. Linearly independent means that no balance equation can be formed by adding together any combination of the others. In batch cultures, the concentration of cells changes with time within the bioreactor bulk liquid. We therefore, have the accumulation or the depletion term, dX/dt in the mass balances. There is no flow of liquid medium into or out of the bioreactor and hence the input and output terms due to mass transfer are zero.

Using the volumetric rates, r, the general material balance of Equation 18. 36 for a batch culture becomes:

$$\frac{dV_y}{dt} = \sum Vr_{gen} - \sum Vr_{cons} \tag{18.36}$$

Where y is the general extensive property, r_{gen} is the volumetric rate of generation (production), r_{cons} is the volumetric rate of consumption, V is the volume of the control regions, which is the bulk liquid for a well-mixed bioreactor. In our case, y can be, for example, one of the following: live biomass concentration (x_v), dead biomass concentration (x_d), total biomass concentration (x_T), substrate concentration (S), product concentration (P), dissolved oxygen concentration (C_o), etc. The sign indicates the summation of the similar terms; for example the summation of all the different volumetric consumption rates of the carbon substrate for biomass formation, for the formation of an excreted product and for the maintenance energy requirements.

For constant volume operation, V can be deleted from Equation 18. 36 to yield

$$\frac{dy}{dt} = \sum r_{gen} - \sum r_{cons} \tag{18.37}$$

Making an individual mass balance for each biological and chemical species of interest, we will have, for a microbial culture in a batch bioreactor the following mass balances:

Live biomass balance:

$$\frac{d(x_v V)}{dt} = + r_x V - r_d V \tag{18.38}$$

Dead biomass balance:

$$\frac{d(x_d V)}{dt} = + r_d V \tag{18.39}$$

Substrate balance:

$$\frac{d(SV)}{dt} = - r_S V \tag{18.40}$$

Product balance:

$$\frac{d(PV)}{dt} + r_P V \tag{18.41}$$

Where x_v = concentration of viable cells; x_d = concentration of dead cells; S = substrate concentration; P = product concentration; V = volume of bulk liquid (culture volume); r_x = volumetric rate of growth; r_S = volumetric rate of substrate consumption; r_P = volumetric rate of product formation; r_d = volumetric rate of cell death.

In a typical batch bioreactor, liquid volume is constant, since no liquid is added to or removed from the bioreactor. Therefore, V can be neglected and the equations that are given above can be rewritten as:

Viable cells:

$$\frac{dx_V}{dt} = + r_x - r_d \tag{18.42}$$

Non-viable cells:

$$\frac{dx_d}{dt} = + r_d \tag{18.43}$$

Substrate:

$$\frac{dS}{dt} = - r_S = - (r_{Sx} + r_{Sm} + r_{SP}) \tag{18.44}$$

Product:

$$\frac{dP}{dt} = + r_P \tag{18.45}$$

The kinetic expressions for various volumetric rates can then be substituted from the previous sections. Assuming Monod kinetics for biomass growth, Luedeking-Piret type of product formation kinetics and maintenance energy concept, we use Equations 18.36 and 18.41 in Equation 18.42~18.45 and obtain the following equations (for convenience, we use " + " sign for the first term after the equality sign as shown later in the chapter):

$$\frac{dx_V}{dt} = + \mu x_V - k_d x_V \tag{18.46}$$

$$\frac{dx_V}{dt} = +\mu_m \frac{S}{K_S + S} x_V - k_d x_V \tag{18.47}$$

$$\frac{dx_d}{dt} = +k_d x_V \tag{18.48}$$

$$\frac{dP}{dt} = \alpha r_x + \beta x_V = \alpha \mu x_V + \beta x_V = \alpha \mu_{max} \frac{S}{K_S + S} x_V + \beta x_V \tag{18.49}$$

$$\frac{dS}{dt} = -\left(\frac{1}{Y'_{\frac{x}{S}}} r_x + \frac{1}{Y'_{\frac{P}{S}}} r_P + m_S x_V\right) \tag{18.50}$$

When the maintenance energy has to be accounted for in the balances, it should appear as a consumption term in the limiting carbon and energy substrate balance. If on the other hand, the endogenous metabolism has to be accounted for instead of the maintenance energy, the endogenous respiration should appear as a consumption term in the viable biomass balance. (Note: The reader should try to write the balances above for the case when endogenous respiration has to be accounted for.)

A typical plot indicates the curves of the concentrations of biomass, carbon-and-energy substrate and nitrogen source over time in a batch culture (Fig. 18.10), assuming that there is no death, no maintenance energy and no complex product formation that is excreted.

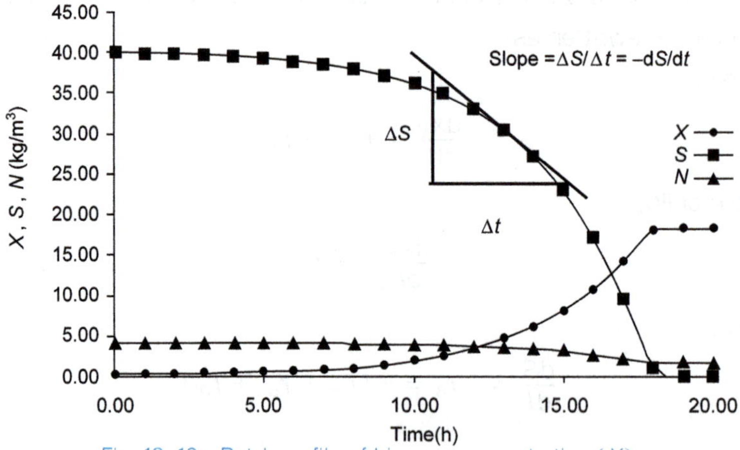

Fig. 18.10 Batch profile of biomass concentration (X), carbon-and-energy substrate (S), and nitrogen source (N)

Advantages and disadvantages of batch culture

Batch fermentation might have some advantages: for example, 1) simplicity of use; 2) operability and reliability; 3) production for secondary metabolites that are not growth-related; 4) fewer possibilities of contamination; 5) it is easy to assign a unique batch number to each run, generating high confidence in the history of each batch of product. This is critically important in a highly regulated environment.

Also there are disadvantages from other aspects: 1) culture ageing; 2) build up of toxic metabolites and the initial substrate concentrations can restrict cell growth and product formation; 3) batch-to-batch variability; 4) the use of batch cultures in industrial systems can lead to an increased nonproductive period; 5) degeneration or differentiation may occur.

18.4.3 Kinetics of continuous fermentation

In the above, appropriate nutrients and other conditions are provided for growth, then an **inoculum** is added and the culture incubated. No further nutrients are added and no waste products are removed. Disadvantages exist in the batch system such as substrate depletion with limited nutrients or product inhibition growth curve and so on. Therefore it will never be able to maintain any steady-state condition. This results in active growth being of limited duration for the reasons outlined above.

The existing limitation and toxic product inhibition can be removed if the system is an open system and the growing culture is in a continuous mode of operation. Sometimes it is desirable to keep the culture in the logarithmic phase. In a continuous culture, nutrient concentrations and other conditions are held constant, and the cells are held in a state of exponential growth. This is achieved by continuously adding fresh culture medium and removing equal volumes of the old. Parameters such as pH can also be monitored and adjusted.

In continuous culture, there are two types of culture mode: chemostat (Fig. 18.11) and turbidostat; both are open systems. Detailed explanations are given below.

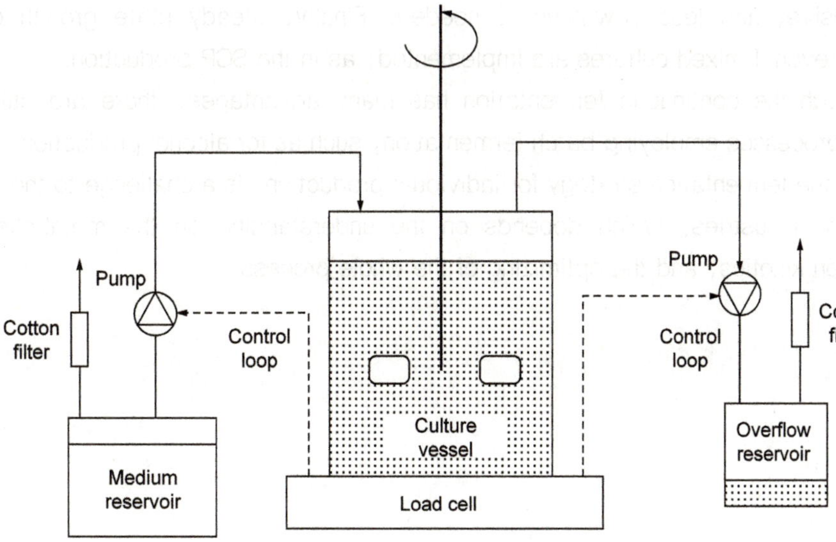

Fig. 18.11 Chemostat with inlet and outlet control loops, feed and product pump with cell loading and recycling

(1) Chemostat [growth rate controlled by dilution rate, D, (h^{-1})];

(2) Turbidostat (constant cell density that is controlled by the fresh medium).

The continuous cultures of chemostat and turbidostat systems have the following criteria:

- The medium and cells are continuously changing;
- The cell density (ρ_{cell}) keeps constant;
- Steady-state growth;
- Open system.

The system is balanced for cell growth by removing old culture and replacing it with fresh medium at the same rate.

Advantages and disadvantages of continuous culture

There are several advantages to continuous culture, where all the problems associated with the batch culture are solved. Firstly, the growth rate is controlled and the cells are well maintained, since the old culture is continously replaced by fresh medium while the dilution is taking place. As a result, the effect of physical and chemical parameters on growth and product formation can easily be examined. The biomass concentration in the cultured broth is well maintained at a constant dilution rate. The continuous process results in substrate-limited growth and cell-growth-limiting nutrients. The composition of the medium can be optimized for maximum productivity; in addition secondary metabolite production can also be controlled. The growth kinetics and kinetic constants are accurately determined. The process leads to reproducible results and reliable data. High productivity per unit volume is achieved. The continuous culture is less labor-intensive, and less downtime is needed. Finally, steady state growth can be achieved, even if mixed cultures are implemented, as in the SCP production.

Although the continuous fermentation has many advantages, there are still many industrial processes employing batch fermentation, such as for alcohol production. How to determine the fermentation strategy for individual production, is a challenge to the modern fermentation industries; which depends on the understanding on the metabolism and fermentation kinetics, and the optimizing of the whole process.

Summary

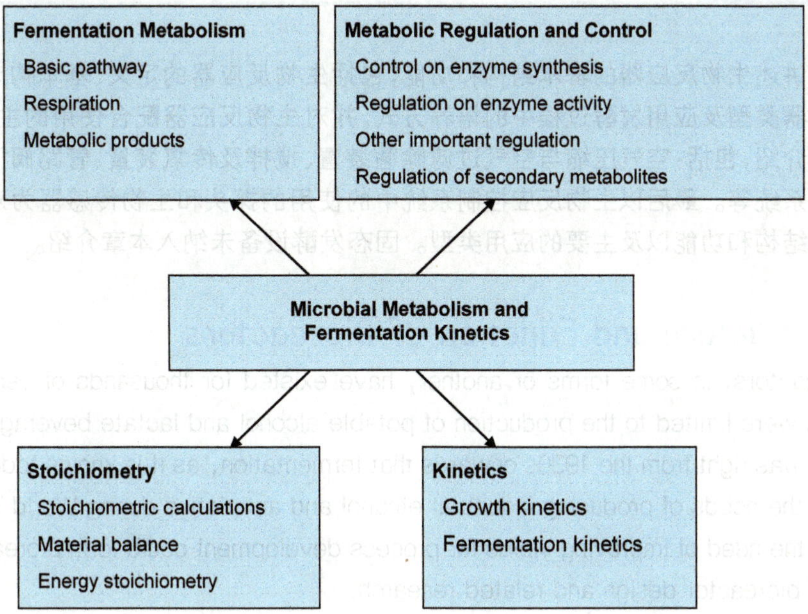
Fermentation Metabolism
Basic pathway
Respiration
Metabolic products
Metabolic Regulation and Control
Control on enzyme synthesis
Regulation on enzyme activity
Other important regulation
Regulation of secondary metabolites
Microbial Metabolism and Fermentation Kinetics
Stoichiometry
Stoichiometric calculations
Material balance
Energy stoichiometry
Kinetics
Growth kinetics
Fermentation kinetics

Chapter 19 Bioreactors and Auxiliary Equipments

本章讲述生物反应器的基本结构和功能，包括生物反应器的定义、基本构成要素、常见的反应器类型及应用发酵过程中的培养方式，并对生物反应器配合使用的主要附属设备进行了介绍，包括：空气压缩与空气过滤除菌装置、搅拌及传氧装置、管路阀门、取样装置及冷却系统等。最后以生物反应控制系统中的使用的探头和生物传感器为对象，介绍了它们的结构和功能以及主要的应用类型。固态发酵设备未纳入本章介绍。

19.1 Structure and Function of Bioreactors

Bioreactors, in some forms or another, have existed for thousands of years though their uses were limited to the production of potable alcohol and lactate beverage until the 1900s. It was right from the 1930s onwards that fermentation, as it is known today, began to satisfy the needs of producing industrial ethanol and antibiotics during World War II. At this point the need of improving yields for process development could form bioreaction that drove the bioreactor design and related research.

Normally, a bioreactor is a **vessel** in which a biochemical process, involving organism cultivation, enzyme or biochemical active substance, derived from an organism is carried out. The reactor is the heart of any fermentation or enzyme conversion process. Current bioreactors are grouped into fermentation vessels, reactors using biocatalysts (e. g., immobilized enzymes/microorganisms), and special reactors for cultivation of plant and animal tissues.

The bioreactor should meet the following requirement as a minimum:

- With function of agitation to mix organism, medium and sterilized air in the form of suspension and maintained the **dissolved O_2**. All this should be achieved without damage to the organisms.
- With capability **of sterile sampling** and aseptic fermentation for extended periods of time.
- Monitoring and controlling the following parameters: pressure, concentration of dissolved O_2, temperature, pH (adjustment by either acid/base addition), and some time controlling of **subtract feeding**.
- Meeting the containment regulations.

Tab. 19.1 shows a number of aerobic fermentation systems which are schematically classified into 1) **internal mechanical agitation reactors**, 2) **external or circulation reactors**, and 3) **bubble column and air-lift loop reactors**. This classification is based on both agitation and aeration as it relates to oxygen supply. In Tab. 19.1, reactor (a) 1 is often used at the industrial level and reactors (a)2, (b)2, (c)2, and (c)3, can be fitted with draught tubes to improve both mixing and oxygen supply efficiencies.

Tab. 19.1 Classification of aerobic fermentation systems

(a) Internal mechanical agitation reactors
1. Turbine-stirring installation
2. Stirred vessel with draft tube
3. Stirred vessel with suction tube
(b) External circulation reactors
1. Water jet aerator
2. Forced water jet aerator
3. Recycling aerator with fritted disc
(c) Bubble column and air-loop reactors
1. Bubble column with fritted disc
2. Bubble column with a draft tube for gyration flow
3. Air lift reactor
4. Pressure cycle reactor
5. Sieve plate cascade system

As mentioned in Chapter 18, culture techniques can be classified into **batch, fed-batch**, and **continuous operation**. In batch fermentation, all the nutrients required for cell growth and product formation are present in the medium prior to cultivation. Oxygen is supplied by sterilized aeration. The cessation of fermentation reflects the exhaustion of the limiting substrate in the medium.

Fed-batch fermentation is the operation in which one or more nutrients are added continuously or intermittently to the initial medium after the start of cultivation or from the halfway point through the batch process. The fed-batch operation can be divided into two basic models either with or without **feedback control**. Fed-batch processes have been utilized to avoid substrate inhibition, glucose effect, and **catabolite repression**, as well as for **auxotrophic mutants**.

Continuous fermentation is an open system in which sterile nutrient solution is added to the bioreactor continuously and an equivalent amount of converted nutrient solution with bio-cells is simultaneously removed from the system. Two basic types of continuous fermentations can be distinguished as: 1) **Homogeneously Mixed Bioreactor**. In the **chemostat** in the steady state, cell growth is controlled by adjusting the concentration of one substrate. Any required substrate (carbohydrates, nitrogen compounds, salts, O_2) can be used as a limiting factor. In the turbidostat, cell growth is kept constant by using turbidity to monitor the biomass concentration and the rate of feed for nutrient solution is appropriately adjusted. 2) **Plug Flow Reactor**. In this type of continuous fermentation, the culture solution flows through a tubular reactor without back mixing. The composition of the nutrient solution, the number of cells, mass transfer, and productivity vary at different locations within the system. At the entrance to the reactor, cells must be continuously added along with the nutrient solution.

Basically, there are three groups of bioreactor currently use for industrial production: 1) Non-stirred, non-aerated vessels; 2) Non-stirred, aerated systems; 3) Stirred and aerated systems. Bioreactors of the first group are usually used in anaerobic fermentation or sewage digestion. Non-stirred, aerated vessels are used in the process for traditional products such as wine, beer and cheese production. Most of the newly established bioprocesses require microbial growth in an aerated and agitated system and many new products require growth of microorganisms in the aerated and agitated vessels, so the stirred and aerated vessels show importance in modern fermentation industry. The performances of various bioreactors are compared as follow:

Stirred tank reactor

This is the most common type of bioreactor used in industry which has the advantages of low capital and low operating costs. Fig. 19. 1 is the mechanical diagram for such a reactor.

Stirred tank reactor essentially consists of a vessel with an aspect ratio of around 3 : 1. It has an agitation system typically driven through the head plate. Although with steaming in-place systems, the mixing could be driven through the base. The head plate will have specific designed ports that allow for the addition of probes, reagents and gas as well as the removal of samples. Vessels for laboratory experiments of volume up to 20L are normally made of glass. For larger volumes, the vessel is made of stainless steel. The height/diameter ratio of the vessel can vary between 2 : 1 and 6 : 1, depending largely on the amount of the heat to be removed. Mixing and bubble dispersion are achieved by mechanical agitation, and the stirrer may be top or bottom driven. **Baffles** are fitted on the inner wall of most vessels,

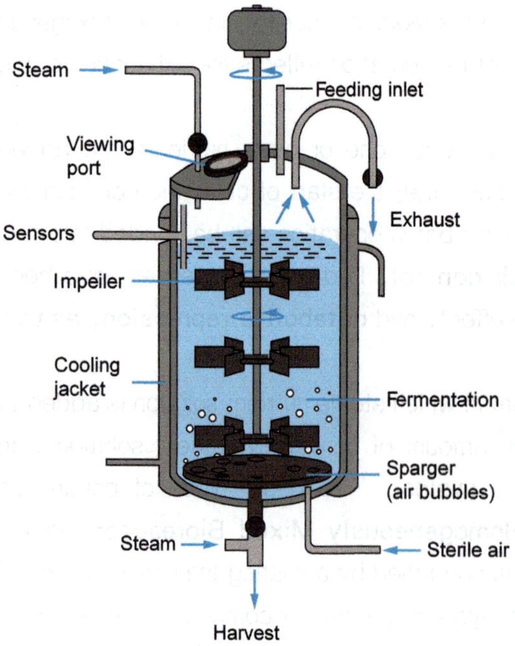

Fig. 19. 1 Stirred tank reactors

which prevent a large central vortex being formed as well as improving the suspension of all the contents. 4 baffles are normally used in the vessels less than 3 meters in diameter, and 6～8 baffles are used in larger vessels. Temperature control and heat transfer in stirred vessels can be accomplished using internal or external **cooling coils**. Typically, only 70%～80% of the volume of stirred reactors is filled with liquid. This allows adequate headspace for disengagement of droplets from the exhaust gas and to accommodate any foam which may develop. If foaming is a problem, a supplementary impeller called a foam breaker should be installed, but the **de-foaming agent** is needed to control the foam bursting for some time.

Airlift bioreactors are the types of non-stirred, aerated fermenters. In an **airlift fermenter**, mixing is accomplished without any mechanical agitation, but by air lifting force. Airlift bioreactors are used for tissue culture because the tissues are sensitive and can't stand the shearing force from strong agitation. Generally, it can be divided into two main reactors on the basis of their structure (Fig. 19.2): 1) **External-loop vessels**, in which circulation takes place through separate and distinct conduits; and 2) **Internal-loop vessels**, in which the baffles are placed strategically in a single vessel creating the channels required for the circulation.

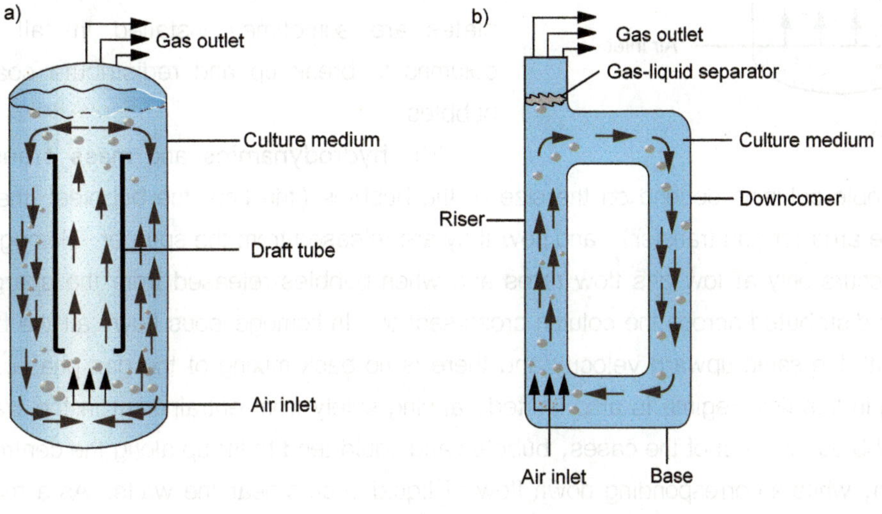

Fig. 19.2 Airlift Bioreactors with internal-loop(a) and external loop(b)

In a commonly used airlift bioreactor, air is fed into the bottom of a central draught tube through **a sparger ring** for reducing the apparent density of the liquid in the tube that is relative to the annular space within the bioreactor. The flow passes up through the draught tube to the head space of the bioreactor, where the excess air and CO_2, disengaged. The degassed liquid then flows down the annular space outside the draft to the bottom of the bioreactor. Cooling can be provided by either making the draught tube an internal heat exchanger or with a heat exchanger in an external recirculation loop.

The advantages of airlift bioreactor are usually considered as: 1) In low shearing force, there is mild mixing which means the airlift bioreactor can be used for cultivation of plant or animal cells; 2) Since there is no agitation, sterility is easily maintained; 3) In a large vessel, the height of liquid layer can be as high as 60m, and the pressure at the bottom of the vessel will increase the oxygen solubility, and also the value of K_{ia} will increase; 4) Extremely large vessels can be constructed. In one single cell protein plant, the reactor had a total volume of $2300m^3$.

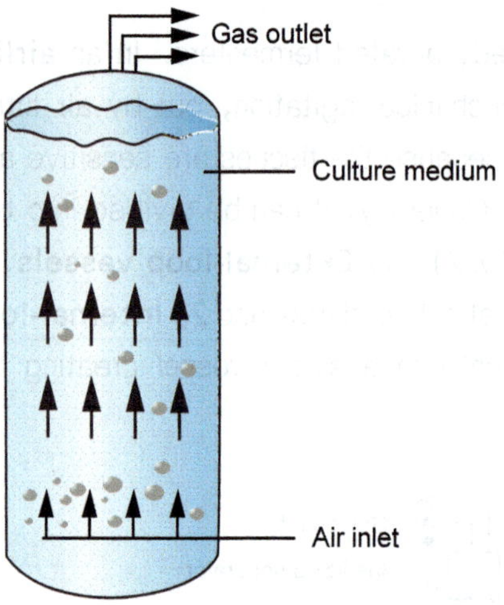

Fig. 19.3 Bubble column fermentor

Bubble column fermentor: For the production of baker's yeast, beer and vinegar, and bubble column fermenters are used. As shown in Fig. 19.3, they are also often used for less severe aeration, for example in treatment of wastewater. In designing such a bioreactor, a common ratio of H : D is also about 3 : 1, and the height of liquid level to tank diameter (H : D) is about 2 : 1. In bakers' yeast production, the ratio of H : D is about 6 : 1. Perforated horizontal plates are sometimes installed in tall bubble columns to break up and redistribute coalesced bubbles.

The **hydrodynamics** and **mass transfer** in the bubble columns depend on the size of the bubbles (the finer the bubbles, the larger surface area for gas transfer), and how they are released from the sparger. Homogeneous flow occurs only at low gas flow rates and when bubbles released from the sparger are evenly distributed across the column cross-section. In homogeneous flow, all the bubbles rise with the same upward velocity and there is no back mixing of the gas phase. Liquid mixing in this flow regime is also limited, arising solely from entrainment in the wakes of the bubbles. In most of the cases, bubbles and liquid tend to lift up along the centre of the column, while a corresponding down flow of liquid occurs near the walls. As a matter of fact, liquid circulation helps the air bubbles friction and decreases the total surface area of bubbles, so back mixing of gas would be inevitably occur.

Packed-bed bioreactors: As shown in Fig. 19.4, **Packed-bed bioreactors** are used with immobilized cells or particulate biocatalysts. The reactor consisting of a tube, usually vertical, is packed with catalyst particles. Medium can be fed either at the top or bottom of the column and forms a continuous liquid phase between the particles. Damage due to particle attrition is minimal in packed beds compared with stirred reactors. Packed-bed reactors have been used commercially with immobilized cells and enzymes for production of biochemicals. Examples are aspartic acid and fumaric acid, which are used for conversion of penicillin to 6-aminopenicillanic acid and resolution of amino acid isomers.

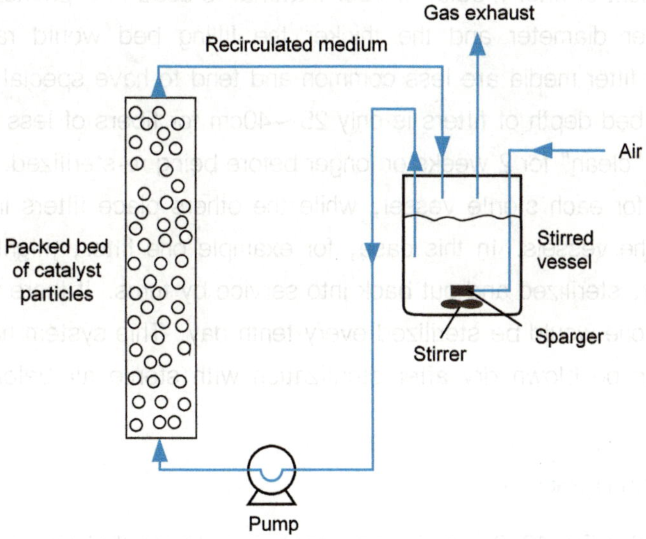

Fig. 19.4 Packed-bed bioreactor with medium recycles

Fluidised bed: Packed beds are operated in upflow mode with catalyst beads of appropriate size and density, and the bed expands at high liquid flow rates due to upward motion of the particles. The basis for operation of **fluidised-bed reactors** is illustrated in Fig. 19.5. Because particles in fluidised beds are in constant motion, channeling and clogging of the bed are avoided and air can be introduced directly into the column. Fluidised-bed reactors are used in waste treatment with sand or similar material supporting mixed microbial populations. They are also used with **flocculating cells** in brewing and for production of vinegar.

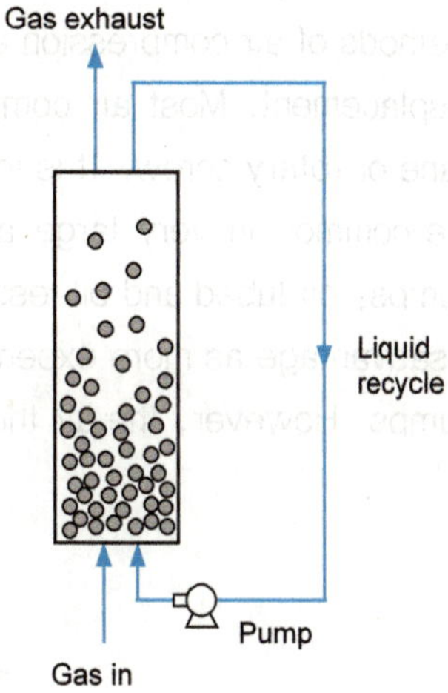

Fig. 19.5 Fluidised-bed reactor

19.2 Auxiliary Equipment

19.2.1 Sterile air filters

Sterile air filtration is used to get rid of any impurities including the germs from inlet of air. The essential method to obtain sterile air, whether packed-bed or cartridge filters are used, is also involved with reduction of the humidity of the air after compression, so that the filter material always remains dry. Larger plants install instrumentation with alarms set at about 85% relative humidity. A well designed packed-bed filter consists of a set of filter units, and it can be operated in excess of three years without replacement of filter media. If fiber material is used in a packed-bed type filter, the finer the fiber diameter and the thicker the filling bed would raise the filtration efficiency. Other filter media are less common and tend to have special problems and/or shorter life. The bed depth of filters is only 25~40cm for fibers of less than 10 microns. These filters run "clean" for 2 weeks or longer before being re-sterilized. Some plants use a separate filter for each sterile vessel, while the others place filters in a central group which feeds all the vessels. In this case, for example one filter, might be taken out of service each day, sterilized and put back into service by turns. If there were ten filters in the group, each one would be sterilized every tenth day. This system has the advantage that the filter can be blown dry after sterilization with sterile air before being put into service again.

19.2.2 Air compressors

As illustrated in Fig. 19.6, an air compressor is a device that converts power (usually from an electric or diesel or gasoline engine) into kinetic energy by pressurizing and compressing air, which is later released in quick bursts in fermentor. There are numerous methods of air compression and they can be divided into two types: positive and negative displacement. Most air compressors are either reciprocating piston type, or with rotary vane or rotary screw. It is ideal to have oil-free air compressors. Centrifugal compressors are common in very large applications. There are two main types of air compressor's pumps: oil lubed and oil-less. The oil-less system is more developed, but still have some disadvantage as more expensive, louder and last for less time than that of the oiled lube pumps. However, the air they delivered has better quality.

Fig. 19.6 Air compressor

19.2.3 Valves

Most fermentation outside of the traditional food and beverage industries are carried out using pure cultures and aseptic conditions. Keeping the reactor free of unwanted organisms is especially important for slow-growing cultures which can be quickly over-run by contaminants. Obviously, valves are very necessary to maintain sterility in fermentation. They must be easily sterilized, maintained, and occasionally replaced. In general, diaphragm (shown in Fig. 19.7) and ball valves require considerable maintenance, but tend to be popular in batch sterilizing operations, while plug type valves are more typical for continuous fermentation where continuous sterilizers are used. Plug or diaphragm valves are also commonly used for inoculum transfer and feed piping sterilization.

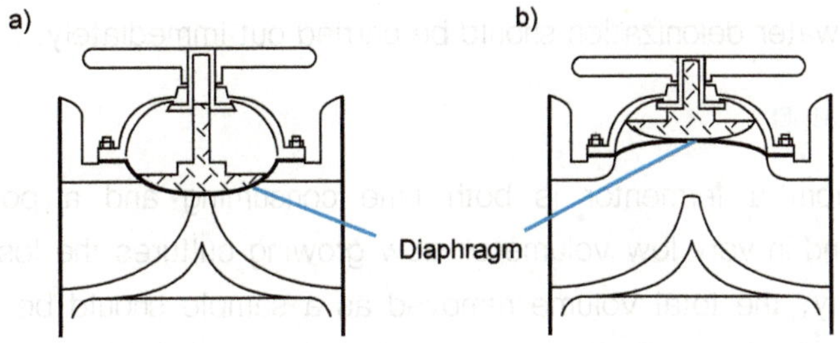

Fig. 19.7 Diaphragm valve in (a) closed and (b) open positions

19.2.4 Pumps

Apart from the continuous sterilizers, pumps are drawing minor attention in the fermentation system, but still the very useful auxiliary apparatus. A simple way to transfer inoculum from a large laboratory flask to a seed fermentor without reducing the back pressure in the vessel is to use a **peristaltic pump**. Connect the **sterile adapter** (which is attached to the flask) to the seed fermentor by sterile technique first, install the gum rubber tubing in the pump, open the hose clamp and start the pump, the inoculum and sterile feeds can be easily transferred to the fermentor under air pressure. **Centrifugal pumps** are used to pump non-sterile raw materials, slurries, harvested broth, etc. The centrifugal pumps and piping should be cleaned immediately after every batch of the transfer has been completed.

19.2.5 Cooling system

Cooling equipments are required in order to cool down the temperature of sterilized media, remove the exothermic heat of fermentation, cool the final broth before harvesting, and also to cool the compressed air. Some portion of the heat can be reclaimed to produce hot water for the preparation of new substrate, or for general cleaning of equipment, platforms and floors; however, the excess heat must be disposed to the environment. Cooling water is provided from tap water or underground water, but chilled water (5~15℃) is produced by steam vacuum or refrigeration units. In any case, the fermentation department should always concern about the cooling water supply with temperature and the chloride content. When chloride ions in water is above 150×10^{-6} and temperature is above 80℃ (while sterilizing even higher), this will cause stress corrosion cracking of stainless steel piping. The chloride content should be determined analytically every two weeks to control the chloride to less than 100×10^{-6}. This is done by draining water from the cooling tower and adding fresh water. Further more, a probe for detecting **conductivity** in the cooling water should be recommended for use. When the concentration of dissolved ions (salts) gets too high and temperature raises up quickly, the conductivity will increase dramatically, which indicates a risk of pipe line corrosion, and implies that water deionization should be carried out immediately.

19.2.6 Sampling

Sampling from a fermentor is both time consuming and a potential source of contamination, and in very low volume or slow growing cultures the loss may distort the process. Typically, the total volume removed as a sample should be kept to less than 10% of the process volume. So if a parameter is to be measured in a bioreactor it is better if it can be measured online— in the bioreactor— without the need for the taking of a sample. As standard in a bioreactor, there is a pH probe, a dissolved oxygen probe and a temperature probe.

19.3 Probes and Sensors

The fermentation process has more than one continuous variation of the variables. To control fermentation process, on-line monitoring devices are very necessary. In order to achieve this goal, scientists invent the probes and biosensors that detect, record, and transmit information regarding a physiological and chemical change, or the presence of various biological materials in the environment.

19.3.1 Commonly used probes

Temperature probe can be generally classified into **thermocouples, thermistors,** and **thermometers miniature integrated circuit devices**. Thermocouples are cheap and simply executed, but they are rather low in resolution and they require a cold junction. Thermistors are semi-conductors, which exhibit a change in electrical conductivity with temperature. They give a highly nonlinear output and are very sensitive and inexpensive. Modern miniature integrated circuit devices display a more linear output which represents the future trend of temperature probes.

pH probe The pH probe measures pH as the activity of **hydrogen cations** surrounding a thin-walled glass bulb at its tip. The probe produces a small voltage (about 0.06 volt per pH unit) that is measured and displayed as pH units by the meter. The indication of fermentation pH is so critical to the operation, which ensures that acid or alkaline is fed accurately through the inlet into broth to keep the optimal condition for microbial metabolism.

Dissolved oxygen probe The Dissolved Oxygen (DO) Probe is a galvanic measuring element which produces a milli-volt output proportional to the oxygen present in the medium it is placed in. It consists of an upper part with cathode, anode and cable, and a cap with membrane and electrolyte. Based on the DO measurement, the automatic system can adjust the speed of stirring and the strength of aeration to ensure the dissolved oxygen in broth is always kept in a decent level.

19.3.2 Biosensors

The biochemical constituents of fermentation broth have been developed by a wide range of biosensors. A biosensor consists of two main elements. These two elements, the biocatalyst and a transducer, are combined as a single detecting probe, in which the transducer and biocatalyst are held together in a very close contact. The biosensor acts as a device as flow passes. It can detect penetration of flow through biocatalysts and measure the biochemical transformation of a given substance, for example change in the concentration of a specific amino acid. The function of the transducer is to detect such change and to produce an output signal, which is related to the concentration of the measured substance. In fact, a biosensor is a combination of biological sensor, attached to transducer which is a simple device that acts specifically, with a high sensitivity in measurements.

Characteristics of biological sensors

The sensor in a bioreactor provides knowledge and information on the state of the process and also supplies suitable operational data for the process variables. Some of the physical and chemical effects on the bioreactor have to be translated to electrical signals, which can be amplified and then displayed on a monitor or recorder and be used as an input signal for a controlling unit. The common characteristics of biosensors are: 1) highly specific; 2) independent of physical parameters such as temperature, pH; 3) accurate; 4) biocompatible; 5) durable; and, 6) allow rapid continuous control.

The principal components (Fig. 19.8) of a biosensor are: 1) a biologically sensitive element as enzymes, antibodies, nucleic acids or whole cells, usually adsorbed on solid substrates; 2) a transducer of the signal that is generated when the sensitive element specifically reacts with the analyte to be determined; and 3) an electronic circuit for signal elaboration. The immobilization of the biological molecule is a particularly important aspect of the fabrication of biosensors as the immobilization procedure must reproducibly keep the target molecule close to the surface of transducer while retaining its biological activity.

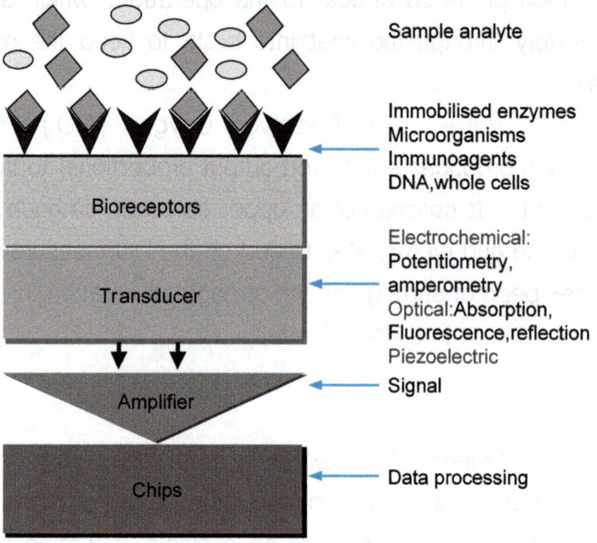

Fig. 19.8 Biosensor principle diagram

Types of biosensors

Enzyme electrode The enzyme electrode is a miniature chemical transducer which functions by combining an electrochemical procedure with immobilized enzyme activity. An enzyme is immobilized on the surface of an electrode, and when the enzyme catalyses its reaction, electrons are transferred from the reactant to the electrode. Moreover, a current is generated, which can then be measured.

Optical biosensors Colorimetric for color Measure change in light adsorption Photometric for light intensity Photon output for a luminescent or fluorescent process can be detected with photomultiplier tubes or photodiode systems. Only the marked macromolecules can be detected, for example, the fluorescent marked DNA fragment can be sequenced under the detection of optical biosensors.

Electrochemical biosensors (ECB) Those are biosensors which utilize immobilized bioactive compounds (such as enzymes, antigen, antibody, DNA, etc.) for the conversion of the target analytes into electrochemically detectable products. ECB is one of the most widely used detection methods and has become an area that draws a wide range of research. The advances in biocompatible nano technology make it possible to develop even more sensitive electrochemical biosensors.

Summary

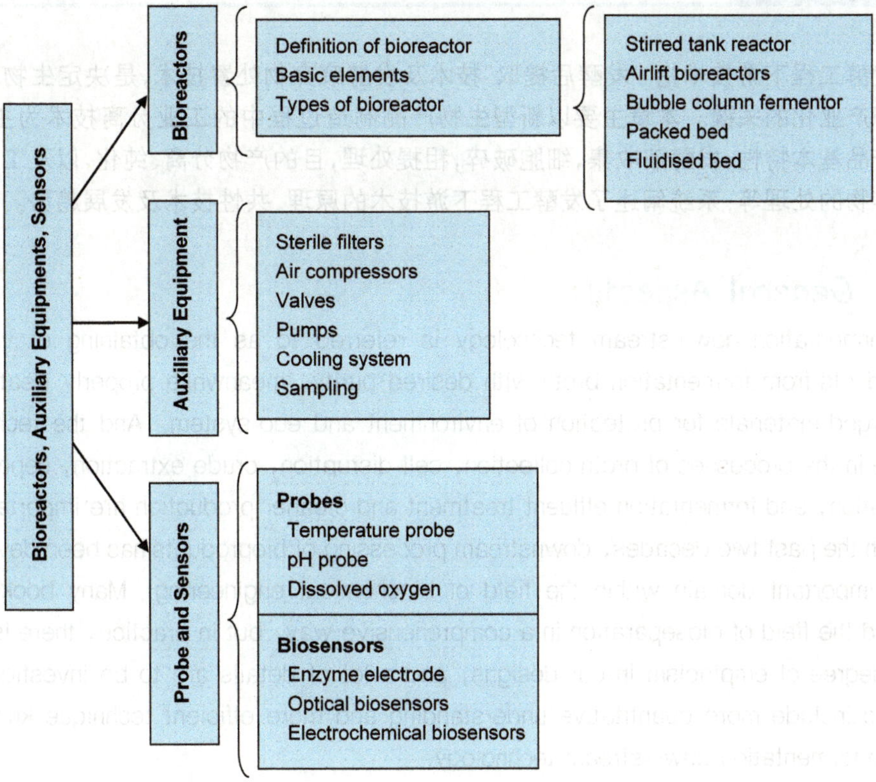
Bioreactors, Auxiliary Equipments, Sensors
Bioreactors
Definition of bioreactor
Basic elements
Types of bioreactor
Stirred tank reactor
Airlift bioreactors
Bubble column fermentor
Packed bed
Fluidised bed
Auxiliary Equipment
Sterile filters
Air compressors
Valves
Pumps
Cooling system
Sampling
Probe and Sensors
Probes
Temperature probe
pH probe
Dissolved oxygen
Biosensors
Enzyme electrode
Optical biosensors
Electrochemical biosensors

Chapter 20 Fermentation Down-Stream Technology

发酵工程下游技术指“发酵后提取”技术及发酵废弃物处置技术，是决定生物工业产品能否产业化的关键。本章主要以新型生物产品制造过程中的工业分离技术为主线，从发酵产品基本特性，发酵液收集，细胞破碎，粗提处理，目的产物分离、纯化，以及工业废水与废弃物的处理等；系统阐述了发酵工程下游技术的原理、共性技术及发展趋势。

20.1 General Aspects

Fermentation down-stream technology is referred to as the obtaining of valuable bioproducts from fermentation broth with desired purity, meanwhile properly treating the discharged materials for protection of environment and eco-system. And the techniques involve in the processes of broth collection, cell disruption, crude extraction, separation, purification, and fermentation effluent treatment and cleaner production are important part of it. In the past two decades, downstream processing of bioproducts has been developed as an important domain within the field of biochemical engineering. Many books have covered the field of bioseparation in a comprehensive way, but in practice, there is still a large degree of empiricism in our designs, and a lot of details are to be investigated in order to include more quantitative understanding and more efficient technique know-how into the fermentation down-stream technology.

20.1.1 Strategy of bio-products recovery

The bio-products from fermentation might cover all the three matter forms: gas, liquid and solid. As for the gas form, methane and hydrogen released from anaerobic fermentation or digestion can be easily harvested, compressed, and stored in the pressure cylinders. Solvents with rather low boiling points, such as ethanol and acetone, are usually the metabolites from yeasts or bacteria. Those products can be crude-distilled from the fermentation broth firstly, then **rectifying-distilled** to exclude the impurities and finally concentrated through the well designed distillers. Some of the fermented organic acids can be extracted by adding calcium into the clear broth first, to precipitate the product in a salt form, then dissolved it again and purified the organic acid through the ion exchange column. The recovery of bio-products in a solid form is normally involved with filtration/centrifugation, isolation and purification.

But for productions of primary and secondary metabolites, the bio-products can also be classified into two categories: extracellular or intracellular products, based on whether they are secreted. For the secreted primary and secondary metabolites, also enzymes and proteins, the bio-products recovery starts from the collection of fermentation broth. But for the intracellular products, collection of microbial cells would be the first step (illustrated as in Fig. 20.1).

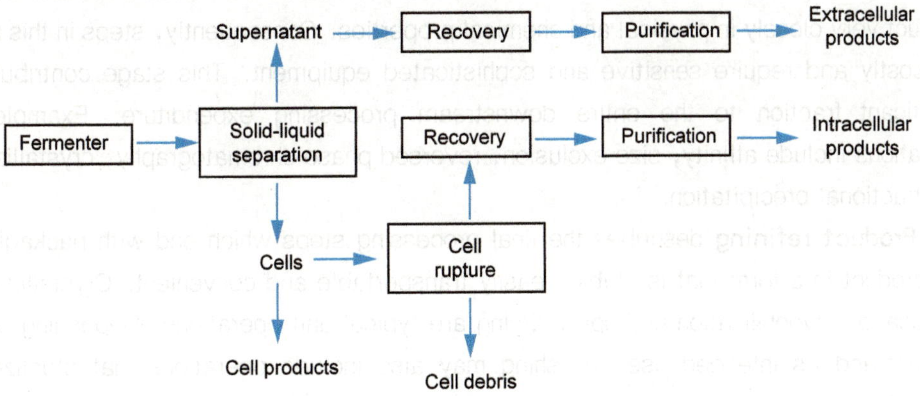

Fig. 20.1 Common process steps in downstream processing

Be aware of the fact that the novel bio-products are more complicate, sensitive to stress, and the purification must be subordinated to the restricted regulations, that's why the more strict downstream technologies are needed for implement of specific bio-products recovery and purification, particularly for the pharmaceuticals made from natural sources such as animal and plant tissues or fermentation broth, also for recycling of **salvageable** components and the proper treatment and disposal of waste.

20.1.2 Stages in downstream processing

Another widely recognized heuristic for categorizing downstream processing operations divides them into four groups which are used in order to bring a product from its natural state as a component of a tissue, cell or fermentation broth through progressive improvements in purity and concentration.

Harvest and pretreatment involve collection of fermentation broth, removal of most impurities and capture of the product as a solute in a particulate-free liquid, for example the separation of cells, cell debris or other particulate matter from fermentation broth that contains an antibiotic. Typical operations to achieve this purpose are filtration, centrifugation, sedimentation, precipitation, **flocculation**, electro-precipitation, and gravity settling. When the desired products are intracellular substances, additional operations such as cell grinding or homogenization are required to recover the products from the ruptured cells, or plant and animal tissue.

Product isolation involves the removal of those components with the properties varied from that of the desired product. For most final products, water is also the chief impurity. Isolation steps are designed to remove most of the water, reducing the volume of material to be handled and concentrating the product. Solvent extraction, adsorption, ultrafiltration, and precipitation are some of the unit operations involved.

Product purification is done for separation of those contaminants that resemble the product very closely in physical and chemical properties. Consequently, steps in this stage are costly and require sensitive and sophisticated equipment. This stage contributes a significant fraction to the entire downstream processing expenditure. Examples of operations include affinity, size exclusion, reversed phase chromatography, crystallization and fractional precipitation.

Product refining describes the final processing steps which end with packaging of the product in a form that is stable, easily transportable and convenient. Crystallization, desiccation, lyophilization and spray drying are typical unit operations. Depending on the product and its intended use, polishing may also include operations that sterilize the product and remove or deactivate trace contaminants which might compromise product safety. Such operations might include the removal of viruses or depyrogenation.

It's encouraged to combine two or more stages into one integrated step, for example, **expanded bed adsorption** may accomplish the removal of insolubles and product isolation in a single step, and **affinity chromatography** often merges the product isolation and purification into a single procedure.

20.1.3 Theoretical basis of fermentation downstream technology

Many bioproducts are soluble and exist either in the clarified extracellular culture fluid or in the intracellular soluble fraction. These products may be further processed and purified by using molecular separation techniques which exploit property differences such as size, charge, solubility, volatility, biological affinity, etc.

The objective requires specification, while achieving maximum product yield at minimum recovery costs. The nature of the down-stream process will be influenced by the type of fermentation, physical and chemical properties of the product and the unwanted by-products or contaminants, product concentration, product location (intracellular, extracellular), scale of operation, waste treatment implication, product stability and desired specification.

On the other hand, it is better to couple the fermentation with down-stream process as an integral part in overall process. The nature of the fermentation process can significantly influence the downstream processing, for example, the cost savings or yield improvements achieved by using raw materials or more complex medium must take into account of the possible implications for down-stream processing. For products which are difficult to separate from contaminating cell or medium materials, the specific activity of the desired product (activity of product per unit weight of biological material) may be more important than the absolute product yields. Attempts should be made to eliminate or control production of undesirable by-product, by employment of industrial strains with superior traits or running fermentation under optimal conditions. Fermentation run-to-run variations with consequential production of variable broth or cell batches often cause problems in down-stream processing. Contaminated fermentations are particularly problematic in this regard. The broth harvesting time is also critical to the stability of cells or its bio-products.

20.2 Harvest and Pretreatment

20.2.1 General characteristics of fermentation broth

Before working on harvest and pretreatment of fermentation broth, it's necessary to learn the general characteristics of the fermentation broth and the chemical and biological properties of the desired bio-products, as follows:

(1) Dilute concentration in an aqueous medium;

(2) Usually temperature sensitive, pH sensitive and sometimes proteases sensitive;

(3) The products to be separated exist in small proportion;

(4) Might be intracellular, often as insoluble inclusion bodies in the recombinant *E. coli* cells;

(5) Some time the physical and chemical properties of the desired products are similar to the contaminants existing in broth;

(6) Easy to be contaminated again, and lost of bioactivity, thus should be extracted promptly;

(7) Required in extremely high purity and homogeneity for human health care, needs costly purification.

20.2.2 Collection and pretreatment

Collection would be the first step in a downstream process, which involves the separation of the cells and other insoluble particles from the fermentation broth. Shapes of the particles may be sphere, ellipsoid, rod, filament, or flocculent. Typical sizes for various cells vary widely such as: bacterial cells (0.5~1μm), yeast cells (1~7μm), fungal mycelia (5~15μm in diameter, more than 50μm in length). The separation of solid particles from the fermentation broth can be accomplished by filtration or centrifugation.

The purpose of pretreatment is not only to isolate bacteria and other particles, but also to remove some soluble impurities and change the nature of the filtrate. The pretreatment methods of various fermented products are diverted because of the strains and fermentation characteristics are all different. The majority of the fermentation products exists in the fermentation broth, a small number present in the body of the bacteria, or exist in both bacteria and broth. For extracellular products, we should try our best to transfer the products to the liquid phase by pretreatment, and then remove the solid phage by solid-liquid separation. For intracellular products, we should collect bacteria or cells first, after cell disruption, the product enters the liquid phase, and then separate the cell fragments.

Fermentation broths tend to be very dilute and contain complex mixtures of inorganic or organic substances. The pretreatment of a product (MW: 500～2500Da) such as an antibiotic, organic acid or animal vaccine from fermentation broth takes several processing steps. The first step is the clarification of broth to separate the low molecular weight soluble product from microorganisms and other particulate matter such as cells, cell debris, husks, colloids and macromolecules in the broth medium. In this step, either microporous membrane filters (MWCO: 10,000 to 500,000Da), pre-coat vacuum filter or centrifuge can be selectively used.

Centrifugation

Centrifugation is used to separate materials of different densities when a force greater than gravity has been implemented, known as the centrifugal forces. Centrifugation may be used to remove cells from fermentation broth; yeast, for example, can be easily harvested in a centrifuge unit. For dilute suspensions each cell may be treated as a single particle in an infinite fluid. In the concentrated fluid with suspended solids, the particle's motion is influenced by neighboring particles. A continuous centrifugation is commonly used in industrial separation of solid particles from fermentation broth (Fig. 20.2).

Fig. 20.2 Industrial centrifuges

In some cases, the decanter centrifuge with large throughput capacity of the nozzle is used for recovery of mould pellets or large particles of impurities. Another type of centrifuge is known as a scroll conveyer centrifuge. By comparing the types of continuous centrifuges, the scroll type of centrifuge has an important feature similar to decanter, which is being able to handle large solid particles without any clogging, and the settled solid is easily removed with a scroll centrifuge.

Filtration

Filtration is also an important unit operation in downstream technology. Solid particles are separated from a fluid-solid mixture by forcing the fluid through a filter medium or filter cloth. As a result, solids are retained by filter media and the filtrate is obtained, which is a clear solution without any particles. The solid particles deposited on the filter form a layer known as filter cake. The depth of the filter cake gradually increases as more solids are retained and create more resistance to further filtration. Filtration can be performed by using either vacuum or positive-pressure equipment (Fig. 20.3). The exerting differential pressure across the filter separates fluid from solid and is called the **filtration-pressure-drop**, and also the viscosity of the fluid may create filtration resistances. In many cases, the fermentation broths are troublesome and are hard to filter, because of the non-Newtonian behavior of the broth. Most microbial filter cakes are compressible. When the filter pores are clogged by cell bodies, the increasing pressure results in major problems in the filtration of fermentation broth. Finally, the filter cake should be removed and the cleaned cloth can be used again.

Fig. 20.3 Plate and frame type filter press

Plate filters are suitable for filtration of batch fermentation broth; accumulated biomass must be cleaned periodically. A rotary drum vacuum filter is used for a continuous treatment and is often used for removal of *Penicillium* and *Streptomyces* mycelia in the production of antibiotics. In these processes, the rotary drum filter is used with a pre-coated cloth filter with filter aid; by a knife blade which scrapes the solid accumulation from the rotating drum, the filter cake is continuously removed.

Cross-flow filtration (**CFF**) also known as tangential flow filtration, is not of recent origin. It began with the development of reverse osmosis (RO) more than three decades ago. Industrial RO processes include desalting of sea water and brackish water, and recovery and purification of some fermentation products. The cross-flow membrane filtration technique was then applied to the concentration and fractionation macromolecules commonly recognized as ultrafiltration (UF) in the late 1960's. Major UF applications include recovery of enzyme or protein molecules, enzyme or protein molecules, and UF is also used for removal of **pyrogens**.

In the past ten to fifteen years, the applications sphere of crossflow filtration has been extended to include microfiltration (MF) which primarily deals with the filtration of colloidal or particulate suspensions with a size ranging from 0.02 to about 10 microns. Microfiltration applications are rapidly developing and range from sterile water production to clarification of beverages, fermentation products, concentration of cell mass, yeast, *E. coli* and other media in biotechnology are also the related applications.

Tab. 20.1 shows the types of separations achievable with MF, UF and RO membranes when operated in cross-flow configuration. For MF or UF application, the choice of filter materials includes ceramics, metals or polymers, whereas for RO only polymer membranes are predominantly used at the present. Although cross-flow filtration is practiced in all the above three types of membrane applications, the description of membrane characteristics, operational aspects and applications will be limited to MF and UF, and here the cross-flow mode shows the greatest impact on filtration performance compared with dead end filtration. The schematic of cross-flow filtration is shown in Fig. 20.4, which includes the critical issues and operational modes for clarification or concentration using a semi-permeable polymeric or inorganic membrane.

Tab. 20.1 Separation spectrum of different filtration

Average size	Separation examples	Process
100～500Da	Organic acid, acetic acid, citric acid, amino acids	UF RO
200～2,000Da	Penicillin, cephalosporin	MF/UF
10,000～200,000 Da	Proteins, polysaccharides	UF
0.01～0.3μm	Viruses	UF
0.1～1μm	Colloidal sillica	MF/UF
0.1～10μm	*E. coli*, *Pseudomonas diminuta*, mammalian cells, air-borne microbes	MF MF
1～100μm	Oily emulsions, bacteria cells, yeasts, molds	MF/UF MF

Despite the increasing use in a wide range of applications, cross-flow filtration still remains a semi-empirical science. Mathematical models and correlations are generally unavailable or unapplicable under very specific and well-defined conditions, owing to the complex combination of hydrodynamic, electrostatic and thermodynamic forces that affect flux and/or retention. Membrane fouling is not fully understood yet and perhaps is the

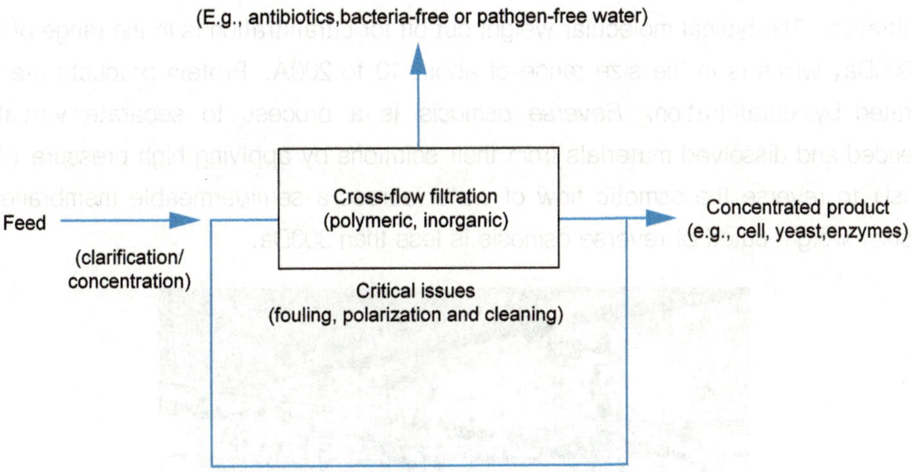

Fig. 20.4 Schematic of cross-flow filtration

biggest obstacle to widespread use of CFF in solid-liquid separations. The success of a membrane based filtration process depends on its ability to obtain a reproducible performance in conformance with the design specifications over a long period of time with periodic (typically once a day) membrane cleaning.

Ultrafiltration

Filtration separates particles by forcing the fluid through a filtering medium on which solids are deposited; the conventional filtration canvas is made from synthetic fabric, or glass fiber. By using polymeric membranes, we can apply the same filtration principle for the separation of smaller particles down to small size of the molecular level. Depending upon the size range of the particles separated, membrane separation processes can be classified to three categories based on the ranges of molecular weight cutoff (summarized in Tab. 20.2): microfiltration, ultrafiltration, and reverse osmosis.

Tab. 20.2 Pressure driven membrane separation processes

Process size	Cutoff	Molecular Wt. cutoff	Pressure drop (psi)	Material retained
Micro-filtration	0.02～10μm		10	Suspended material including microbes
Ultra-filtration	10～200Å	300Da～3000kDa	10～100	Biochemicals, colloids, macromolecules
Reverse osmosis	1～10Å	<300	100～800	All suspended and dissolved material

Microfiltration refers to the separation of suspended material such as bacteria by using a membrane with pore sizes of 0.02 to 10μm. If the pore size is further decreased so that the separation (Fig. 20.5) can be achieved at the molecular level, it is called ultrafiltration. The typical molecular weight cut off for ultrafiltration is in the range of 300 to 300,000Da, which is in the size range of about 10 to 200Å. Protein products are easily separated by ultrafiltration. *Reverse osmosis* is a process to separate virtually all suspended and dissolved materials from their solutions by applying high pressure (100 to 800 psi) to reverse the osmotic flow of water across a semipermeable membrane. The molecular weight cutoff of reverse osmosis is less than 300Da.

Fig. 20.5 Ultrafiltration device

The ultra filters are usually made by casting a polymer solution on a surface and then let the liquid film gelled slowly by exposing it to humid air. The pore size in the membrane can be varied by altering the composition of the casting solution or the gelation condition. Another common technique is to irradiate a thin polymeric film by exposure to radiation then chemically etching of the film to produce well-defined pores.

20.2.3 Cell disruption

The extraction of biologically-interesting molecules from cloned or cell engineered cultures would be more complicated. Except of excreted molecules, the cells lacking secretion capability must be disrupted to let the intracellular molecules released. Several factors must be considered before working on cell disruption, for example: 1) average size of cells; 2) basic structure of cell envelope; 3) cell density in the sample; 4) proportion of cell being disrupted; 5) sensitivity of cells to the heat produced from disruption; 6) whether there is a need to add protective factors to maintain the activity of the desired products; 7) whether there is safety problems.

For easily disrupted cells such as insect and mammalian cells grown in culture media, a mild osmosis-based method for cell disruption (lysis) is commonly used. Quite frequently, lowering simply the ionic strength of media will cause the cells to swell and burst. But for cells that are more difficult to disrupt, such as bacteria, yeast, and algae, hypotonic shock alone is insufficient to break the cell. Due to the presence of cell walls, which must be broken to allow access to intracellular components, stronger methods must be employed. These stronger methods are discussed below.

Two categories of well-defined methods for cell rupturing are: 1) mechanical grinding with abrasives, high-speed agitation, high pressure pumping and ultrasound; 2) non-mechanical methods such as osmotic shock, freezing and thawing, enzymatic digestion of cell walls and treatment with solvents and detergents.

A widely used cell disruption technique is high-pressure **homogenization**. Pressing and shearing forces generated in this treatment are sufficient to completely disrupt many kinds of cells. In the homogenization system, a high-pressure pump incorporates an adjustable valve with a restricted orifice through which the cells are forced at a pressure of up to 550 atmospheres. This kind of homogenizer is of general applicability for cell disruption. We should be aware that the homogenizing value would be blocked when it is used for disrupting the highly filamentous cells. Keeping the sample in cold is an important protective way to maintain the activity of macromolecules. And the method might not be encouraged for treating the cells which contains DNA or RNA as desired bio-products, because those long-sequence macromolecules are very fragile under the shearing force.

20.3 Isolation and Purification

20.3.1 General aspect

The fermentation products can be the cells themselves (biomass), components secreted by cells into the fermentation broth (extra cellular), or those trapped within the cells (intracellular). If the product of our interest is the cell, it is easily separated from the fermentation broth and is then washed and dried. In the case of extracellular products, after the cells are separated, products in the dilute aqueous medium need to be recovered and purified. When the intracellular products are needed, the first step would be the harvest of cells, and the next step is rupturing the cells to release the desired products, and finally is recovering and purifying of the end products. The downstream processing for exo-enzyme will be an example similar to the process for extracellular products. Varieties of techniques are used for isolation and purification of bio-products, only few of them are introduced as follows.

20.3.2 Isolation by solvent extraction

Solvent extraction is a method for separation of fermentation products by using a solvent to separate out the particular biochemicals. Solvent in liquid form is commonly used, but in rare case gases and solids can also be employed. In this method, a solvent is introduced to a mixture, and if some of the components are more soluble than the others, the sample starts to separate out, allowing people to remove the separated components individually.

Solvent extraction is popular for recovery of fermentation products downstream. Many antibiotics are dissolved in an organic solvent, which may be the precipitated by converting antibiotic to salt form and separating it from the organic solvents. Ethanol, acetone, propanol, butanol and ketones are commonly used in pharmaceutical industries.

Many products such as antibiotics have excellent solubility in organic solvents and they are water immiscible. A multistage extraction separates the aqueous phase from the organic phase. Extraction can provide concentrated and purified products. A typical penicillin broth contains 30～50mg/L of antibiotic. Filtration is used to remove mycelial biomass from fermentation broth. The filtration may be subjected to filter aided polymers. Neutralization of penicillin at pH 2～3 is required. Amyl acetate or butyl acetate is used as an organic solvent to remove most of the products from the fermentation broth. Finally, penicillin is removed as sodium penicillin and precipitated by a butanol - water mixture.

Isolation of erythromycin using pentyl or amyl acetate is another example of solvent extraction. Solvent extraction is also used for recovery of steroids and purification of vitamin. However, organic solvents are undesirable for the isolation of proteins, for the reason that protein molecules are easily denatured in the solvent extraction. The purification of DNA or RNA molecules can be performed by solvent treatment, for example, proteins become denatured when phenol is added to the sample, but nucleotides remained in the aqueous phase and are separated from the junk protein components.

20.3.3 Chromatographic isolation and purification

Surface adsorption chromatography

In general, adsorption is a surface phenomenon, where gas or liquid is concentrated on the surface of solid particles or fluid interfaces. There are many adsorption systems. Fixed-bed adsorption may give a higher adsorption area per unit volume than any other types of adsorbent. The point of saturation of the bed is called the breakthrough point. By knowing the value of specific breakthrough point, one can determine operation schedules. In designing fixed-bed adsorption, the quantity of resin and the time required for adsorption of a given quantity of solute must be well estimated. Adsorption chromatographic techniques are based on four different sorption mechanisms: surface adsorption, partition, ion exchange and size exclusion. Those techniques can be used for purification of fermentation products.

The separation mechanism depends upon differences in polarity between the different feed components. The more polar a molecule is, the more strongly it will be adsorbed by a polar stationary phase. Similarly, the more non-polar a molecule is, the more strongly it will be adsorbed by non-polar stationary phase. During a surface adsorption chromatography process, there is competition at stationary phase adsorption sites, between the materials to be separated and the mobile phase. The feed molecules of low polarity spend proportionally more time in the mobile phase than those molecules having highly polarity, and retained well in the column, therefore the components of a mixture are eluted in order to increase polarity.

Ion exchange chromatography (*IEC*)

In this process, the stationary phase consists of an insoluble porous resinous material containing fixed charge-carrying groups. When the organic compounds carrying counter-ions are running through the resin bed, the biochemicals would be retained onto the surface of resin, and the unwanted components with the same charge or without any change would be excluded and drained out of the column. As a very popular down-stream separation technology, IEC has been used for extraction of organic acids and amino acids, since they have different charged residues. The disadvantages of IEC involve the pollution from the resin regeneration step, in which acid or alkaline is used to treat the column, and large amount of water is used to rinse the regenerated resin. Now, the glutamate industry is exploiting a new process for replacement of IEC step by developing twice-crystallization of glutamic acid directly from the broth.

Size exclusion chromatography (*SEC*)

Except of enzyme extraction (already mentioned in Part III), the size exclusion chromatography can be applied to purify specific peptide molecules or other macromolecules in a feed material, according to the difference of their size or molecular weight. Many pharmaceutically important biochemicals, such as proinsulin or other hormones produced from engineered *E. coli*, are purified through size exclusion chromatography.

Partition chromatography

Unique to chromatography is the liquid-supported or liquid-bonded solids, where the mechanism is absorption into the liquid, also referred to as a partition mode of separation or partition chromatography. With mobile liquid phases, there is a tendency for the stationary liquid phase to be stripped or dissolved. Therefore, the stationary liquid phase has to be chemically bonded to the solid bonding support.

20.4 Product Polishing

20.4.1 Crystallization and recrystallization

Crystallization is the (natural or artificial) process of formation of solid crystals precipitating from a solution, melting or more rarely depositing directly from a gas. Crystallization is also a chemical solid-liquid separation technique, in which mass transfer of a solute from the liquid solution to a pure solid crystalline phase occurs. In a unit operation of chemical engineering, crystallization occurs in a crystallizer, which incubates the crystal nuclei to grow into pretty crystals. Crystallization is therefore an aspect of precipitation, which is obtained through a variation of the solubility conditions of the solute in the solvent, as compared to precipitation due to chemical reaction.

Crystallization separates a product from a liquid **feed stream**, often in extremely pure form, by cooling the feed stream or adding precipitants which lower the solubility of the desired product so that it forms crystals. Well formed crystals are expected to be pure because each molecule or ion must fit perfectly into the lattice as it leaves the solution. Impurities would not fit as well in the lattice normally, and thus remain in solution preferentially. Hence, molecular recognition is the principle of purification in crystallization. However, there are instances when impurities incorporate into the lattice, hence decreasing the level of purity of the final crystal product. Also, in some cases, the solvent may incorporate into the lattice and form a solvate. In addition, the solvent may be 'trapped' (in liquid state) within the crystal formed, and this phenomenon is known as inclusion. Fermentation products such as amino acids, organic acids and some other biochemicals can be purified by crystallization. Mono sodium glutamate crystal as an example is shown in Fig. 20.6.

Fig. 20.6 Mono sodium glutamate crystals (5X)

Recrystallization

In biochemistry, recrystallization is a procedure for further purifying compounds. The most typical situation is that a desired "compound A" is contaminated by a small amount of "impurity B" with pretty closed properties. Recrystallization is involved with dissolving of the primary crystals as the first step. By readjusting very carefully the dissolving solution and incubating condition, the small amount of impurity B dissociates with the desired product A and detains in the solution, thus the purity of A is up graded. Recrystallization process has been applied to purify antibiotics, such as meleumycin, to ensure those fermentation products with higher purity to meet the strict quality management.

20.4.2 Desiccation

Desiccation is referred to as the state of extreme dryness, or the process of extreme drying. After purification the purified bio-products (in solid form) are usually desiccated in an extreme dry and stable form, and some of end products must be kept at low temperature. There are several desiccation methods, among which, the processes of freeze-drying and spray-drying are commonly employed. In practice, the option depends on the quality management of the bio-product and the processing cost.

Freeze-drying

More often, bioproducts are freeze-dried and easily stored. Freeze-drying (also known as lyophilisation, lyophilization or **cryodesiccation**) is a dehydration process typically used to preserve a perishable material or make the material more convenient for transport. Freeze-drying works by freezing the material and then reducing the surrounding pressure to allow the frozen water in the material to sublimate directly from the solid phase to the gas phase.

Fig. 20.7 Industrial freeze-drier

Pharmaceutical companies often use freeze-drying (Fig. 20.7) to increase the shelf life of products, such as vaccines and other **injectables**. By removing the water from the material and sealing the material in a vial, the material can be easily stored, shipped, and later reconstituted to its original form for injection. Another example from the pharmaceutical industry is the use of freeze drying to produce tablets or wafers, the advantage of which is less excipient as well as a rapidly absorbed and easily administered dosage form.

In bio-separation processes, freeze-drying can be used also as a late-stage purification procedure, because it can effectively remove solvents. Furthermore, it is capable of concentrating substances with low molecular weights that are too small to be removed by a filtration membrane.

But, freeze-drying is a relatively expensive process, for the equipment is expensive and the high energy demands lead to high energy costs. Furthermore, freeze-drying also needs a long processing time, because the avoidance of heat can prevent the material from melting or structural deformations. Freeze-drying is often applied to the materials that are high value added products but heat-sensitive, such as proteins, enzymes, bioactive factors, and blood plasma. The low operating temperature of the process leads to minimal damage of these heat-sensitive products.

Spray drying

Spray drying is a method of producing a dry powder from a liquid or slurry by rapidly drying the bioproducts with hot air. This is the preferred method of drying for many thermally less-sensitive biochemicals and foods. When the liquor medium (water or solvent) containing bioproduct is sprayed into the drying chamber, the medium evaporates promptly and the vapor is carried away by drying air, and then the dried powders are collected right away. A consistent particle size distribution is a reason for spray drying some industrial products such as catalysts. Air is the routinely heated drying media, but if the bioproduct is dissolved in a flammable solvent such as ethanol, or the product is oxygen-sensitive, then the drying media should be replaced by nitrogen gas.

20.4.3 Sterilized packaging

Product polishing would be the last step which provides a high quality and stable final product. Depending on the product and its intended use, polishing may also include operations to sterilize the product and remove or deactivate trace contaminants which might compromise product safety. Such operations might include the removal of viruses or depyrogenation, that's why the packaging technique should be designed for joint processing in order to make the product in a form that is stable, easily transportable and convenient.

20.5 Waste Treatment and Cleaner Production

20.5.1 Aspect of discharge from fermentation

Fermentation plants produce ethanol, alcoholic drinks, organic acids, amino acids, antibiotics and enzymes etc. Most of the fermentation industries are running in large scale, which means the amount of effluent discharged by those industries is huge. It's almost impossible to achieve the conversion rate of 100%, from substrate to product, which means that the substrate remained incompletely utilized or has been converted partially into some other by-products. Meanwhile, the cells of industrial strain running into death phase are partially broken, which adds more waste contents to the effluent. For example, the effluent form ethanol fermentation contains unused-up starch and dead yeast cells and the BOD value would be around 20,000. In the effluent of glutamic acid fermentation, not only the carbohydrate, but also the nitrogen and phosphate components are very rich.

There are many factors that affect the volume and strength (BOD value) of wastewaters discharged from fermentation plants. The amount of wastes depends upon how much of the syrup that is not converted into a quality product. It is a well known fact that the efficient plants generate less waste than poorly running plants, for the higher level of the techniques and management are employed, which afford higher productivity and lower waste discharge. Management attitude and precise control would be the primary factors. Modern measurement on plant efficiency which can be used to calculate plant efficiency is now used to calculate the daily production of BOD_{5D} in terms of pounds or kilograms. According to the strict environmental regulations, all kinds of fermentation discharge must be well treated to lower the BOD_{5D} value and meet the regulation standards.

20.5.2 Basic techniques for effluent treatment

Wastewaters discharged from fermentation plants normally contain no toxic organics or heavy metals. The primary contaminants are mainly composed of the residual of substrates, fermentation by-products and the cell debris. That's why in most of the treatment processes, the biological procedures are predominated technology, which would not cause the secondary pollution (compared to chemical treatment). For small size production, the effluent will be pumped to local treatment plant, while for large scale fermentation in situ treatment would be more practical and is the necessary domain in the plant. Firstly, the fermentation effluent is usually treated with anaerobic digestion (Fig. 20.8) in order to allow the component with high BOD value to be degraded into methane. Then the sludge produced from anaerobic microbes would be removed by precipitation and dehydrated procedure. Both methane and sludge can be utilized as fuel and fertilizer. The treated waste water might not meet right away the discharging regulation, so the aerobic treatment (Fig. 20.9) would be employed to lower the remained BOD value, which is operated simply by exposing the waste water to fresh air and let the organic compounds being further oxidized.

An alterative technique is using algae to treat fermentation effluent, whose organisms

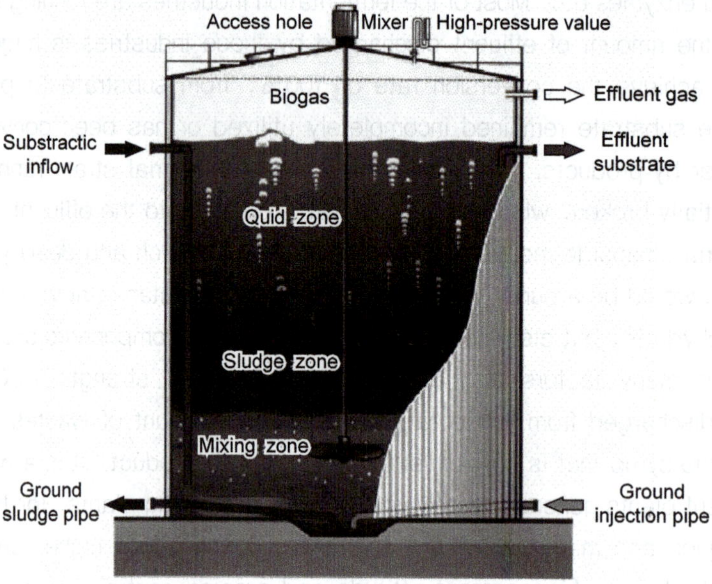

Fig. 20.8 Anaerobic digester

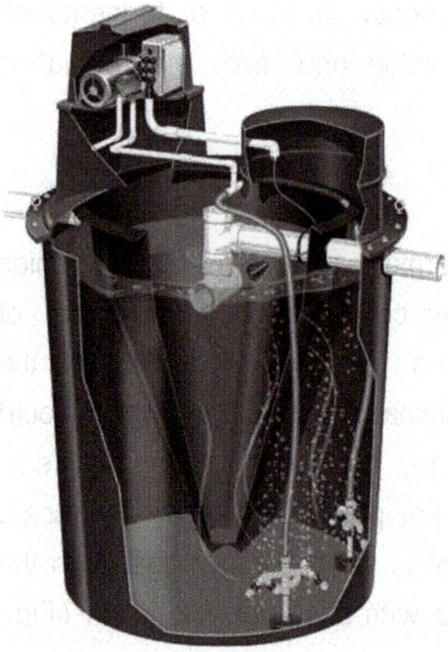

Fig. 20.9 Aerobic treatment unit

take sunlight as energy and grow well in the elements C, N, and P that contain waste water. As by-products from bio-treatment, the single cell protein of algae can be used as feed supplement, and the oil synthesized by algae would be a potential biofuel resource.

20.5.3 Cleaner production and "green biorefining"

Cleaner production is a preventive, company-specific environmental protection initiative. It is intended to minimize waste and emissions while maximize product output. By analyzing the flow of materials and energy consumption in a production system, one tries to identify options to minimize waste and optimize the industrial processes through source reduction strategies. Improvements of organization and technology help to reduce energy cost, offer better suggestions in use of materials, as well as avoid solid waste and effluent generation, gaseous emissions, and waste energy and noise.

Examples for cleaner production options are:

• Documentation of consumption (as a basic analysis of material and energy flows, e. g., with a Sankey diagram);

• Use of indicators and controlling (to identify losses from poor planning, poor education and training, mistakes);

• Substitution of raw materials and auxiliary materials (especially renewable materials and energy);

• Increase of useful life of auxiliary materials and process liquids (by avoiding drag in, drag out, contamination);

• Improved control and automation;

• Reuse of waste materials and water;

• Employment of novel technologies for cleaner production.

For example, the cleaner production would be a major task to the glutamic acid fermentation, and the integrated procedure might involve the readjustment of fermentation media, application of superior master strain, enhancement of process control, substitution of ion exchange procedure by dual-crystallization, and separation between solid and liquor from waste for material and water recycling.

A biorefinery is a facility that integrates biomass conversion processes and equipment to produce fuels, power, heat, and value-added chemicals out of biomass. The biorefinery concept is analogous to today's petroleum refinery, which produces multiple fuels and products from petroleum.

The **International Energy Agency Bioenergy Task** 42 **on Biorefineries** has defined biorefining as the sustainable processing of biomass into a spectrum of bio-based products (food, feed, chemicals, and pharmaceuticals) and bioenergy (biofuels and power and/or heat produced from bio-production).

By producing multiple products, a biorefinery takes advantage of the various components in biomass and their intermediates, therefore maximizing the value derived from the biomass feedstock. A biorefinery could, for example, produce one or several **low-volume but high-value** chemicals, pharmaceutical products and a **low-value but high-volume** liquid transportation fuel such as biodiesel or alcohol fuel, and generate electricity and process heat through combined heat and power (CHP) technology, for its own use and perhaps enough for sale of electricity to the local utility. The high-value products increase profitability, while the high-volume fuel helps meet energy needs, and the power production helps to lower energy costs and reduce greenhouse gas emissions from traditional power plant facilities. Although some facilities that can be called bio-refineries exist, the bio-refinery has not been fully realized yet. Future biorefineries may play an important role in producing chemicals and materials that are traditionally dominated by petroleum. Several potential biorefinery examples have been proposed, from flax straw and the residues from the bio-production to produce the alcohol fuel.

Summary

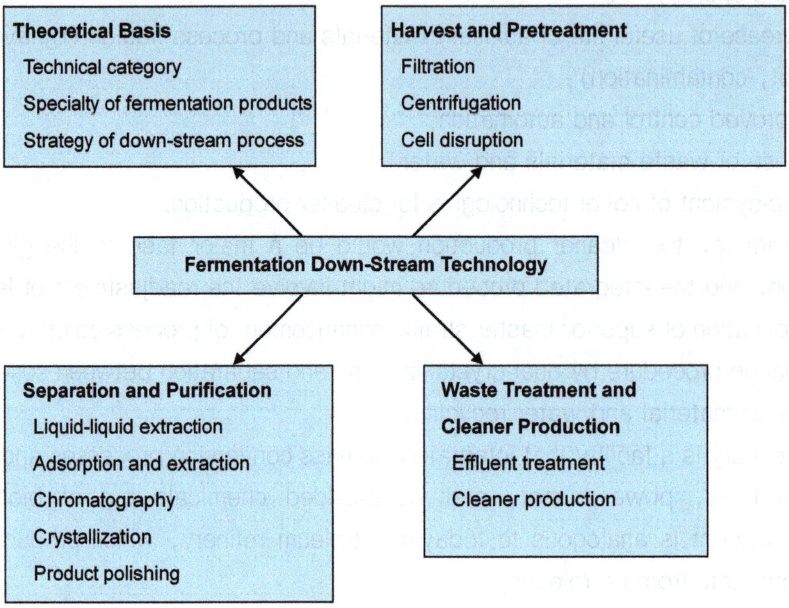
Theoretical Basis
Technical category
Specialty of fermentation products
Strategy of down-stream process
Harvest and Pretreatment
Filtration
Centrifugation
Cell disruption
Fermentation Down-Stream Technology
Separation and Purification
Liquid-liquid extraction
Adsorption and extraction
Chromatography
Crystallization
Product polishing
Waste Treatment and
Cleaner Production
Effluent treatment
Cleaner production

Chapter 21 Progress in Fermentation Engineering

发酵工程作为各类生物技术实现生产力转化的重要平台，其自身的技术升级至关重要。近些年来，基因组学、克隆技术、新材料研究、计算机辅助及节能技术等方面的研发成果，对发酵的各环节产生着重要的影响，促进了工业发酵的可持续发展。

21.1 Approaches of Up-stream Technology

In this section, the Fermentation Up-stream Technology is referred to as the technique applied previously to the conduction of fermentation in a bioreactor. Up-stream technology involves the construction, selection and improvement of novel industrial strains; the extending of resource utilization; and reconsideration of non-submerged fermentation mode in particular circumstance for high efficient and sustainable production.

21. 1. 1 Exploitation of novel trains

Extremophiles

The term **extremophile** usually refers to the microbes living in extreme environments. These microbes may live in the Arctic or Antarctic, the volcanic vents on land, oceanic floor, extremely dry places, the place deep inside the Earth, the severe chemical environments (acid, alkali and salt) which are harmful to most life forms, and even the high-radiation environments such as on the control rods of nuclear powerplants. The species living in these environments can survive for a long time. It was not until the 1970's that such creatures were recognized, and then the exploitation of extremophilic industrial strains started. The areas of employing extremophilic microorganisms involve microbial biodiversity and systematic studies, techniques of extremophile cultivation, isolation of extremozymes and bioleaching. There are increasing reports on the discovery of those microorganisms for exploitation in new antibiotic and therapeutic agents, probiotics, fine chemicals, novel enzymes and polymers. Different kinds of extremophiles were constructed for uses in different domains such as scientific and industrial applications, bioremediation of polluted environments, bioleaching and recovery of minerals, as well as making defensive agents against exotic and emerging pathogens of humans, animals, and plants.

The thermophilic strains which are tolerated in high temperature are the first exploited industrial extremophiles. From those microbials, thermophilic enzymes, for example, thermo-amylases, proteases and cellulases for special utilization are produced. Not only the fermentation itself benefits from the shortening of processing cycle and saving of energy, but also the thermo-enzymes which normally catalyze chemical reactions at rather high temperature gains higher efficacy in the enzymatic conversions. Uses of the extremophilic microbials adapting to tougher chemical conditions might create new biocatalytic techniques, which may replace the chemical catalytic process that is carried out under tough condition requiring a high concentration of acid or alkaline.

Breeding of extremophilic microbes usually undergoes several steps: 1) Field-sampling of extremophiles with specific traits based on pre-research results. For example the DNA polymerase Taq is commonly used for the 2nd strand DNA synthesis under high temperature and in high speed, which was discovered from a thermophilic bacterium living in a spa port; 2) Selection of the individual colonies which express superior extremophilic traits; 3) Breeding the strains by various mutagenesis treatment to increase the genetic stability and productivity. Conducting of *in vitro* mutagenesis to alter the molecular sequence of an extremophilic enzyme (only one or few amino acid residuals in the peptide chain might cause the 3-D structure to change, as mentioned in Part I); 4) Transform the extremophilic traits from the wild type to a suitable host through cloning, which aims to construct the engineered strains for industrial purpose. 5) Design a specific screening procedure for picking up the most ideal individuals for industrial uses. 6) All the bred extremophiles should be tested for biosafety, genetic stability and fermentation capacity.

Metabolic engineered strains

Metabolic engineering is generally referred to as the targeted and purposeful alteration of metabolic pathways found in an organism for better understanding and utilizing the cellular pathways for chemical transformation and energy transduction. Therefore, it is a new field with applications to the production of chemicals, fuels, materials, pharmaceuticals, and medicine at the metabolic level. The field's novelty lies in the application of engineering principles on metabolic pathway design and analysis. The application helps to achieve a particular goal involving the increase of process productivity, as in the case of fermentation of antibiotics, biosynthetic precursors or polymers, and also a goal of extending metabolic capability by the addition of extrinsic activities for chemical production or degradation. Previous strategies to attain these goals seem to be more of an art with experimentation by trial-and-error, but the metabolic engineering allows rational selection of targets for genetic modification through measurements and control of metabolic fluxes. In the early 1990s, the switch from this artistic approach to a more systematic and rational approach was advocated by Bailey, Stephanopoulos and Vanillo. Later on, the *Sacchromyces* was engineered to make it fermented pentose (originally only use hexose as substrate) to produce biofuel. The successful approaches in metabolic engineering are mostly involved with gene cloning and they also require a better understanding of cellular physiology to modify a particular metabolic pathway on purpose.

The interest in metabolic engineering is stimulated by potential commercial applications where improved methods for developing strains can increase production of useful metabolites. Recent endeavors have focused on using biologically derived processes as alternatives to chemical processes. Such manufacturing processes pursue goals related to "sustainable developments" and "green chemistry" as well as positioning companies to exploit advances in the biotechnology field. For example, when running recombinant *E. coli* cells for producing intercellular protein products, the biomass yield in the broth is expected to be as high as possible, but acetic acid accumulated as byproduct occurs in the stationary phase might inhibit cell growth. By genetically weakening the acetate pathway, the growth inhibition can be partially removed. Some other examples of metabolic engineering also include the microbial production of indigo (Genencor) and propylene glycol (DuPont), others involving improvements in the more traditional areas of antibiotic, and amino acid production by a large number of firms.

Natural products and their derivatives play an important role as **frontline** treatments for many diseases and as inspiration for chemically synthesized therapeutics in modern healthcare. With advances in metabolic engineering, many of the biosynthetic pathways have been elucidated, which is responsible for the production of these chemically complex but yet valuable compounds. With an ever-expanding toolkit of biosynthetic components, metabolic engineering is an increasingly powerful method to improve natural product titers and generate novel compounds. The extension of metabolic engineering for producing the desired compounds in plant tissues and provide better understanding of human metabolic disorders has broadened the interests beyond the fermentation industry.

Novel producer from metagenomic blast

Thousands of genes from industrial strains have been sequenced and recorded in different gene libraries, and varieties of software are developed for blasting their genomes and many industrial microbials have been identified. But it was reported that more than 99% of the microbial species are non-cultured or un-cultured yet by the traditional screen and cultivation approaches, which limits the discovery of more useful microorganisms from natural environment. "Metagenome" is referred to as the genomic library of the total microbes found in nature, which is constructed by directly extracting the DNA from environmental sample, amplifying and transforming to the surrogate host. The libraries were screened for novel genes or unknown molecules based on bio-informatics study. These techniques enormously amplify the space of microbial resource utilization and enhance the opportunity of obtaining novel bioactive compounds from nature. Faster and cheaper sequencing technologies applied to those uncultured microbes sampled directly from their habitats are broadening and transforming our views of the microbial world. Distilling meaningful information from millions of new genomic sequences leads to a challenge for the bioinformaticians. In the cultured microbes, the genomic data only come from a single clone, making sequence assembly and annotation tractable. But in metagenomics, the data come from heterogeneous microbial communities, sometimes containing more than 10,000 species, with the sequence data being noisy and partial. From sampling to assembly, gene calling and function prediction, bioinformatics faces new demands in interpreting voluminous, noisy, and often uncompleted sequence data. Although metagenomics is a relative newcomer to science, the past few years have seen an explosion in computational methods applied to metagenomic-based research. We've notice that, at least, some of the microbes with novel antifungal activities or toxic-degradation capacity have been screened out from the metagenomic libraries. And this threw a light to modern fermentation industry, for the sustainable progress of this industry always relies on the endless fountainhead of massive novel industrial strains.

Application of "synthetic biology"

Based on the engineering methodology, a new sub-discipline named "Synthetic Biology" has emerged in recent years. Aimed on improvement of bio-species, synthetic biology focuses on assembling of artificial sequences or recombination of ideal sequence segments that already exist (for example, operon, promoter, reporter, or other core components for specific expression) and building a new "bio-facility" with desired bio-functions. Since some of the microbial species are well studied and their genomes are throughout sequenced, it's not difficult to synthesize total genomic sequence of viruses or bacteria (for example, the successful case of *Mycoplasma genitalium*). Synthetic biology might lead to the incredible "Artificial Evolution" of the target organism in a comparatively short period and of high efficiency. Scientists are now able to treat cells as a Nanosystem. Specific DNA subunits can be put together to form an artificial genome to meet the user' specification. Synthetic biology has been applied to produce lysine, which is made by a bacterial strain with various modified genes and ideal expression sequence in its genome, and the lysine has over secreted from the "synthesized strain". The same strategy is also implemented in the production of various secondary metabolites from the recombinant yeasts.

Another purpose to reconstruct industrial super-strains would be the assembly of "**Minimal Genome**". Blattner group has made precise deletion from the genome of *E. coli* MG 1655 which get rid of many unnecessary segments from the genome. It was proved that by 15% genomic deletion, the growth rate of the bacterium was not affected. Nevertheless, the genetic stability of the chromosome and episome has been improved, though the electric transformation efficiency was higher than that of the wild type. The work of constructing industrial strains with "**Minimal Genome**" points out a promising future to the fermentation industries.

21.1.2 Extended bio-resource

The benefits of exploiting bio-resources are manifold. These resources do not rely on fossil resources, and are more energy efficient. Their substrates and wastes are biologically degradable, which helps decrease their environmental impact. By using alternative substrates and energy sources, many innovations were made in chemical, textile, food, packaging and health care fields.

Utilization of celluloid resources

Wood is one of the most abundant resources in the bio-based industry. Yet it is also one of the most complex materials. Woody plant species are so diverse and are composed of different kinds of polymers, such as cellulose, hemicellulose, lignin etc. **Cellulose** is a polymer composed of glucose (sugar) chains. It makes up nearly 50 percent of woody plant mass. Cellulose can be hydrolyzed into glucose by cellulases and added to the fermentation medium for production of single cell protein, ethanol, cellophane, and cellulose ethers. Over last 20 years, the increased population and growing economics demanded more and more wood products, and the climate situation is also not optimistic, so cellulose is no longer the biomass as abundant as we thought. But still there are quite amount of waste from the wood industries that are waiting to be used up reasonably, such as **Hemicellulose** which consists of pentose, mainly xylose, and comprises 25 to 35 percent of the dry weight of wood residues, being the second only to cellulose in abundance. The use of hemicellulose is currently limited due to manufacturing related costs. However, quantities of hemicellulosic pectins and various other plant polymers are abundant and have great potential in the production of chemicals and materials. Also **lignin**, as phenylpropane polymer which holds together the cellulose and hemicellulose components of woody biomass, can be exploited as new resource to fermentation industry. Lignin constitutes about 15 to 25 percent of the weight of woody biomass, but has not been used yet as a raw material for industrial purposes in large quantities. This reflects the chemical complexity of lignin. In addition, attempts by pulp and paper research laboratories were made to develop by-product markets for lignin. Many fungal species were found to have enzymatic decomposition ability on lignin. Mixed with cellulases and semi-cellulases together, the wood waste can be degraded into a saccharine containing resource.

The complex components existed in the celluloid materials usually limit them to be used as the substrate for fermenting food stuff and pharmaceuticals, but fermentation of wood waste for producing solvents or bio-fuel would be a reasonable choice. The industrial strains can also be metabolically engineered to make them the utilized inexpensive substrate, such as using wheat straw to produce ethanol. The agricultural celluloid material consists of a large percentage of pentose in addition to hexose. However, in conventional ethanol fermenting, yeast can only use hexose as substrate. There were successful cases that engineered yeast strains convert both hexose and pentose into ethanol after the pentose metabolic pathway has been reinforced by cloning. Celluloid materials mixed with waste water containing nitrogen, phosphate and other nutrients can also be utilized to cultivate single cell protein(SCP) to produce feed protein.

Novel biopolymers

An entire branch of biotechnology, known as "white biotechnology", is devoted to the use of living cells (from yeast, moulds, bacteria and plants) and enzymes to synthesize polymer materials that are easily degradable, require less energy and create less waste. One of the first goals on white biotechnology's agenda has been the production of **novel biopolymers**. Over the past 20 years, efforts have been concentrated mainly on **biodegradable plastics**, such as polyesters of 3-hydroxyacids (PHAs), which are naturally synthesized by a wide range of bacteria as an energy reserve and carbon source. These compounds have the properties similar to the synthetic thermoplastics and elastomers from propylene to rubber, but are completely and rapidly degraded by bacteria in soil or water. The most abundant PHA is poly 3-hydroxy-butyrate (PHB), which is systhesized by bacteria from acetyl-CoA. Growing in glucose medium, the bacterium *Ralstonia eutropha* can amass up to 85% of its dry weight in PHB, which makes this microorganism a miniature bioplastic factory. Much effort has therefore made on reducing production costs through the construction of superior bacterial producers. More microbial producers were bred to produce various white biopolymers to replace the synthetic polymers. For example, some species of *corynebacteria* produce poly-lactic acid, which can be used to manufacture items such as clothing fiber, packaging materials and food containers. Under certain culture condition, *Bacillus subtilis* produces poly glutamic acid (PGA) which is not only safer and degradable, but also possesses strong water absorption capacity, and is suitable to be processed as drug delivery carrier. Recently, a potentially more economic and environmentally friendly alternative method emerges, namely the transgenic plants to synthesize novel biopolymers.

Photosynthesis

To the earth, the only external energy source is the sun, so life has to cope with utilizing sun light to fuel all the metabolic demands. Not only can the plants convert energy from sunlight to chemical forms, but also some microbes such as *Cyanobacteria* (blue algae) and their relatives that responsible for a major part of photosynthesis in oceans and on wet land. All these organisms convert CO_2 (carbon dioxide) to organic material by reducing this gas to carbohydrates in a rather complex set of bioreactions. Electrons for this reduction reaction ultimately come from water, which is then converted to oxygen and protons. During the process of photosynthesis, cyanobacterial cells are propagated, and the protein, lipid, and other organic components contained in the cells are valuable to our human being. Utilizing fermentation effluent and the daily waste water that contain nitrogen and phosphate to grow those photosynthesis bacteria should be encouraged for sustainable development.

21.1.3 Alterative fermentation strategy

Solid state fermentation (SSF)

Although submerged fermentation is the predominant pattern applied in modern biotechnology field as mentioned in previous chapters, the strategies of **solid state fermentation (SSF)** and **shallow tray fermentation** are also reconsidered by entrepreneurs for many reasons. SSF has been practiced since prehistorical time, and is still applied for production of liquors, food additives, enzymes, and nutrient enriched animal feeds. Advantages of SSF over the submerged fermentation are as follows: higher volumetric productivity, simpler with lower energy requirements, easier to meet aeration requirements, resembles the natural habitat of some fungi and bacteria, and in some cases easier downstream processing. Utilization of **renewable materials** is somehow a challenging task to industrial fermentation. Huge amount of agriculture wastes are waiting to be converted into feedstock, methane, ethanol, and fertilizer. Probably it is worth to develop novel SSF process to fulfill this goal. Alone with the application of modern biotechnical knowledge, redesigning of SSF bioreactors and performance of process control aided by computer, one could offers this traditional technique a promising future.

Shallow tray fermentation

The technique is still under performance for production of citric acid by growing *Aspergillum niger* in very high concentration of glucose broth filled in shallow trays. Automatic operation ensures the process being controlled precisely. At the end of fermentation, the broth containing high concentration of citric acid is easily harvested. The mould mycelia is further rinsed and under *plate and frame filter* pressed, to recover the remained citric acid. Shallow tray fermentation was also performed in the case of **nattokinase** biosynthesis. Compared with the other fermentation pattern, *Bacillus subtilis* produces higher amount of nattokinase in shallow tray fermentation. The advantage of this process would be as follows: higher product concentration, lower water consumption, less loss of enzyme activity, less overall cost, and easier waste treatment.

Fermentation with high cell density

Fermentation with high cell density is also called as High cell density culture (HCDC). It is developed for the production of very important industrial strains in the specifically designed fermentors. The yeast, *Saccharomyces cerevisiae*, *Pichia pastoris*, *Kluyveromyces fragilis*, and *Candida utilis* have been grown in bench-scale fermentors at cell densities of over 120g/L, while the bacteria, *Escherichia coli*, *Bacillus megaterium*, *Methylomonas* sp. and *Pseudomonas putida* bacteria have been cultivated to cell densities of more than 110g/L. Various HCDC techniques have been developed for growing recombinant *E. coli* strains in fed-batch cultures at the concentrations greater than 100 grams (dry cell weight) per liter for most of the strains amass high valued bioactive substances within their cells, and the productivity of HCDC process seems quite important to the industrialists. Yeast cell productivities [g cells/(L • h)] greater than 25 have been achieved in both bench-scale and 150 Liter fermentors, and values as high as 55 have been achieved with bacteria in the bench-scale fermentor. The fermentor, capable with high O_2 and heat transfer rates, was operated at constant volume with continuous feed and product discharge. The problem of acetic acid accumulation in the stationary phase has been partially overcome by metabolic engineering. The advantages of performing HCDC process are obvious: smaller bioreactor size that is required, reduce in energy consumption, and less amount of wastes and recycle streams.

21.2 Process Optima

21.2.1 Advanced biosensors

More biosensors with different functions have been developed in recent years. And some of them can be introduced into industries for detection of parameters in fermentation process, if the techniques for aseptic installation of those biosensors become matured.

Calorimetric biosensors If the enzyme catalyzed reaction is exothermic, two thermistors may be used to measure the difference in resistance between reactant and product, and hence, the analyte concentration can be precisely measured. A microfluidic device was then designed which is capable of measuring real-time enthalpy changes of biochemical reactions and thermal properties of biological fluids. This device consists of a freestanding microthermopile integrated with a glass chamber for microfluidic reaction. The heat of reaction from the catalytic action of glucose oxidase, catalase, and urease on glucose, hydrogen peroxide, and urea, respectively, are reported. Calorimetric Biosensors were performed in open air using liquid batch testing and in enclosed fluidic reaction chamber by continuous flow experiments. The aim of this work is to demonstrate the potential of the integrated calorimetric microfluidic device for fundamental thermodynamic studies in biochemical reactions. Using arrays of such devices with immobilized enzymes, multi-analyte detection can be accomplished and the effects of interfering factors from competing substrates can be compensated.

Piezoelectric biosensors Piezoelectric sensors utilize crystals which undergo an elastic deformation when an electrical potential is applied to them. A circulating-flow piezoelectric biosensor, based on an Au nanoparticle amplification and verification method, was used for real-time detection on food borne pathogen, for example *E. coli* O157:H7. A synthesized thiolated probe (Probe 1; 30-mer) specific to the eaeA gene of *E. coli* O157:H7 was immobilized onto the piezoelectric biosensor surface. Hybridization was induced by exposing the immobilized probe to the *E. coli* O157: H7 eaeA gene fragment (104bp) amplified by PCR, resulting in a mass change and a consequent frequency shift of the piezoelectric biosensor. A second thiolated probe (Probe 2), complementary to the target sequence, was conjugated to the Au nanoparticles and was used as a "mass enhancer" and "sequence verifier" to amplify the frequency change of the piezoelectric biosensor. A linear correlation was found when the cells of *E. coli* O157:H7 detected were between the range between $10^2 \sim 10^6$ CFU/mL.

Surface plasmon resonance (SPR) biosensors A commercial example of a SPR biosensor employs a photoelectric principle, called surface plasmon resonance, for detection of changes in concentration of macromolecular reactants, which relates the energy transferred from photons bombarding a thin gold film at the resonant angle of incidence to electrons in the surface of the gold. This loss of energy results in a loss of reflected light at the resonant angle. The resonant angle is affected by changes in the mass concentration in the vicinity of the metal's surface which is directly correlated to the binding and dissociation of interacting molecules. The SPR array with sub-wavelength nanoholes in a gold film was first used to monitor the binding of organic and biological molecules to a metallic surface. Now SPR device has been developed with rapid, simple and specific immunoassay for detection of any particular molecules or microbials in the bioreaction process. SPR biosensor would be useful in food safety monitoring.

21.2.2 Computer control and expert system for online process control

As mentioned in Chapter 18, bioprocessing requires tight control of key nutrients and byproducts to optimize growth of cells in culture and maximize the yield of products. Online control allows industrialists the freedom to sample automatically, adjust both nutrient and byproduct concentrations and regulate all kinds of physical parameters. The techniques of online process control are progressed along with the IT technology. The online analyzers and monitors are available for industry choices by providing fast and accurate measurements of multiple parameters and feedback of the critical signals which keep the process running along the right track. The Carbon Dioxide Monitor measures real-time *in situ* dissolved carbon dioxide levels in bioreactors. Varieties of sensors are developed to meet the control requirement for individual process. In recent years, bioreactor suppliers have developed sophisticated software modules for the control of fermentation processes. However, the available systems can be further improved by incorporating more advanced algorithms. In particular, it is important to address the issue of process optimization, which is only marginally dealt with by firmware currently.

Expert system applied in fermentation

Recently, some of the fermentation industries are employing Expert system to optimize the process. Expert system is a computer system that emulates the decision-making ability of a human expert. Expert systems are designed to solve complex problems by reasoning about knowledge and experience like an expert. An expert system has a unique structure which consists of two parts: one fixed, independent of the system, i. e. the inference engine; and one variable, i. e. the data base. To run an expert system, the engine reasons about the data base like a human, and users can communicate with the system through a graphic user interface. Building, maintaining and development of expert systems are sometimes referred to as knowledge engineering.

For some of the individual fermentation has already been run into the state of art, as in the case of beer fermentation, the experts' knowledge and experience have become very dear to the industry. A good expert system must have the idea of looking fermentation process as a whole. In combination with the functions of advanced decision-making, statistical analysis, database sorting and other methods of optimization (including fault diagnosis) and post-analysis, an expert system can record all aspects of the fermentation process including raw materials, microbial strains, apparatus and the operating environment as the main data categories, as well as optimize the fermentation condition, control the process and predict the outcome of fermentation. A good expert system allows operator, when necessary, to readjust the programming during the fermentation, helping the industrialists to up grade the product quality and improve the overall production efficacy. For running expert system in fermentation, the cross-disciplinary talents are most welcomed by fermentation industry.

21.2.3 Energy saving

Lowering down the energy consumption is not only a practical method for cutting down the production cost, but also a responsible attitude to the society. There is room for energy saving in different sections of fermentation process:

(1) Controlling the amount of air input and increasing the oxygen transfer efficiency by reducing power supply form air compressor. Redesign of air nozzles and baffles would be helpful to increase the total surface area of air bubble. And supplying the amount of air and

providing the stirring power exactly matched with the fermentation dynamics would be intellectual means to save energy.

(2) Reducing energy supply from steam, heat water and cooling water. It is known to us that the energy consuming of mashing workshop occupies 50% energy in whole brewery. Recently, decreasing the energy consumption of mashing workshop is performed in many breweries by adoption of new boiling technology and reuse of waste steam and heat water. It's known that in order to run fermentation at the optimal temperature, the cooling water is always needed to exchange the energy released from the bioreaction. Using thermophilic strains and allowing the bioreaction being performed at rather high temperature would definitely reduce the cooling energy. Also, using sectional heat exchanger might benefit both heating and cooling sections.

(3) Well control of agitation power with the aid of computer in the way of just matching the needs of cell growth and metabolism, in avoidance of electricity over supply and excessive heat generation.

(4) Saving energy from the starch gelatinization by using novel enzymes. Different kinds of amylases and **glucoamylases** have been screened or constructed with the capability of raw-starch liquefaction, which help save the energy for cooking starch into gelatinized form to allow the conventional amylases to digest.

21.3 Development of Down-stream Technology

Along with the diversification of fermentation processes, the fermentation down stream treatment is under challenge. It's well known that the biotechnology industry today employs recombinant bacteria, mammalian cells, and transgenic animals for the production of high-value therapeutic proteins, and the techniques for recovery of those cloning products need some *case to case* works. The problem of how to improve the efficiency of down stream processing requires the engineers to bringing over the advanced techniques from other disciplines. Only a few approaches are introduced in this section, but it should be remembered that the thorny problem needs comprehensive solutions, since the fermentation industries still produce large amount of effluent and solid waste.

21.3.1 Affinity purification for cloned products

Significant efforts are made for the design of large-scale purification processes of proteins due to great demands of cost efficiency and safety. In order to design an effective purification scheme, the unit operations need to be reduced to a minimum volume. Affinity purification of enzymes has been introduced in Part III, and the method can be further designed for purification of engineered protein products with the expectation of offering purities >95% in one step.

E. coli are by far the most widely used type of bacterium. The recombinant proteins produced from *E. coli* increase greatly in recent years, many sophisticate techniques have been invented for their construction, fermentation and purification. The more that is known about the characteristics of a protein, the more easily it can be isolated and purified. Since a donor sequence is linked to an already known peptide segment, a conventional separation pattern can be applied to extracting the fused proteins even with only one single purification step, for example, through a commercially available affinity chromatography column is sufficient. One of the practical methods is the application of Glutathione **Stransferase** (GST) Gene Fusion System on purification and detection of GST fusion proteins in *E. coli*. The mechanism of this system involves the affinity of glutathione to the enzyme. Glutathione is attached to Sepharose by coupling to the oxirane group on its surface, and the structure of glutathione is complementary to the bind site carried by glutathione S transferase. Therefore, any target peptide fused with GST section will be easily attracted by Sepharose beads, and then be simply eluted. In the other affinity systems, similar mechanisms have been arranged. For example, recombinant protein molecules with **his-tag** may bind to the Ni Sepharose, therefore the peptide can be easily separated by Ni Sepharose chromatography. Purification of natural protein molecules which possess surfaced histidine residuals might also be optimized based on the similar principle.

21.3.2 Novel separation

Membrane technology emerges as one of the most important separation methods, which have captured the attention of industries, biochemical, pharmaceutical, biomedical, and food industries. Due to the current intensive development and expansion, membrane technology is also gaining market recognition and application rapidly in biotechnology.

Microfiltration is used extensively for medium exchange and harvest, and ultrafiltration can be found in virtually many modern bioprocesses. A significant number of mammalian cell processes use filtration as an integral part of the overall strategy for viral clearance. Depth filters have also seen the widespread use for the clarification of both mammalian and bacterial feed streams. Improvements in membrane technology are now focused on high-resolution applications, including improved protein-virus separation, protein purification by high-performance tangential flow filtration and enhanced membrane chromatography.

Membrane separation for aqueous-organic bioconversions

Hydrophilic and hydrophobic membranes were used to separate multiphasic organic solvent/water mixtures produced from fermentation. It was found that one phase could be separated from the mixture as long as the breakthrough pressure, which derived from the Laplace law, was not exceeded. The breakthrough pressure was found to be a function of the pore diameter and the interfacial tension. The flux through the membranes was found to be function of the volume fraction of water. Between a water volume fraction of 0.4 and 0.8, both hydrophilic and hydrophobic membranes were found to give high fluxes in the absence of surface active material. Yeast and *Pseudomonas putida* were used as examples of microbial material present in liquid/liquid mixtures downstream of a multiphasic biocatalytic reactor. The fluxes of both membranes decreased upon addition of the cells to the mixture. While the hydrophilic membranes were reversibly fouled and retained acceptable fluxes, in contrast, the hydrophobic membranes were fouled irreversibly. Subsequently, the hydrophobic membranes were modified and the flux through the hydrophobic modified membrane remained high. Both hydrophilic and hydrophobic membranes could also be used at high pressure when both liquids permeated the membrane.

21.3.3 Approaches to Zero-waste-discharge

Large quantities of industrial waste waters/effluents with strong contaminations of different compositions are generated in the production and processes of almost all fermentation sectors. Although many strict regulations on waste treatment have been issued by different countries, there is still a long way to achieve the goal of **Zero-waste-discharge**.

In the end of the 1980's, together with the awareness on the increased environmental fragile within society and industry, the conception and action of "Zero waste discharge" were ignited and soon became the long-term goal within industries including fermentation field. The principle of "Zero-waste-discharge" is to recycle all the industrial effluent and solid wastes and not allow the exhaust gas being released into air, the waste water being drained into water body or the solid waste being discarded. A pattern of "comprehensive process of corn alcohol" was practiced in some of the alcohol industries. Corn cereal was **degermed** without major damage to germ integrity and germ oil quality. Corn husk was removed from the grains and milled to make dietary fiber additive for improvement of human health. Corn kernel was smashed into small granules but not powder, for the raw starch glucoamylase to convert corn granules into glucose, by this process the starch cooking temperature can be lowered down. Finally, alcohol fermentation was carried out by yeast genus *Saccharomyces*. By sacrifying the highest conversion rate of alcohol, industries are able to produce byproducts, such as corn oil, dietary fiber, and distiller grains, which could be more valuable than ethanol itself. Probably the most significant advantage of this comprehensive process is that, form non-gelatinized cooking, the solid substance remaining in the distillation wastewater could be easily separated and used as feedstock and nitrogen & phosphate enhanced fertilizers, and the final effluent can be easily clarified and recycled in the next run of alcohol fermentation.

Still it is rather difficult to promote the "Zero discharge process" in all the fermentation industries. As fermentation engineers, it's their duty to perform the environmental friendly processes if possible.

Summary

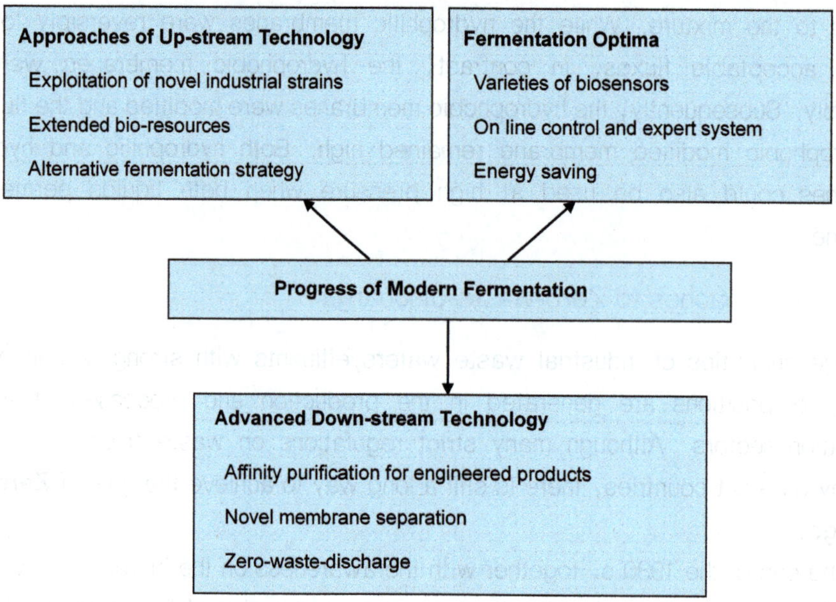

Appendix

Terms Translation

A haemophilia	A 型血友病
acetyl-CoA	乙酰辅酶 A
Acidithiobacillus ferrooxidans	嗜酸氧化亚铁硫杆菌
adaptor	平黏接头
adenosine triphosphate	三磷酸腺苷
affinity CG	亲和层析
agar plate	琼脂平板
agarose gel electrophoresis	琼脂糖凝胶电泳
agarose	琼脂糖
Agrobacterium tumefaciens	根癌农杆菌
agrolistic	农杆枪法
alkaline denaturation	碱变性处理法
alkaline phosphatase	碱性磷酸酶
alkaptonurea	黑尿酸尿症
allosteric enzyme	变构酶
alpha interferon	α-干扰素
amidinotransferase	脒基转移酶
amino acid auxotrophic strain	氨基酸营养缺陷株
aminotransferase	转氨酶
ampicillin	氨苄青霉素
amylase	淀粉酶
anabolism	合成代谢
anamorphic stage	变形阶段
antibiotics resistant marker	抗生素限制性标记
antifoam	消泡剂
antisense technology	反义技术
anti-cytokine	抗细胞素
anti-herbicide	抗除草剂
aqueous-organic monophasic system	水相-有机相单相系统
artificial seed	人工种子
Aspergillus niger	黑曲霉
atrial natriuretic peptide	心房钠尿肽

auxin	生长素
auxotrophy	营养缺陷型
avidin	抗生物素蛋白
Bacillus amyloliquefaciens	解淀粉枯草芽孢杆菌
bacterial artificial chromosome(BAC)	细菌人工染色体
bacterial conjugation	细菌接合
bacteriophage T4	细菌噬菌体 T4
base mutation	碱基变异
binary fission	有丝分裂
biocatalyst	生物催化剂
biofuel	生物燃料
bioleaching	生物浸矿
biolistic transformation	基因枪转殖技术
bioremediation	生物修复
blastocyst	囊胚
blendor experiment	匀浆试验
blood clotting factor VIII	凝血因子 VIII
brain natriuretic peptide	脑钠素
bromophenol blue	溴酚蓝
Bt insecticidal crystal protein gene	苏云金杆菌杀虫伴晶胞蛋白
bud scar	芽痕
budding	芽殖
by-product	副产物
callus	愈伤组织
calorimetric	量热
Candida yeast	假丝酵母
capillary zone electrophoresis (CZE)	毛细管条带电泳
carcinogenic effect	致癌效应
catabolism	分解代谢
catalase	过氧化氢酶
cell entrapment	细胞包埋
cell fusion	细胞融合
cellular senescence	细胞衰老
cellulase	纤维素酶
centromeric	着丝粒
cephalosporin	头孢菌素
chemiluminescence	化学发光
chemostat	恒化器
chimeric antibody	嵌合抗体
chiral	手性
chitinase	几丁质酶
cholesterol oxidase	胆固醇氧化酶

chromosomal theory of inheritance	染色体遗传理论
citrate	柠檬酸
coenzyme	辅酶
cohesive (sticky) end	黏末端
colicin	大肠杆菌素
colicinogenic factor	大肠杆菌毒素因子
column based electrophoresis	柱式电泳
combinatorial screening	组合筛选
constitutive enzyme	组成酶
Corynebacterium glutamicum	谷氨酸棒杆菌
cosmid	黏粒
cross-flow filtration	横流过滤
crown gall	冠瘿
cryodesiccation	冷冻干燥
custom-made animal	定制动物
cyclooxygenase	环氧合酶
cymbium	新生组织
cystic fibrosis	囊性纤维化
cytokine	细胞激素
cytokinesis	胞质分离
cytokinin	细胞分裂素
cytoplasm	细胞质
dedifferentiation	脱分化
degermed	脱胚的
denaturation	变性
deoxynucleoside triphosphate	脱氧核苷三磷酸腺苷
deoxyribonuclease I (DNase I)	脱氧核糖核酸酶 I
deoxyribose	脱氧核糖
desiccation	干燥
diauxic growth	二次生长
differentiation	分化
dimer	二联体
diploid	二倍体
division cycle	分裂周期
DNA fingerprint	DNA 指纹鉴定
DNA ligase	DNA 连接酶
DNA polymerase I	DNA 聚合酶 I
DNA restriction endonuclease	DNA 限制性内切核酸酶
DNA shuffling	DNA 分割重组
domain	结构域
electrofusion	电融合
electroporation	电转化

embryogenesis	胚胎发生
encapsidation	噬菌体粒子包装
endocytosis	内吞作用
endoenzyme	胞内酶
endogenous	内源性的
endotoxin	内毒素
entropy	熵
enzyme inhibitor	酶抑制物
enzymes catalyze	酶催化
epidermal growth factor	表皮生长因子
episomal genome	染色体外基因组
epitope	表位
equilibrium centrifugation in a density gradient	密度梯度平衡离心
error-prone PCR	易错 PCR
erythropoietin	促红素
ethidium bromide (EB)	溴化乙锭
eukaryotic genome	真核基因组
exoenzyme	胞外酶
exon	外显子
exonuclease activity	外切核酸酶活力
expanded bed adsorption	扩张床吸附
explant	分离块
extracellular	细胞外的
extremophiles	极端微生物
facultative anaerobe	兼性厌氧
feedback inhibition	反馈抑制
feedstream	供水液流
fibrinolysis	血纤溶解
fission	分裂
flocculation	絮凝
foot-mouth disease	口蹄疫
frontline treatment	一线治疗
fructose-glucose isomerase	果葡糖异构酶
fruit fly *Drosophila*	果蝇属名
fumarate	富马酸
fusant	融合子
fusion expression system	融合表达系统
galactoside	半乳糖苷
gel filtration	凝胶过滤
gene cluster	基因簇
gene knockout	基因敲除
gene locus	基因位点

gene remains	基因遗骸
gene therapy	基因治疗
genetic modified organism	遗传修饰生物体
genetic parasite	遗传寄生体
genetically modified organism(GMO)	转基因生物
genomic fingerprinting	基因指纹鉴别
genomic library	基因文库
germ-cell gene therapy	繁殖细胞基因治疗
glucoamylase	糖化酶
gluconeogenesis	糖元再生
glycolysis	糖酵解
haemophilia	血友症
hammerhead ribozyme	锤头状核酶
haploid	单倍体
hemicellulase	半纤维素酶
hepatitis B vaccine	乙肝疫苗
hereditary disorder	遗传缺陷症
hereditary factor	遗传因子
hereditary trait	遗传性状
heuristic	启发式的
hexose	己糖
histone	组蛋白
His-tag	组氨酸标签
holoenzyme	全酶
homogenization	均质
homologous tailing technique	同聚物加尾技术
horseradish peroxidase	辣根过氧化氢酶
human commensal flora	人体共生微生物群落
human growth hormone(HGH)	人类生长激素
human insulin	胰岛素
human pituitary cell	人脑垂体细胞
hybrid cell	杂交细胞
hybridoma	杂交瘤
hybrid-arrest translation (HART)	杂交分子中止翻译
hydrodynamics	水动力学
hydrogel	水凝胶
hydrolase	水解酶
hydrophilicity/hydrophobicity	亲水性/疏水性
hydrophobic CG	疏水层析
hygroscopic	吸湿性的
hypertonic	高渗
hyphae	菌丝

icosahedral	正二十面体
icosahedral tailless	无尾二十面体
idiophase	繁殖期
immortalized cell line	永生细胞系
immunodiagnostic agent	免疫诊断试剂
immunoglobulin	免疫球蛋白
in vitro mutagenesis	体外诱变技术
in vivo	体内
indigo	靛蓝
inducible	诱导的
inducible enzyme	诱导酶
inhibitor	抑制剂
injectable	血管注射剂
inoculum	种子液
inosine	肌苷
insertional inactivation	插入失活
interferon	干扰素
interleukin-2	白细胞介素-2
intervening sequence	间隔序列
intron	内含子
IS element	插入因子
isocitrate	异柠檬酸
isomerase	异构酶
isoprenoid	类异戊二烯
isotonic	等渗
jumping gene	跳跃基因
Klenow fragment	Klenow 片段
lac operon	乳糖操纵子
lactate dehydrogenase	乳酸脱氢酶
lactoferrin	乳铁蛋白
lag phase	延滞期
law of independent assortment	独立分配律
law of segregation	分离定律
ligand	配体
ligninase	木质素酶
linker	衔接片段
lyase	裂解酶
lyophilization	冻干
lysate	溶菌液
lysozyme	溶菌酶
lysozyme-heat treatment	溶菌酶速热处理
lytic phage	烈性噬菌体

major surface antigen	主要表面抗原
malate	苹果酸
mass transfer	传质
matrix	基质
Maxam-Gilbert (chemical degradation Maxam-Gilbert)	化学降解法测序
medullary plate	髓板
meiosis	减数分裂
membrane enzyme	膜酶
meristem	分生组织
metabolism	代谢
metagenome	宏基因组
microaqueous system	微水相系统
microinjection	显微注射
microthermopile	微型热电
milkweed	乳草
mini-satellite probe	微卫星 DNA 探针
mitosis	有丝分裂
monarch butterfly larva	黑脉金斑蝶幼虫
monoclonal antibody	单克隆抗体
monosaccharide	单糖
morphological mechanism	形态学机制
Mucor	毛霉
multigene family	多基因家族
multi-antigenitic vaccine	多效抗原疫苗
Mycoplasma genitalium	支原体
nattokinase	纳豆激酶
nested PCR	巢式 PCR/嵌套 PCR
nick translation	带缺口翻译
obligate aerobe	严格好氧
oligonucleotide	寡聚核苷酸工具
oncogene	癌基因
opines	冠瘿碱
orbital steering	轨道转向
organelle transplantation	器官移植
organogenesis	器官发生
osmolarity	渗透压
osmoticum	渗压剂
overlapped gene	重叠基因
oxaloacetate	草酰乙酸
oxidative phosphorylation	氧化磷酸化
oxidoreductase	氧化还原酶
paclitaxel	紫杉醇

paratope	配位
Parkinson's symptom	帕金森症
partition CG	分配层析
pathogenicity	致病性
pentose	戊糖
perfusion pump	灌注泵
permeability and surface area	渗透性与表面积
phenylketonuria	苯丙酮尿症
phosphodiester bond	磷酸二酯键
phosphoramidite	亚磷酰胺
piezoelectric	压电
plantule	胚芽
plaque	噬菌斑
plasmid	质粒
plasminogen activator	纤维蛋白溶酶原激活因子
plasmolyticum	质壁分离剂
plug flow reactor	恒流器
pluripotent	多能性的
polar mutation	极性变异
polishing	精制
polyacrylamide-based gel electrophoresis	聚丙烯酰胺凝胶电泳
polyanionic	多聚负离子
polycistron	多顺反子
polymerase chain reaction (PCR)	聚合酶链反应
polynucleotide kinase	多核苷酸激酶
polysaccharide capsule	多糖荚膜
polysaccharide	多糖
porcine endogenous retrovirus	猪内源反转录病毒
potential pathogenicity	潜在致病性
primary metabolite	初级代谢产物
primer	引物
probiotic	益生素
product polishing	产品修饰
proinsulin	前胰岛素
prokaryote	原核生物
prokaryotic cell	原核细胞
proline	脯氨酸
promotor	启动子
pronuclei	原核仁
pronucleus	前核
proof-reading function	证读功能
protease	蛋白酶

proteolysis	蛋白水解作用
protists	原生生物
protoplast	原生质体
proximity	紧邻性
pseudo rabies vaccine	狂犬疫苗
pseudogene	假基因
Pseudomonas	假单胞菌
purine	嘌呤
pyrimidine	嘧啶
pyrogen	致热源
pyruvate	丙酮酸
random hexamer primer	随机六聚体引物
random mutagenesis	随机突变
recrystallization	重结晶
rectifying-distilled	精馏
redox	氧化
replication fork	复制叉
replication origin (ori)	复制原点(复制子)
restriction fragment length polymorphisms	限制性片段多态性(RFLP)
reverse phase CG	反相层析
reverse transcriptase (RTase)	反转录酶
reversed micelle system	反相胶束系统
Rheological property	流变学性质
ribonuclease	核苷核酸酶
ribozyme	核酶
RNA interference	RNA 干涉
RNA retrovirus	RNA 反转录病毒
rolling-circle replication	滚环复制
rough-edged (R) mutant	粗糙型变异株
Ruppy Puppy	荧光红狗仔
Saccharomyces	酿酒酵母属
salvageable	可回收的
Sanger-Coulson (dideoxy or enzymatic)	Sanger 酶法/双脱氧链终止法测序
scattered highly repeated sequence	散在的高度重复序列
Schizosaccharomyces	裂殖酵母
secondary metabolite	次级代谢产物
segmented genome	片段式基因组
self-cleaving ribozyme	自切核酶
sequenase	高保真测序酶
shallow tray fermentation	浅盘发酵
shikonin	紫草素
sickle cell anemia	镰状细胞贫血症

single cell colony stimulating factor	单细胞集落刺激因子
site directed mutagenesis	定点诱变
slab gel electropheresis	板式凝胶电泳
slot	进样孔
smooth-edged colony	光滑型菌落（S）
sodium dodecyl sulphate	十二烷磺酸钠
somatic-cell gene therapy	体细胞基因治疗
somatic	体细胞胚
somatostatin	生长抑素
Soybean rhisobia	大豆根瘤菌
splicing	拼接作用
spore	孢子
stationary phase	平稳期
sterile sampling	无菌取样
stoichiometry	化学计量学
streptavidin	抗生蛋白链菌素
Streptococcus pneumoniae	肺炎球菌
streptomyces	链霉素
subculture	继代培养
substrate	底物
subtilisin	枯草杆菌蛋白酶
succinate	琥珀酸
succinyl-CoA	琥珀酰辅酶 A
supercritical fluid	超临界流体
superoxide dismutase	超氧化歧化酶
surface antigenic protein	表面抗原蛋白
surface antigenic vaccine	表面抗原疫苗
synkaryon	核子
S-transferase	S-转移酶
taxol	紫杉醇
telomere	端粒
terminase	末端酶
Tetrahymena thermophila	嗜热四眼虫
thalassemia	地中海贫血症
thermal denaturation	热变性
Thermus acquaticus	耐热性水生细菌
toad *Xenopus*	蟾蜍属名
topoisomerase	拓扑异构酶
totipotent	全能性的
toxins guide	毒素导向
transduplication	转复制作用
transferase	转移酶

transferrable plasmid	可转移质粒
transformation	转化
transgenic farm animal	转基因饲养动物
transition state	过渡态
transposable element	可转座因子
transposon	转座子
tricarboxylic acid cycle	三羧酸循环
trichloroethylene	三氯乙烯
Trichoderma	木霉
triple codon	三联体密码子
trypsin	胰蛋白酶
tumor necrosis factor	肿瘤坏死因子
ultracentrifuge	超速离心
ultrafiltration	超滤
urbidostat	恒浊器
vaccines against HBV	抗 HBV 疫苗
vector or carrier molecule	载体
vegetative seed	新鲜种子液
viral coat protein	病毒外壳蛋白
Western blotting	Western 印迹
wild type	野生型
xylanase	木聚糖酶
yeast artificial chromosome (YAC)	酵母人工染色体
yeast-derived vaccine	重组酵母疫苗
Yunnan minipig	云南微型猪
zero-waste-discharge	零污染排放
zygotic embryos	受精卵的胚胎
"Central Dogma" theory	中心法则理论
α-complementation	α-互补
α-ketoglutarate	α-酮戊二酸
β-galactosidase	β-半乳糖苷酶
β-immune interferon	β-免疫干扰素

Related Nobel Laureates

A 诺贝尔生理学或医学奖获得者

获奖年	科学家	主要贡献
1910	Albrecht Kossel(German)	Determining the chemical composition of nucleic acids, the genetic substance of biological cells
1923	Frederick Banting (Canada); John James Richard Macleod (Canada)	The discovery of insulin
1933	Thomas Hunt Morgan (American)	The role the chromosome plays in heredity
1935	Hans Spemann(German)	Embryonic induction that directs the development of groups of cells into particular tissues and organs
1939	Gerhard Domagk(German)	The discovery of sulfonamidochrysoidine
1945	Alexander Fleming(Britain); Ernst Boris Chain(German-born British); Howard Walter Florey(Australian)	The discovery of Penicillin
1946	Hermann Joseph Muller(American)	The discovery that mutations can be induced by X-rays
1952	Selman Waksman(Ukrainian-American)	Discovery of "streptomycin" the first antibiotic active against tuberculosis
1953	Hans Adolf Krebs(German-born British); Fritz Albert Lipmann(German-American)	Citric acid cycle
1955	Hugo Theorell(Swedish)	Discovering the oxidation enzyme and its effects
1958	George Wells Beadle (American); Edward Lawrie Tatum (American)	The role of genes in regulation biochemicalevents within cells
1959	Arthur Kornberg(American); Severo Ochoa(American)	The mechanisms in the biological synthesis of deoxyribonucleic acid (DNA)
1962	Francis Crick(Britain); James D. Watson(American); Maurice Wilkins(Britain)	The molecular structure of nucleic acids and its significance for information transfer in living material
1965	François Jacob (French); André Michel Lwoff (French); Jacques Monod (French)	Discoveries concerning genetic control of enzyme and virus synthesis
1968	Robert W. Holley(American); Har Gobind Khorana(American); Marshall Warren Nirenberg(American)	Describing the structure of alanine transfer RNA linking DNA and protein transfer RNA linking DNA and protein

续表

获奖年	科学家	主要贡献
1969	Max Delbrück(German-American); Alfred Hershey(American); Salvador Luria(American)	Discovery on the repli synthesis cation of viruses and their genetic structure
1972	Gerald Edelman(American); Rodney Robert Porter(Britain)	Discovery of the structure of antibody molecules
1974	Albert Claude(Belgian); Christian de Duve(Belgian); George Emil Palade(American)	Discoveries concerning the structural and functional organization of cells
1975	David Baltimore(American); Renato Dulbecco(American); Howard Martin Temin(American)	Discoveries concerning the interaction between tumor viruses and the genetic material of the cell
1978	Werner Arber(Swiss); Daniel Nathans(American); Hamilton O. Smith(American)	The discovery of restriction endonucleases
1980	Baruj Benacerraf(American); Jean Dausset(French); George Davis Snell(American)	Characterisation of the genes making the major histocompatibility complex
1983	Barbara McClintock(American)	The discovery of genetic transposition
1984	Niels Kaj Jerne(Danish); Georges J. F. Köhler(German); César Milstein(Argentina, United Kingdom)	Control of the immune system and the discovery of the principle monoclonal for production of antibodies
1986	Stanley Cohen(American); Rita Levi-Montalcini(Italian)	Nerve growth factor
1987	Susumu Tonegawa(Japanese)	The genetic mechanism that produces antibody diversity
1989	J. Michael Bishop(American); Harold E. Varmus(American)	Discovery of the cellular origin of retroviral oncogenes
1990	Joseph Murray (American); E. Donnall Thomas(American)	Organ and cell transplantation
1991	Erwin Neher(German); Bert Sakmann(German)	The function of single ion channels in cells
1992	Edmond H. Fischer(American); Edwin G. Krebs(American)	Describing how reversible phosphorylation works as a switch to activate proteins and regulate various cellular
1993	Richard J. Roberts(British); Phillip Allen Sharp(American)	The discovery of introns in eukaryotic DNA and the mechanism of gene-splicing

续表

获奖年	科学家	主要贡献
1995	Edward B. Lewis(American); Christiane Nüsslein-Volhard(German); Eric F. Wieschaus(American)	G control of embryonic development
1997	Stanley B. Prusiner(American)	Prion research
2001	Leland H. Hartwell(American); Tim Hunt(British); Paul Nurse(British)	Discoveries of protein molecules that control the division (duplication) of cells
2002	Sydney Brenner(South African); H. Robert Horvitz(American); John E. Sulston(British)	Genetic regulation of organ development and programmed cell death
2006	Andrew Fire(American); Craig Mello(American)	The discovery of RNA interference (RNAi)
2007	Mario Capecchi(American); Martin Evans(British); Oliver Smithies(American)	A method to create mice in which a specific gene is turned off (knocked out)
2009	Elizabeth Blackburn(American); Carol W. Greider(American); Jack W. Szostak(American)	Discovery of telomerase and its mechanism
2010	Robert G. Edwards (British);	For the development of *in vitro* fertilization

B 诺贝尔化学奖获得者

获奖年	科学家	主要贡献
1970	Eduard Buchner(German)	Cell free fermentation
1915	Richard Martin Willstätter(German)	The structure of plant pigments, chlorophyll included
1929	Arthur Harden(British); Hans von Euler-Chelpin(Sweden)	The fermentation of sugar and fermentative enzymes
1937	Walter Haworth(British); Paul Karrer(Switzerland)	Investigations on carbohydrates and vitamin C
1946	James B. Sumner(American); John Howard Northrop(American); Wendell Meredith Stanley(American)	Isolation, crystallization, and study of enzymes, proteins, and viruses
1948	Arne Tiselius(Sweden)	Protein analysis by electrophoresis and adsorptional analysis
1952	Archer John Porter Martin(British); Richard Laurence Millington Synge(British)	The invention of partition chromatography

续表

获奖年	科学家	主要贡献
1953	Vincent du Vigneaud(American)	Isolation, structural identification, and total synthesis of the cyclic peptide, oxytocin
1954	Linus Pauling(American)	Elucidating the nature of chemical bonds and the structures of molecules
1957	Alexander R. Todd; Baron Todd (British)	Structure and synthesis of nucleotides, nucleosides, and nucleotide coenzymes
1958	Frederick Sanger(British)	The structure of proteins, especially that of insulin
1959	Jaroslav Heyrovský(Czechoslovakia)	Polarographic method
1961	Melvin Calvin(American)	Discovering the Calvin cycle
1962	Max Perutz(British); John Kendrew(British)	The structures of hemoglobin and globular proteins
1972	Christian B. Anfinsen(American); Stanford Moore(American); William Howard Stein(American)	Structure of the enzyme ribonuclease and the connection between chemical structure and catalytic activity of the ribonuclease
1980	Paul Berg(American); Walter Gilbert(American); Frederick Sanger(British)	Determination of base sequences in nucleic acids
1984	Robert Bruce Merrifield(American)	The invention of solid phase peptide synthesis
1988	Johann Deisenhofer (German); Robert Huber (German); Hartmut Michel (German)	Determination of the first crystal Structure of an integral membrane protein
1989	Sidney Altman(Canada & American); Thomas Cech(American)	The catalytic properties of RNA
1993	Kary Mullis(American); Michael Smith(Canada)	Polymerase chain reaction (PCR) technique Oligonucleotide-based site-directed mutagenesis
2002	John Bennett Fenn(American); Koichi Tanaka(Japan); Kurt Wüthrich(Swiss)	Mass spectrometry Nuclear magnetic resonance spectroscopy
2004	Aaron Ciechanover(Israel); Avram Hershko(Israel); Irwin Rose (American)	Ubiquitin-mediated protein degradation
2006	Roger D. Kornberg (American)	Genetic information from DNA is copied to RNA The molecular basis of eukaryotic transcription
2008	Osamu Shimomura(Japan); Martin Chalfie (American); Roger Y. Tsien (American)	Development of green fluorescent protein (GFP)
2009	Venkatraman Ramakrishnan (American); Thomas A. Steitz (American); Ada E. Yonath(Israel)	The structure and function of the ribosome

Further Reading

1. Jocelyn E. K., Elliott S. G. and Stephen T. K. *Lewin's Genes X*. London: Jones and Bartlett Publishers, LLC. 2011. ISBN: 076376320.

2. Sandy B. P. and Richard T. *Principles of Gene Manipulation and Genomics*. Hoboken: Wiley-Blackwell. 2006. ISBN: 1405135441.

3. Brown T. *Gene Cloning and DNA Analysis: An Introduction*, 6th Edn. Hoboken: Wiley-Blackwell. 2010. ISBN: 1444334077.

4. Tamara L. R. *Genetic Engineering*. San Diego: ReferencePoint Press. 2008. ISBN: 1601520388.

5. Jeremy D. and Malcolm von S. *From Genes to Genomes: Concepts and Applications of DNA Technology*, Chiehester: Wiley-Interscience. 2007. ISBN: 0470017340.

6. Mitsuo G., Cheng R. and Lin L. *Network Models and Optimization Multiobjective Genetic Algorithm Approach*. London: Springer-Vedag, Ltd. 2008. ISBN: 1848001800.

7. Desmond S. T. N. *An Introduction to Genetic Engineering*, 3rd Edn. New York: Cambridge University Press. 2008. ISBN: 0521615211.

8. Jane S. *Genetic Engineering Principles and Methods*, New York: Plenum Press. 2007. ISBN: 0387258558.

9. Chris J. M. *Engineering genetic circuits*. Boca Raton: CRC Press, LLC. 2010. ISBN: 9781420083.

10. Nair A. J. *Intro Biotechnology and Genetic Engineering*. Massachusetts: Jones & Bartlett Learning. 2010. ISBN: 0763773751.

11. Mark L. S. *Encyclopedia of Biotechnology and Genetic Engineering*. New York: Facts On File. 2010. ISBN: 0816067985.

12. Stille D. R. *Altering the Biological Blueprint: The Science of Genetic Engineering*. Mankato: Compass Point Books. 2010. ISBN: 0756540494.

13. Sylvia E. *Genetic engineering*. New York: Greenhaven Press. 2007. ISBN: 0737732555.

14. Lisa Y. *Biotechnology and genetic engineering*. New York: Facts On File. 2008. ISBN: 0816072170.

15. Wexler B. *Genetic and Gentic Engineering*. Farmington Hills: Thomson Gale. 2008. ISBN: 9781414407.

16. Linda T. *Genetic Engineering: Modern Progress or Future Peril*. Minneapolis: Twenty-First Century Books. 2009. ISBN: 9780761340812.

17. Balgir P. *Genetic Engineering: Principles and Applications*. New York: Oxford University Press. 2008. ISBN: 195686272.

18. Claire H. C. *Uncertain Peril: Genetic Engineering and the Future of Seeds*. Boston: Beacon Press. 2009. ISBN: 0807085812.

19. Robert M. B. *The ethics of genetic engineering*. London: Routledge. 2007. ISBN: 0415769945.

20. Feng Z. and Qu S. *Biochemistry and Molecular Biology*. Beijing: People's Medical Publishing House. 2007. ISBN: 117082716.

21. Cornel M. *The Experimenter Series: Molecular and Genomics*. Burlington: Elsevier Inc. 2007. ISBN: 0120885466.

22. David P. C. and Nanette J. P. *Molecular Biology*. Burlington: Elsevier Inc. 2012. ISBN: 0123785947.

23. Burton E. T. and David F. *Molecular Biology: Genes to Proteins*, 3rd Edn. Canada: Jones and Bartlett Publishers, Inc. 2008. ISBN: 0763709167.

24. Dr. P. S. and Namita S. *Molecular Biology principles and Practices*. New Delhi: Laxmi Publications. 2007. ISBN: 8131802700.

25. Helen K. and Adrianne M. *Molecular Biology and Biotechnology: A Guide for Teachers*. Washington: ASM Press. 2008. ISBN: 1555814717.

26. Martin V. and Limsoon W. *Research in Computational Molecular Biology*. Heidelberg: Springer Verlag Berlin Heidelberg. 2008. ISBN: 3540788387.

27. Pierre C. *Pseudomonas Genomics and Molecular Biology*. Norfolk: caister Academic Press. 2008. ISBN: 904455196.

28. Henry W., Edmund M., Ronald L. and Bryce P. *Enzymology and Molecular Biology of Carbonyl*. West Lafayette: Purdue University. 2007. ISBN: 1557534477.

29. Frank H. S. *Calculations for Molecular Biology and Biotechnology: A Guide to Mathematics in the Laboratory*, 2nd Edn. Burlington: Elsevier Inc. 2010. ISBN: 0123756909.

30. Sharma R. K. and Sangha S. P. S. *Basic Techniques in Biochemistry and Molecular Biology*. New Delhi: I. K. International Publishers. 2009. ISBN: 81907462.

31. Peter G. *Bacillus Cellular and Molecular Biology*. Norfolk: caister Academic Press. 2007. ISBN: 1904455127.

32. Harvey L., Arnold B., Chris A. K. and Monty K. *Molecular Cell Biology*. W. H. Freeman: W. H. Freeman & Co Ltd. 2007. ISBN: 0716776014.

33. James D. W., Tania A. B. and Stephen P. B. *Molecular Biology of the Gene*. Alexander Gann: Pearson Education. 2008. ISBN: 0321507815.

34. Michael B. *Cell Culture and Upstream Processing*. London: Taylor & Francis Group. 2007. ISBN: 0415399692.

35. Gerhard M. A., Stephen M. and Jurgen H. *Stem Cell Engineering: Principles and Applications*. London: Spinger. 2010. ISBN: 3442118647.

36. Satya P. *Artificial Cells, Cell Engineering and Therapy*. Cambridge: Woodhead. 2007. ISBN: 1845690362.

37. David S., Joseph D. B. and Donald R. P. *Stem Cell Engineering: Principles and Practices*. London: Taylor & Francis Group. 2012. ISBN: 1439872048.

38. Stephen K. *Cell*. London: Hodder & Stoughton. 2007. ISBN: 1848941013.

39. Geoffrey M. C. and Robert E. H. *The cell: a molecular approach*. Herndon: ASM Press. 2009. ISBN: 0878933006.

40. David B. *Mechanics of the Cell*, 2nd Edn. New York: Cambridge University Press. 2012. ISBN: 0521130691.

41. Stephen R. B., Elizabeth A. S., Hugh A. W. and Jeremy S. H. *Cell Biology: A Short Course*, 3rd Edn. New Jersey: John Wiley & Sons, Inc. 2011. ISBN: 0470526996.

42. René F. K. *Molecular & Cell Biology for Dummies*. Indiana: Wiley Publishing, Inc. 2009.

ISBN: 0470430668.

43. Lars I. L. *Cell Fusions: Regulation and Control*. London: Springer. 2011. ISBN: 9048197712.

44. John M. D. *Animal Cell Culture: Essential Methods*. West Sussex: John Wiley & Sons Ltd. 2011. ISBN: 0470666586.

45. Frederic P. M., Agnes F. V. and John M. B. *Cell Culture*. Saarbrücken: VDM Publishing House Ltd. 2009. ISBN: 6130226428.

46. Jan T. S. and Kee W. N. *A Manual for Primary Human Cell Culture*, 2nd Edn. London: World Scientific Publishing Co. Pte. Ltd. 2010. ISBN: 9812834775.

47. Brian S. M. *Cellular Engineering Strategies for Improvement of Mammalian Cell Phenotypes in Commercial Applications*. Ann Arbor: ProQuest LLC. 2009. ISBN: 0549945291.

48. Alane E. W. *Combinatorial Cellular Engineering for Improved Biopharmaceutical Production*. Ann Arbor: ProQuest Information and Learning Company. 2007. ISBN: 0549383406.

49. Lynne C., Vishwanath R. L. and George P. *Lewin's Cells*, 2nd Edn. London: Jones and Bartlett Publishers, LLC. 2011. ISBN: 0763766641.

50. Bruce A. *Molecular Biology of the Cell*, 5th Edn. New Jersey: Garland Publishing Inc. 2007. ISBN: 0815341067.

51. Wu Y., HsiuChiung Y and James R. M. *Enzyme Technologies: Metagenomics, Evolution, Biocatalysis and Biosynthesis*. Hoboken: John Wiley & Sons Inc. 2009. ISBN: 0470286245.

52. Robert J. W. and Maarten V. *Enzymes in Food Technology*. Hoboken: Wiley-Blackwell. 2009. ISBN: 1405183666.

53. WuKuang Y., HsiuChiung Y. and James R. M. *Enzyme Technologies: Metagenomics, Evolution, Biocatalysis and Biosynthesis*. Hoboken: John Wiley & Sons, Inc. 2010. ISBN: 0470286245.

54. Klaus B., Volker K. and Uwe T. B. *Biocatalysts and Enzyme Technology*. Hoboken: Wiley-VCH Verlag GmbH & Co., KGaA, Weinheim. 2005. ISBN: 3527304975.

55. Schomburg D. and Stephan D. *Enzyme Handbook* 15, Berlin: Springer-Verlag. 1998. ISBN: 3540641165.

56. Andres I. *Enzyme Biocatalysis: Principles and Applications*, Berlin: Springer Science + Business Media B. V. 2005. ISBN: 1402083600.

57. Robert A. C. *Enzymes: A Practial Introduction to Structure, Mechanism, and Data Anlysis*. New York: John Wiley & Sons, Inc. 2000. ISBN: 0471359297.

58. Ashok P., Colin W., Carlos Richardo S. and Christian L. *Enzyme technology*. New York: Springer Science + Business Media, Inc. 2006.. ISBN: 0387292942.

59. Herbert A. K., Wu-Kuang Y. And Milton J. Z. *Enzyme Technologies for Pharmaceutical and Biotechnological Applications*, New York: Marcel Dekker, Inc. 2001. ISBN: 0824705491.

60. Bayindirli A. *Enzymes in Fruit and Vegetable Processing: Chemistry and Engineering Applications*, Boca Raton: CRC Press, Inc. 2010. ISBN: 1420094336.

61. Svendsee A. *Enzyme Functionality: Design, Engineering, and Screening*. Boca Raton: CRC Press, Inc. 2004. ISBN: 0824747091.

62. Kwan-Hwa P. and Yang-Do C. *Enzymes for carbohydrate engineering*, Amsterdam: Elsevier, Inc. 1996. ISBN: 0444824081.

63. Jose M. G. *Immobilization of Enzymes and Cells*, 2nd Edn. New York: Humana press, Inc. 2008. ISBN: 1588292908.

64. Linqiu C. and Rolf D. S. *Carrier-bound Immobilized Enzymes: Principles, Application and Design*. Hoboken: John Wiley & Sons, Inc. 2006. ISBN: 3527312320.

65. Julio P. and Andrew P. M. *Industrial Enzymes: Structure, Function and Applications*. Bolin: Springer, Inc. 2007. ISBN: 1402053762.

66. Trevor P. and Philip L. R. B. *Enzymes: Biochemistry, Biotechnology, Clinical Chemistry*. London: Horwood, Inc. 2007. ISBN: 1904275273.

67. Otto C. *Enzymes*, 2nd Edn. Charleston: BiblioBazaar, Inc. 2009. ISBN: 1113706155.

68. Fuyuhiko T. *The Enzymes: Protein Methyltransferases*, 2nd Edn. Amsterdam: Elsevier, Inc. 2006. ISBN: 0121227251.

69. Brian M. and Linda H. *Practical Fermentation Technology*. Chiehester: John Wiley & Sons, Ltd. 2009. ISBN: 0470014342.

70. Ashok P., Carlos R. and Christian L. *Current Developments in Solid-state Fermentation*. New York: Springer. 2008. ISBN: 9780387752129.

71. Annadurai B. *Handbook of Biochemical Engineering and Fermentation*. Cambridge: Woodhead Pub. 2011. ISBN: 9780857091109.

72. Antoine B. *The Mystery of Fermentation*. Montana: Kessinger Pub LLC. 2010. ISBN: 1161588422.

73. Miller A. K. *Practical Studies in Fermentation-Being Contributions to the Life History of Micro-Organisms*. Philadelphia: Wharton Press. 2008. ISBN: 9781409767015.

74. Henry V. and Celeste T. *Fermentation and Biochemical Engineering Handbook*, 3rd Edn. Cambridge: Westwood. 2012. ISBN: 1455725536.

75. Mansi E. and Charles B. *Fermentation Microbiology and Biotechnology*. Boca Raton: CRC Press. 2011. ISBN: 0748407340.

76. Peter S., Stephen H. and Allan W. *Principles of Fermentation Technology*, 2ed Edn. Oxford: Elsevier. 2007. ISBN: 0750645016.

77. Sundar M. and Sukumar M. *Fermentation*. New York: LAP Lambert Academic Publishing. 2010. ISBN: 3843363285.

78. Bhavbhuti M., Afaf E. and Robert Z. *Fermentation: Effects on Food Properties*. London: Taylor & Francis. 2012. ISBN: 1439853347.

79. Sandor K. *The Art of Fermentation: An In-Depth Exploration of Essential Concepts and Processes from Around the World*. Chelsea: Green Publishing Company. 2012. ISBN: 160358286X.

80. David P. C. and Nanette J. P. *Biotechnology: Academic Cell Update Edition*. London: ACACL. 2011. ISBN: 0123850630.

81. Thieman W. and Palladino M. A. *Introduction To Biotechnology*, 2nd Edn. US: Pearson Education. 2008. ISBN: 0321589033.

82. Molly F. and Frieda R. *DNA and Biotechnology*, 3rd Edn. London: Academic Press. 2009. ISBN: 0120489309.

83. Altman A. and Hasegawa P. M. *Plant Biotechnology and Agriculture: Prospects for the 21st Century*. London: Academic Press. 2011. ISBN: 0123814669.

84. Stewart C. N. *Plant Biotechnology and Genetics: Principles, Techniques and Applications*. Hoboken: Wiley-Interscience. 2008. ISBN: 0470043814.

85. Shigeo K. and Fumitake Y. *Biochemical Engineering: A Textbook for Engineers, Chemists and Biologists*. Weinheim: Wiley-VCH. 2009. ISBN-10: 3527325360.

86. Ghasem N. *Biochemical Engineering and Biotechnology*. Oxford: Elsevier Science. 2006. ISBN-10:0444528458.

87. Miichael S. I. *Downstream Processing of Natural Products: A Practical Handbook*. Chiehester: Wiley. 1996. ISBN-10:0471963267.

88. Miguel B. and Victoria A. *Protein Purification*. Nova Science Pub Inc. 2012. ISBN-10:1614700982.

89. Richard R. B. and Murray P. D. *Guide to Protein Purification*, 2nd Edn. London: Academic Press. 2009. ISBN: 0123749786.

90. Jan-Christer J. *Protein Purification: Principles, High Resolution Methods, and Applications*. London: Wiley. 2011. ISBN: 0471746614.

91. Scott M. W. *Protein Purification: Design and Scale Up of Downstream Processing*. London: Wiley-Interscience. 1991. ISBN: 0471037230.

后　记

当我们合上最后一页书稿时，动笔已历时两年，本科双语教材 *Essentials of Biotechnology* 一书终于呼之欲出。现代生物技术领域实在过于宽泛，发展变化又如此之快，很难全面反映所有的内容。考虑到作为生物技术或生物工程专业本科生的基础教材，既要帮助学生掌握该领域内最基本的理论知识，又要反映本领域的技术前沿，作者在书中分别介绍了**基因克隆**，**细胞工程**，**酶技术**和**工业发酵**四大板块的科技进展，各板块既相对独立，又有所关联。附录中列出了20世纪以来有关的诺贝尔奖获得者的名单，旨在提醒学生关注这些世界级顶尖人才与成果对科学发展和社会进步的贡献。

Part I(基因克隆)由王武教授和张玲编撰；Part II(细胞工程)由唐蕾副教授编撰；辛瑜副教授编写了 Part III(酶技术)，而 Part IV(工业发酵)由夏小乐副教授与杨海麟副教授负责。研究生巩波参与了 Part IV 中一章的初稿整理，李奇泽为本书的插图承担了绘制和修改工作，其他几位研究生也参与了素材的收集。外语教师杨昊为 Part I & II、李珊珊与李莉分别为 Part III & IV 进行了英语润色。王武教授作为本书的主编，承担了全书章节的策划，各章内容的调整、衔接、修改，以及最终的定稿；马里兰大学王南新教授和长江学者王小元教授分别对书稿提出了重要的修改意见。本书有幸邀请到中国工程院伦世仪院士作序，这是对编者极大的鼓励。

对于编者而言，英语并非母语，撰写本科专业英文教材的难度可想而知。所幸大多数编者拥有国内、外博士学位，或海外研修经历。经过了十余年教学素材的积累和教学实践，现在将书稿整理出来，为各院校相近专业本科生的双语教学服务，此乃我们心愿也。江南大学和生物工程学院多年来大力支持我们开展本科双语教学实践，使之列为国家双语示范课程建设项目。目前在推广教学成果的阶段，编者希望不同院校的同类专业本科生可以分享我们的成果，也期盼本书有望成为相近学科研究生的参考教材，或作为高职高专院校优秀学生的辅读教材。

本书编撰过程中，内容的疏漏和文句的不畅恐怕在所难免，诚恳希望专家和广大读者予以批评指正，以帮助作者进一步完善这部教材，争取来年再版。

诚挚感谢科学出版社的鼓励与厚爱，贵社的负责人和编辑通过“走基层”的方式来到江南大学，与编写组成员直接交换意见，指导本书的编撰。没有科学出版社的支持，我们不敢冒昧提笔。现在本书最终完稿了，亦作为对科学出版社鼎力支持的答谢。

江南大学生物工程学院教授

2012年6月5日